Technology Development for Army Unmanned Ground Vehicles

Committee on Army Unmanned Ground Vehicle Technology
Board on Army Science and Technology
Division on Engineering and Physical Sciences
NATIONAL RESEARCH COUNCIL
OF THE NATIONAL ACADEMIES

THE NATIONAL ACADEMIES PRESS
Washington, D.C.
www.nap.edu

THE NATIONAL ACADEMIES PRESS • 500 Fifth Street, N.W. • Washington, DC 20001

NOTICE: The project that is the subject of this report was approved by the Governing Board of the National Research Council, whose members are drawn from the councils of the National Academy of Sciences, the National Academy of Engineering, and the Institute of Medicine. The members of the committee responsible for the report were chosen for their special competences and with regard for appropriate balance.

This study was supported by Contract/Grant No. DAAD 19-01-C-0051 between the National Academy of Sciences and the Department of Defense. Any opinions, findings, conclusions, or recommendations expressed in this publication are those of the author(s) and do not necessarily reflect the views of the organizations or agencies that provided support for the project.

International Standard Book Number 0-309-08620-5

Additional copies of this report are available from National Academies Press, 500 Fifth Street, N.W., Lockbox 285, Washington, D.C. 20055; (800) 624-6242 or (202) 334-3313 (in the Washington metropolitan area); Internet, http://www.nap.edu

Copyright 2002 by the National Academy of Sciences. All rights reserved.

Printed in the United States of America

THE NATIONAL ACADEMIES
Advisers to the Nation on Science, Engineering, and Medicine

The **National Academy of Sciences** is a private, nonprofit, self-perpetuating society of distinguished scholars engaged in scientific and engineering research, dedicated to the furtherance of science and technology and to their use for the general welfare. Upon the authority of the charter granted to it by the Congress in 1863, the Academy has a mandate that requires it to advise the federal government on scientific and technical matters. Dr. Bruce M. Alberts is president of the National Academy of Sciences.

The **National Academy of Engineering** was established in 1964, under the charter of the National Academy of Sciences, as a parallel organization of outstanding engineers. It is autonomous in its administration and in the selection of its members, sharing with the National Academy of Sciences the responsibility for advising the federal government. The National Academy of Engineering also sponsors engineering programs aimed at meeting national needs, encourages education and research, and recognizes the superior achievements of engineers. Dr. Wm. A. Wulf is president of the National Academy of Engineering.

The **Institute of Medicine** was established in 1970 by the National Academy of Sciences to secure the services of eminent members of appropriate professions in the examination of policy matters pertaining to the health of the public. The Institute acts under the responsibility given to the National Academy of Sciences by its congressional charter to be an adviser to the federal government and, upon its own initiative, to identify issues of medical care, research, and education. Dr. Harvey V. Fineberg is president of the Institute of Medicine.

The **National Research Council** was organized by the National Academy of Sciences in 1916 to associate the broad community of science and technology with the Academy's purposes of furthering knowledge and advising the federal government. Functioning in accordance with general policies determined by the Academy, the Council has become the principal operating agency of both the National Academy of Sciences and the National Academy of Engineering in providing services to the government, the public, and the scientific and engineering communities. The Council is administered jointly by both Academies and the Institute of Medicine. Dr. Bruce M. Alberts and Dr. Wm. A. Wulf are chair and vice chair, respectively, of the National Research Council.

www.national-academies.org

COMMITTEE ON ARMY UNMANNED GROUND VEHICLE TECHNOLOGY

MILLARD F. ROSE, *Chair*, Radiance Technologies, Inc., Huntsville, Alabama
RAJ AGGARWAL, Rockwell Collins, Cedar Rapids, Idaho
DAVID E. ASPNES, North Carolina State University, Raleigh
JOHN T. FEDDEMA, Sandia National Laboratories, Albuquerque, New Mexico
J. WILLIAM GOODWINE, JR. University of Notre Dame, Indiana
CLINTON W. KELLY III, Science Applications International Corporation, McLean, Virginia
LARRY LEHOWICZ, Quantum Research International, Arlington, Virginia
ALAN J. McLAUGHLIN, Massachusetts Institute of Technology, Lincoln Laboratory, Lexington
ROBIN R. MURPHY, University of South Florida, Tampa
MALCOLM R. O'NEILL, Lockheed Martin Corporation, Bethesda, Maryland
ERNEST N. PETRICK, General Dynamics Land Systems (retired), Detroit, Michigan
AZRIEL ROSENFELD, University of Maryland, College Park
ALBERT A. SCIARRETTA, CNS Technologies, Inc., Springfield, Virginia
STEVEN E. SHLADOVER, University of California, Berkeley

Board on Army Science and Technology Liaisons

ROBERT L. CATTOI, Rockwell International (retired), Dallas, Texas
CLARENCE W. KITCHENS, IIT Research Institute, Alexandria, Virginia

National Research Council Staff

ROBERT J. LOVE, Study Director
JIM MYSKA, Research Associate
TOMEKA GILBERT, Senior Project Assistant
ROBERT KATT, Technical Consultant

BOARD ON ARMY SCIENCE AND TECHNOLOGY

JOHN E. MILLER, *Chair*, Oracle Corporation, Reston, Virginia
GEORGE T. SINGLEY III, *Vice Chair*, Hicks and Associates, Inc., McLean, Virginia
ROBERT L. CATTOI, Rockwell International (retired), Dallas, Texas
RICHARD A. CONWAY, Union Carbide Corporation (retired), Charleston, West Virginia
GILBERT F. DECKER, Walt Disney Imagineering (retired), Glendale, California
ROBERT R. EVERETT, MITRE Corporation (retired), New Seabury, Massachusetts
PATRICK F. FLYNN, Cummins Engine Company, Inc. (retired), Columbus, Indiana
HENRY J. HATCH, Army Chief of Engineers (retired), Oakton, Virginia
EDWARD J. HAUG, University of Iowa, Iowa City
GERALD J. IAFRATE, North Carolina State University, Raleigh
MIRIAM E. JOHN, California Laboratory, Sandia National Laboratories, Livermore
DONALD R. KEITH, Cypress International (retired), Alexandria, Virginia
CLARENCE W. KITCHENS, IIT Research Institute, Alexandria, Virginia
SHIRLEY A. LIEBMAN, CECON Group (retired), Holtwood, Pennsylvania
KATHRYN V. LOGAN, Georgia Institute of Technology (professor emerita), Roswell
STEPHEN C. LUBARD, S-L Technology, Woodland Hills, California
JOHN W. LYONS, U.S. Army Research Laboratory (retired), Ellicott City, Maryland
JOHN H. MOXLEY, Korn/Ferry International, Los Angeles, California
STEWART D. PERSONICK, Drexel University, Philadelphia, Pennsylvania
MILLARD F. ROSE, Radiance Technologies, Huntsville, Alabama
JOSEPH J. VERVIER, ENSCO, Inc., Melbourne, Florida

National Research Council Staff

BRUCE A. BRAUN, Director
MICHAEL A. CLARKE, Associate Director
WILLIAM E. CAMPBELL, Administrative Officer
CHRIS JONES, Financial Associate
DEANNA P. SPARGER, Senior Project Assistant
DANIEL E.J. TALMAGE, JR., Research Associate

Preface

The Army's strategic vision calls for transformation to a full-spectrum Objective Force that can project overwhelming military power anywhere in the world on extremely short notice. It must be agile, versatile, and lethal, achieving its objectives through the application of dominant maneuver, precision engagement, focused logistics, information superiority, and highly survivable combat systems. The key to transformation is innovative technology, and the future force will be composed of a family of systems that networks advanced air and ground assets, both manned and unmanned, to achieve superiority in ground combat.

Unmanned vehicles, both air and ground, will play a vital role in such a force structure. There are many tasks that unmanned systems could accomplish more readily than humans, and both civilian and military communities are now developing robotic systems to the point that they have sufficient autonomy to replace humans in dangerous tasks, augment human capabilities for synergistic effects, and perform laborious and repetitious duties.

Unmanned ground vehicles (UGVs) have the potential to provide a revolutionary leap ahead in military capabilities. If UGVs are developed to their full potential, their use will reduce casualties and vastly increase combat effectiveness. To achieve this potential, however, they must be capable of "responsible" autonomous operation. Human operators may always be needed to make the critical decisions, even to take control of critical events, but it is impractical to expect soldiers to continuously control the movement of unmanned systems. Technologies needed to enable autonomous capabilities are still embryonic. Given technical success, there will be "cultural" programs as soldiers learn to trust robot counterparts.

Presentations to the committee and the Demo III demonstrations clearly show that the Army has started down that path and is pursuing many of the enabling technologies. However, without specific requirements to focus the technology base and without funding emphasis, the Army's efforts are less likely to translate into tactically significant unmanned ground vehicle systems. It is particularly important that there be high-level advocacy to coordinate the generation of requirements and the evaluation and acceptance of system concepts.

The Deputy Assistant Secretary of the Army (Research and Technology) requested that the National Research Council's Board on Army Science and Technology conduct this study to evaluate the readiness of UGV technologies. The study was specifically tasked to examine aspects of the Army UGV program, review the global state of the art, assess technology readiness levels, and identify issues relating to implementing UGV systems as part of the Future Combat Systems program. In addition, the committee was tasked with projecting long-term UGV developments of value to the Objective Force.

The committee approached its task by organizing its efforts around the specific technologies and specific charges in the statement of task, subdividing into working groups that could proceed in parallel. Because expertise in many disciplines was necessary to effectively cover all of the elements of robotic vehicles, participants representing many fields were picked from academia and industry (see Appendix A for the biographies of committee members). Several of the committee members had relevant experience in the development, acquisition, testing, and evaluation of combat systems. These members played a vital role, given that concepts for the Future Combat Systems and Objective Force imply many capabilities that have not yet been translated into system requirements.

I want to express my personal gratitude to the members who donated their time to this study. They adhered to a demanding schedule, attended numerous meetings and demonstrations, and had to review copious quantities of material necessary to effectively carry out the task. The report is theirs and represents the committee's collective consensus on the current state of technology development for unmanned ground vehicles.

Any study of this magnitude requires extensive logistical and administrative support, and the committee is grateful to the excellent NRC staff for making its job easier.

Millard F. Rose, *Chair*
Committee on Army Unmanned
Ground Vehicle Technology

Acknowledgments

This report has been reviewed in draft form by individuals chosen for their diverse perspectives and technical expertise, in accordance with procedures approved by the National Research Council's Report Review Committee. The purpose of this independent review is to provide candid and critical comments that will assist the institution in making its published report as sound as possible and to ensure that the report meets institutional standards for objectivity, evidence, and responsiveness to the study charge. The review comments and draft manuscript remain confidential to protect the integrity of the deliberative process. We wish to thank the following individuals for their review of this report:

Harold S. Blackman, Idaho National Engineering and Environmental Laboratory,
Johann Borenstein, University of Michigan,
Roger W. Brockett, Harvard University,
Jagdish Chandra, George Washington University,
Paul Funk, LTG, USA, General Dynamics,
Jasper Lupo, Applied Research Associates,
Larry H. Matthies, Jet Propulsion Laboratory, and
Robert E. Skelton, University of California San Diego.

Although the reviewers listed above have provided many constructive comments and suggestions, they were not asked to endorse the conclusions or recommendations, nor did they see the final draft of the report before its release. The review of this report was overseen by Thomas Munz. Appointed by the National Research Council, he was responsible for making certain that an independent examination of this report was carried out in accordance with institutional procedures and that all review comments were carefully considered. Responsibility for the final content of this report rests entirely with the authoring committee and the institution.

Contents

EXECUTIVE SUMMARY   1

1 INTRODUCTION   13
   Background, 13
   Report Organization, 16

2 OPERATIONAL AND TECHNICAL REQUIREMENTS   17
   Operational Requirements, 17
   Technical Requirements for UGV Capabilities, 19
   UGV Configurations, 19

3 REVIEW OF CURRENT UGV EFFORTS   30
   Army Science and Technology Program, 30
   Other Initiatives, 32

4 AUTONOMOUS BEHAVIOR TECHNOLOGIES   42
   Perception, 42
   Navigation, 51
   Planning, 55
   Behaviors and Skills, 58
   Learning/Adaptation, 65
   Summary of Technology Readiness, 68

5 SUPPORTING TECHNOLOGIES   72
   Human–Robot Interaction, 72
   Mobility, 76
   Communications, 79
   Power/Energy, 83
   Health Maintenance, 87
   Summary of Technology Readiness, 91

6 TECHNOLOGY INTEGRATION   94
   Status of Unmanned Ground Vehicle System Development, 94
   Life-Cycle Support, 96
   Software Engineering, 97

Computational Hardware, 99
Assessment Methodology, 100
Modeling and Simulation, 102

7 ROADMAPS TO THE FUTURE 104
Milestones for System Development, 104
Time Lines for Generic UGV Systems, 109

8 FINDINGS AND RECOMMENDATIONS 111
Technology Development Priorities, 111
Focus on Compelling Army Applications, 112
Systems Engineering Challenge, 113
Advocate for UGV Development, 114

REFERENCES 116

APPENDIXES

A Committee Member Biographical Sketches 123

B Meetings and Activities 125

C Autonomous Mobility 127

D Historical Perspective 151

Figures, Tables, and Boxes

FIGURES

ES-1 UGV technology areas, 4
ES-2 Time lines for development of example UGV systems, 10

1-1 Army transformation to the Objective Force, 15

4-1 Areas of technology needed for UGVs, 43
4-2 Autonomous behavior subsystems, 44
4-3 Perception zones for cross-country mobility, 46
4-4 User interface for controlling a formation of robot vehicles, 63
4-5 User interface for perimeter surveillance, 63
4-6 User interface for a facility reconnaissance mission, 64
4-7 Probability of success, 65

5-1 Areas of technology needed for UGVs, 73
5-2 Schematic of typical hybrid electric power train for UGVs, 86
5-3 System mass as a function of mission energy requirements, 87
5-4 Hybrid UGV 50-watt to 500-watt systems, 87

6-1 Life-cycle cost decisions, 98

7-1 Evolution of UGV systems, 105
7-2 Possible evolution of UGV system capabilities, 105
7-3 Notional FCS acquisition program, 106
7-4 Time lines for development of sample UGV systems, 107
7-5 Technology development roadmap for the Searcher, 107
7-6 Technology development roadmap for the Donkey, 108
7-7 Technology development roadmap for the Wingman, 108
7-8 Technology development roadmap for the Hunter-Killer, 109
7-9 Technology roadmap for development of generic "entry-level" systems in capability classes, 110

C-1 Pedestrian detection, 133
C-2 Demo III vehicle and PerceptOR vehicle, 136
C-3 Perception of traversable slope as an object, 136
C-4 Color-based terrain classification, 137

C-5   Tree-line detection, 139
C-6   Geometric challenge of negative obstacles, 141
C-7   Negative obstacle detection using stereo video, 141

D-1   Autonomous land vehicle (ALV), 153
D-2   ALV and Demo II operating areas, 154
D-3   Demo II vehicle and environment, 156
D-4   Stereo obstacle detection results, 157

TABLES

ES-1   Example Systems Postulated by the Committee, 3
ES-2   Estimates of When TRL 6 Will Be Reached for Autonomous Behavior and Supporting Technology Areas, 5
ES-3   Capability Gaps in Autonomous Behavior Technologies, 6
ES-4   Capability Gaps in Supporting Technology Areas, 8

2-1   UGV Capability Classes, Example Systems, and Potential Mission Function Applications, 20
2-2   Relative Dependence of Technology Areas for Each UGV Class, 20
2-3   Searcher: Basic Capabilities for an Example of a Small, Teleoperated UGV, 22
2-4   Donkey: Basic Capabilities for an Example of a Medium-Sized, Preceder/Follower UGV, 24
2-5   Wingman: Basic Capabilities for an Example of a Medium-Sized to Large Platform-Centric UGV, 27
2-6   Hunter-Killer Team: Basic Capabilities for a Small and Medium-Sized Marsupial Network-Centric UGV Team, 29

4-1   Criteria for Technology Readiness Levels, 44
4-2   Perception System Tasks, 45
4-3   Technology Readiness Criteria Used for Perception Technologies, 49
4-4   TRL Estimates for Example UGV Applications: On-Road/Structured Roads, 49
4-5   TRL Estimates for Example UGV Applications: On-Road/Unstructured Roads, 50
4-6   TRL Estimates for Example UGV Applications: Off-Road/Cross-Country Mobility, 50
4-7   TRL Estimates for Example UGV Applications: Detection of Tactical Features, 50
4-8   TRL Estimates for Example UGV Applications: Situation Assessment, 50
4-9   Estimates for When TRL 6 Will Be Reached for Autonomous Behavior Technology Areas, 68
4-10 Capability Gaps in Autonomous Behavior Technologies, 69

5-1   Desired Criteria for a High-Mobility UGV Weighing Less Than 2,000 Pounds, 76
5-2   Current Options for Army UGV Mobility Platforms, 77
5-3   Summary of Power/Energy Systems, 85
5-4   Estimates for When TRL 6 Will Be Reached in UGV Supporting Technology Areas, 91
5-5   Capability Gaps in Supporting Technology Areas, 92

C-1   Sample Environments and Challenges, 129
C-2   Imaging Sensor Trade-offs, 143
C-3   Sensor Improvements, 143
C-4   Impact of Feature Use on Classification, 146

D-1   Performance Trends for ALV and Demo II, 158

BOXES

1-1   A Glimpse of the Future, 14

3-1   Task Statement Question 2.a, 31
3-2   Task Statement Question 2.b, 31
3-3   Task Statement Question 2.c, 39
3-4   Task Statement Question 3.c, 41

4-1   Task Statement Question 4.a (Perception), 51
4-2   Task Statement Question 4.a (Navigation), 54
4-3   Task Statement Question 4.a (Planning), 59
4-4   Task Statement Question 4.b (Tactical Behaviors), 61
4-5   Task Statement Question 4.b (Cooperative Behaviors), 66
4-6   Task Statement Question 4.a (Learning/Adaptation), 68
4-7   Task Statement Question 3.d (Autonomous Behavior Technologies), 71
4-8   Task Statement Question 4.c (Autonomous Behavior Technologies), 71

5-1   Task Statement Question 4.b (Human–Robot Interaction), 75
5-2   Task Statement Question 4.b (Mobility), 79
5-3   Task Statement Question 4.b (Communications), 82
5-4   Task Statement Question 4.b (Power/Energy), 88
5-5   Task Statement Question 4.b (Health Maintenance), 91
5-6   Task Statement Question 3.b, 91
5-7   Task Statement Question 3.d (Supporting Technology Areas), 93

6-1   Task Statement Question 5.c, 99

7-1   Task Statement Question 5.a, 110

8-1   Task Statement Question 3.a, 113
8-2   Task Statement Question 5.b, 115

Acronyms and Abbreviations

AADL	Avionics Architecture Definition Language
ACC	adaptive cruise control
ACN	assign commercial network
ACS	agile combat support
ALN	adaptive logic networks
ALV	autonomous land vehicle
ALVINN	autonomous land vehicle in a neural network
AMCOM	Army Aviation and Missile Command
AMUST-D	Airborne Manned/Unmanned System Demonstration
AOE	automated ordnance excavator
ARL	Army Research Laboratory
ARTS-FP	All-purpose Remote Transport System-Force Protection
ARTS-RC	All-purpose Remote Transport System-Range Clearance
ARV	armed reconnaissance vehicle
ASA (ALT)	Assistant Secretary of the Army (Acquisition, Logistics, and Technology)
ASB	Army Science Board
ASTMP	Army Science and Technology Master Plan
ATD	Advanced Technology Demonstration
ATR	automated target recognition
ATV	all-terrain vehicle
AVRE	Armored Vehicle Royal Engineers
BAST	Board on Army Science and Technology
BDA	battle damage assessment
BLOS	beyond line of sight
BUGS	Basic UXO Gathering System
C2	command and control
CAT	crew integration and automation testbed
CCD	camouflage concealment deception
CECOM	Communications Electronics Command
CET	combat engineer tractor
CIS	communications interface shelter
CJCS	Chairman, Joint Chiefs of Staff
CMU	Carnegie Mellon University
COP	common operation picture
COTS	commercial off-the-shelf

CTA	Collaborative Technology Alliance
CVA	canonical variate analysis
DARPA	Defense Advanced Research Projects Agency
DGPS	differential global positioning system
DOD	Department of Defense
DOE	Department of Energy
DRP	dynamic remote planning
DSP	digital signal producer
DSRC	dedicated short-range communications
DTED	digital terrain elevation data
DUECE	deployable universal combat earthmover
EEA	essential elements of analysis
EOD	explosive ordnance disposal
EWLAN	enhanced wireless local area network
FCC	Federal Communications Commission
FCS	Future Combat Systems
FDIR	fault detection, identification, and recovery
FFN	friend, foe, or neutral
FLIR	forward looking infrared radar
FOC	Future Operational Capabilities
FOLPEN	foliage penetration
FPGA	field programmable gate arrays
FY	fiscal year
GIPS	giga instructions per second
GIS	geographical information systems
GLOMO	global mobile
GOPS	giga operations per second
GPS	Global Positioning System
HAZMAT	hazardous materials
HCI	human–computer interface
HMI	human–machine interface
HMMWV	high-mobility multi-purpose wheeled vehicle
HRI	human–robot interaction
IFF	identification of friend or foe
IFFN	identifying friends, foes, and noncombatants
IFOV	instantaneous field of view
IMU	inertial measurement unit
INS	inertial navigation system
IR	infrared
JAUGS	Joint Architecture for Unmanned Ground Systems
JFCOM	Joint Forces Command
JPL	Jet Propulsion Laboratory
JRP	Joint Robotics program
JTRS	Joint Tactical Radio System
JVB	Joint Virtual Battlespace
LADAR	laser detection and ranging
LAN	local area network

LORAN	long-range navigation
LOS	line of sight
LPD	low probability of detection
LPI	low probability of intercept
LSI	lead system integrator
M&S	modeling and simulation
MARDI	Mobile Advanced Robotics Defense Initiative
MARS	Mobile Autonomous Robot Software
MC2C	multisensor command and control constellation
MDARS-E	Mobile Detection Assessment Response System-Exterior
MDARS-I	Mobile Detection Assessment Response System-Interior
MEP	Mobility Enhancement program
MFLIR+R	monocular forward looking infrared plus radar
MILS	multiple independent levels of security
MIPS	million instructions per second
MNS	mission needs statement
MOE	measure of effectiveness
MOP	measures of performance
MOPS	million operations per second
MOUT	military operations in urban terrain
MOV	measure of value
MPRS	Man-Portable Robotic System
MURI	Multidisciplinary University Research Initiative
MV+R	monocular video plus radar
NASA	National Aeronautics and Space Administration
NBC	nuclear, biological, chemical
NC-AGV	network-centric autonomous ground vehicle
NIST	National Institute of Standards and Technology
NLOS	non–line of sight
NLP	natural language processing
NRL	Naval Research Laboratory
OAR	organic air vehicle
ODIS	Omni-Directional Inspection System
OMG	Object Management Group
OO	object-oriented
OP	observation post
ORD	operational requirements document
OSD	Office of the Secretary of Defense
PC-AGV	platform-centric autonomous ground vehicle
PerceptOR	Perception off-road
PM	program manager
PRIMUS	Program of Intelligent Mobile Unmanned Systems
QoS	quality of service
RACS	robotics for agile combat support
RAIM	receiver autonomous integrity monitoring
RALPH	rapidly adapting lateral position handler
RBF	radial basis function
RCRV	remote crash rescue vehicle
RCSS	Robotics Combat Support System

RDA	Research, Development, and Acquisition
RF	radio frequency
RGB	red, green, blue
RONS	Remote Ordnance Neutralization System
RSTA	reconnaissance, surveillance, and target acquisition
S&T	science and technology
SA	situational awareness
SAE	Society of Automotive Engineers
SAF-UGV	semiautonomous follower unmanned ground vehicle
SAP/F-UGV	semiautonomous preceder-follower
SARGE	Surveillance and Reconnaissance Ground Equipment
SDD	system development and demonstration
SEAD	suppression of enemy air defenses
SFLIR	stereo forward looking infrared
SLOC	source lines of code
SOP	standard operating procedure
SORC	statement of required capabilities
SPC	software process control
SRS	Standardized Robotics System
STO	science and technology objective
STRICOM	Simulation, Training, and Instrumental Command
SV	stereo video
SWAT	special weapons and tactics
SYRANO	Systeme Robotise d'Acquisition pour la Neutralisation d'Objectifs
TACOM	Tank-Automotive and Armaments Command
TARDEC	Tank-Automotive Research, Development, and Engineering Center
TGV	teleoperated ground vehicle
TMR	tactical mobile robot
TRAC	TRADOC Analysis Center
TRADOC	Training and Doctrine Command
TRL	technology readiness level
UAV	unmanned air vehicle
UCAV	unmanned combat air vehicle
UDS	UCAV Demonstration System
UGCV	unmanned ground combat vehicle
UGV	unmanned ground vehicle
UOS	UCAV Operating System
URPR	University Research Program in Robotics
USD-AT&L	Under Secretary of Defense for Acquisition, Technology and Logistics
USDOT	U.S. Department of Transportation
UUV	unmanned underwater vehicle
UWB	ultra-wide band
UXO	unexploded ordnance
VCI	vehicle cone index
VTOL	vertical takeoff and landing
XUV	experimental unmanned vehicle

Executive Summary

The Army has long recognized the potential of robotics on the battlefield. Capitalizing on early work by the Defense Advanced Research Projects Agency (DARPA), it has sponsored basic and advanced research in intelligent systems and led developments in ground vehicle crew automation technology as well as tactical unmanned air vehicles (UAVs). At the same time, the Army has successfully adapted commercial teleoperated ground vehicles for specialized military uses, such as mine clearing and urban reconnaissance, and it has made early progress toward developing semiautonomous mobility capabilities.

The urgent need to transform the Army—from one characterized by heavy armor and firepower into a lighter, more responsive Objective Force that is at once both lethal and survivable—has made development of practical unmanned ground vehicle (UGV) systems a necessity for the future. Concepts for the Army's Future Combat Systems (FCS), which are now being evaluated, include unmanned systems, both ground and air, and thus will be required for fielding with other elements of the FCS as early as 2010.

The Army plans to use UGVs for such things as weapons platforms, logistics carriers, and surrogates for reconnaissance, surveillance, and target acquisition (RSTA), both to increase combat effectiveness and to reduce the number of soldiers placed in harm's way. Congress, too, has recognized the potential of unmanned systems and has mandated that at least one of every three future Army systems be unmanned. The Army UGV technology development program includes unmanned ground vehicles; however, it is unclear whether UGV technologies can be developed rapidly enough to keep up with the accelerated acquisition pace of the Army FCS program.

Unmanned ground vehicle systems are one of the few areas across the entire Department of Defense (DOD) that legitimately qualifies as having "leap-ahead" (revolutionary as opposed to evolutionary) potential for the battlefield. Considering that it has taken more than 40 years to develop rudimentary unmanned air vehicles, which operate in a relatively simple operational environment, the Army faces a daunting challenge to develop UGV systems. The efforts required to organize and to manage the evolution of technologies and integration of systems will be immense.

This study was sponsored by the Assistant Secretary of the Army (Acquisition, Logistics, and Technology), who requested that the National Research Council undertake the following:

- Review Army operational requirements for UGVs, including the Army Future Combat Systems (FCS) baseline program, the Army Research Laboratory (ARL) UGV science and technology objective (STO), and other UGV requirements.
- Review the current Army UGV efforts at ARL and the Army Tank-Automotive Research, Development, and Engineering Center (TARDEC).
- Review the state of the art in unmanned vehicle technologies applicable to UGV systems (e.g., "intelligent" perception and control, adaptive tactical behaviors, human–system interfaces).
- Identify issues relating to technical risks and the feasibility of implementing applicable UGV technologies within the FCS baseline program time frame.
- Document the results of the examination in a study report that will be provided to the Army. The report will contain a recommended roadmap for the development of UGV technology and systems, including topics that could be the subjects of investigations of longer-term (2015 and beyond) UGV technology applications.

Answers to specific questions in each of the last four task areas are highlighted in the study report and provided in this executive summary.

OPERATIONAL AND TECHNICAL REQUIREMENTS

At the time of the study, the Army had not established a plan for integrating UGV and other technologies into an FCS roadmap. The absence of definitive UGV requirements made it difficult to determine where the Army science and technology (S&T) community should place emphasis in technology development. To help resolve this problem, in early 2002 the Army selected a lead system integrator (LSI) for the FCS program, who requested proposals from industry on concepts for three FCS UGV systems, including the following:

- Soldier UGV, a small soldier-portable reconnaissance and surveillance robot
- Mule UGV, a 1-ton vehicle suitable for an RSTA or transport/supply mission
- Armed reconnaissance vehicle (ARV) UGV, a 6-ton vehicle to perform the RSTA mission, as well as a fire mission, carrying a turret with missile and gun systems.

UGVs can be developed in all sizes and outfitted to perform an assortment of military tasks. Aside from scale and function, a major characteristic of any UGV is its level of autonomy, ranging from 100 percent teleoperated (remote operation), through various stages of semiautonomy, to fully autonomous (ideal). Numerous military and civilian applications for small, teleoperated UGVs are in advanced development and have seen wide use. Improvements in the human–robot interface would greatly increase their effectiveness.

Increasing levels of autonomy in future UGVs would greatly expand the list of military uses. Army STOs have been focusing on technologies needed for semiautonomous mobility from point A to point B (A-to-B mobility), which is uniformly recognized as crucial to the ultimate acceptability of autonomous ground vehicles.

To facilitate its work, the Committee on Army Unmanned Ground Vehicle Technology categorized UGVs as belonging to one of four capability classes. These classes are distinguished by the following characteristics:

1. *Teleoperated ground vehicle (TGV)*—In teleoperation, a human operator controls a robotic vehicle from a distance. The operator conducts all cognitive processes. The sensors onboard the vehicle and the communications link allow the operator to visualize the UGV's location and movement within its environment. TGVs come in all sizes.
2. *Semiautonomous preceder-follower (SAP/F-UGV)*—Like the TGV, SAP/F-UGVs can come in all shapes and sizes. Follower UGVs are the focus of current Army development and demonstrations. Preceder UGVs are follower UGVs with advanced navigation capability that minimizes the need for operator interaction to achieve A-to-B mobility. The preceder must have sufficient autonomy to move in advance of its controller, which could be a dismounted soldier or a vehicle. It has sufficient cognitive processes onboard to select the best route to traverse an objective designated by the controller without the need for marking terrain.
3. *Platform-centric autonomous ground vehicle (PC-AGV)*—An autonomous ground vehicle can be assigned a complex task or mission and will then execute it, perhaps acquiring information from other sources as it goes, or perhaps responding to additional commands from a controller, but without requiring further guidance. Military missions demand "responsible" autonomy for PC-AGVs capable of delivering lethal weapons and require fail-safe interrupt mechanisms. PC-AGVs must have autonomous A-to-B mobility and must be able to carry out assigned missions in a hostile environment. As with negotiating difficult terrain, the benchmark here is that the UGV should have survivability and self-defense roughly equivalent to those of a similar manned vehicle sent on the same mission.
4. *Network-centric autonomous ground vehicle (NC-AGV)*—NC-AGVs are PC-AGVs with levels of autonomy sufficient to operate as independent nodes in a net-centric warfare model. They must be able to receive information from the communications network and incorporate it in their mission execution and respond to appropriate information requests and action commands received from the network, including resolution of conflicting commands. Again, a rough benchmark for operational performance is an equivalent manned system, similarly tasked.

On the surface, the four classes represent a progression of increasing levels of autonomy, but each class has distinctive needs for development in the various UGV technology areas. Using communications technologies as an example: TGVs have a high requirement at all times; SAP/F-UGVs have a moderate requirement for mobility (e.g., placing electronic "breadcrumbs") or contingencies (unusual obstacles or enemy attack); PC-AGVs have little or no need for human control (unless due to specific mission function); and NC-AGVs have little or no need for human control but high needs for network connectivity.

Technology readiness levels (TRLs) can provide a uniform measure for the maturity of different technologies against stated requirements. TRL 6 is especially important, because it is defined as the point when a technology component or subsystem has been demonstrated in a relevant environment. In the absence of firm Army requirements, the study postulated four compelling examples of systems with associated technical requirements that would provide "marks on the wall" against which to estimate the TRLs in each

UGV technology area. The systems postulated by the committee for the study are listed in Table ES-1, and each represents one of the capability classes. The examples include systems with capabilities subsequently implied to be needed for FCS by the LSI.

CURRENT DEVELOPMENT EFFORTS

The Army UGV development program consists of two primary STO efforts: the robotic follower Advanced Technology Demonstration (ATD) managed by TARDEC, and the semiautonomous robotics for FCS STO managed by ARL. In addition to several smaller STO efforts in robotics, the Army also participated in the DEMO III UGV program, which is part of the DOD Joint Robotics Program and has funded several Defense Advanced Research Projects Agency (DARPA) projects, including:

- Tactical mobile robot
- Unmanned ground combat vehicle
- Perception off-road, and
- Organic air vehicle.

The overall Army UGV program includes teleoperated UGV operations, soldier-in-the-loop experimentation, and demonstrations of both follower and semiautonomous mobility under controlled environments. It is focused on technology demonstrations but has included limited testing and experimentation.

The study found that technologies developed for the follower UGV ATD will achieve the ATD objectives but do not include relevant supporting technologies for likely FCS system-level requirements. Basic semiautonomous off-road mobility is scheduled for demonstration by FY06 under the ATD. The ATD is being restructured (in consonance with other STOs) to focus on the Mule and Armed Reconnaissance Vehicle capabilities, consistent with the concepts requested of industry by the FCS LSI.

The original ARL STO addressed capabilities that would support autonomous requirements that were not addressed in the follower ATD, including planning, navigation, and human–robot interaction. However, the emphasis on demonstrations and the heavy reliance on laser detection and ranging (LADAR) sensors limited advances in perception state of the art. To the extent that the restructured STOs assist the Army in defining system requirements, they are clearly a step in the right direction. It remains to be seen, however, whether the redirected focus on Mule and ARV capabilities will accelerate development of UGV systems.

The study considered other government UGV efforts, including those sponsored by the National Aeronautics and Space Administration (NASA), the Department of Transportation, the Department of Energy (DOE), and the National Institute of Standards and Technology (NIST). Except for the DARPA efforts, the interrelationships between other government UGV efforts and those of the Army are informal and unstructured. But the small size of the UGV industry and the small number of robotics experts tend to encourage technical interchange and collaboration. For example, the Jet Propulsion Laboratory supports NASA, DARPA, and Army robotics programs, so collaboration among these programs can be very high. Similarly, NIST experts in intelligent systems participated with others in the Demo III program. There is concern, however, that collaboration in some technology areas may be inhibited by intense competition for a limited number of UGV-related contracts.

Foreign research is on a par with U.S. research in some fields relevant to future UGV technology. Based on information available to the committee, however, there are no foreign UGV technology applications that are significantly more advanced than those in this country.

STATE OF THE ART

A UGV consists of a mobility platform with sensors, computers, software (including modules for perception,

TABLE ES-1 Example Systems Postulated by the Committee

Example System	Capability Class	Other Possible Applications
Small robotic building and tunnel searcher ("Searcher")	Teleoperated ground vehicle	Mine detection, mine clearing, engineer construction, EOD/UXO materials handling, soldier-portable reconnaissance/surveillance
Small-unit logistics mover ("Donkey")	Semiautonomous preceder/follower	Supply convoy, medical evacuation, smoke laying, indirect fire, reconnaissance/surveillance, physical security
Unmanned wingman ground vehicle ("Wingman")	Platform-centric autonomous ground vehicle	Remote sensor, counter-sniper, counter-reconnaissance/infiltration, indirect fire, single outpost/scout, chemical/biological agent detection, battle damage assessment
Autonomous hunter-killer team ("Hunter-Killer")	Network-centric autonomous ground vehicle	Deep RSTA, combined arms (lethal direct fire/reconnaissance/indirect fire for small unit defense or offense), static area defense, MOUT reconnaissance

EOD/UXO = explosive ordnance disposal/unexploded ordnance; RSTA = reconnaissance, surveillance, and target acquisition; MOUT = military operations in urban terrain.

navigation, learning/adaptation, behaviors and skills, human–robot interaction, and health maintenance), communications, power, and a separate mission package depending upon the UGV's combat role. The committee's study assumed that technologies needed for the mission-function packages would be developed independently.

Autonomous Behavior Technology Areas

All of the UGV system components are illustrated in Figure ES-1. The committee assessed the state of the art in UGV-specific and supporting technology areas to uncover capability gaps and estimate technology readiness. Table ES-2 provides committee estimates for when TRL 6 will be reached in each of the technology areas for the four postulated examples. Tables ES-3 and ES-4 outline capability gaps by technology area for each example along with the committee's assessment of difficulty and/or risk.

Perception

The greatest uncertainties in perception technologies, including sensors and software for mobility and situational awareness, are in describing UGV performance and in determining the effect of perception (and other technology subsystems) on overall UGV system performance. Metrics do not exist, and there are no procedures for benchmarking algorithms, so there is considerable uncertainty as to whether the current algorithms are the best available. In the absence of metrics and data, there is little basis for system optimization and a corresponding uncertainty about performance losses due to system integration issues. There is no systematic way to determine where improvements are required and in what order. The uncertainties exist because perception technologies other than basic sensors have not been emphasized in the Army program, and adequate resources have not been applied.

Navigation and Planning

Achieving nearly fully autonomous UGV navigation will require the integration of perception, path planning, communication, and various navigation techniques. This integration of multiple technologies is the largest technology gap in autonomous navigation. UGVs must be able to detect when they are lost and then react appropriately. The ability to detect navigation errors will have to be developed. Further integration between navigation and communication technologies will help to create more robust positioning solutions.

While path planning for a single UGV is relatively mature, algorithm developments for multiple UGV and UAV

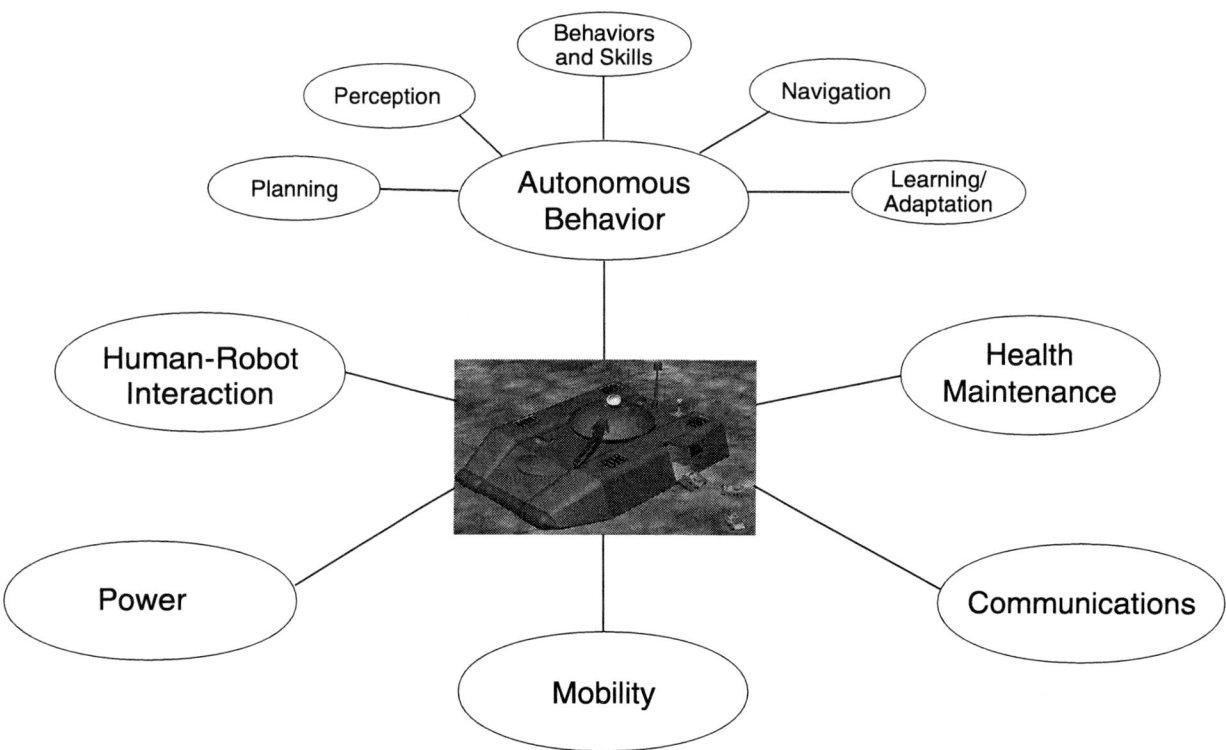

FIGURE ES-1 UGV technology areas.

TABLE ES-2 Estimates of When TRL 6 Will Be Reached for Autonomous Behavior and Supporting Technology Areas

Technology Areas	Searcher	Donkey	Wingman	Hunter-Killer
Perception				■
Navigation				■
Planning				■
Behaviors and skills			■	■
Learning/adaptation				■
Human–robot interaction				■
Mobility				
Communications			■	■
Power/energy				
Health maintenance				

Near term
Mid term (2006–2015)
Far term (2016–2025)

planning (both path planning and mission planning) are relatively immature. When planning the path of multiple UGVs and UAVs, the bandwidth of communications between vehicles is a very important factor. Trade studies need to be performed to determine how much bandwidth is available and how the requirements will vary for specific missions. Once this is determined, it will be possible to develop the appropriate path planners for multiple coordinated vehicles.

Tactical Behaviors and Learning/Adaptation

Technologies for UGV behaviors (both tactical and cooperative) are still in their infancy. Development of modules to enable complex tactical behaviors, such as difficult terrain negotiations or stealthy operations, will not happen in the near term. Although some simple cooperative control strategies have been demonstrated at universities and at the national laboratories, there is still no basic understanding of how to design cooperative behaviors for multiple robots.

There is significant uncertainty about the degree to which methods from machine learning will ultimately provide software solutions for complex, real-world problems like UGVs. Critical missing elements are measures for the complexity of the environment in which UGVs will have to operate that enable comparison with the levels of complexity that high-level control algorithms (i.e., current decision-making algorithms) can effectively handle. Funding for research is less of a factor for advances in this area given the amount and intensity of research devoted to machine-learning paradigms, particularly in academia.

UGVs capable of adaptive behaviors sufficient to deal with complex and changing battlefield environments are far from reality. Uncertainty exists on where to draw the line between adaptive control solutions and artificial intelligence solutions. This applies to all of the software-based components of the autonomous system, including perception, navigation, planning, and behaviors.

Supporting Technology Areas

Developments in several supporting technology areas, not totally unique to unmanned ground vehicles, are essential to the development of UGV systems. These include human–robot interaction, mobility, communications, power/energy, and vehicle health-maintenance technologies.

Human–Robot Interaction

Human–robot interaction (HRI) requirements involve more than basic man–machine interfaces and will be much more demanding than those for commercial robots. Multimodal interfaces, including distinctive sounds and gestures for operations in combat, must be developed. The Army must consider requirements for secure, natural-user interfaces. Studies are needed to determine optimum means for all facets of soldier–robot and robot–robot interactions. HRI technology is resource-intensive and will require many experiments and tests under realistic conditions to achieve acceptable levels of reliability.

Mobility

The mobility platform is highly application-dependent. Platforms for different mission applications will have to be designed based on differences in the mission requirements. While a UGV has the advantage of not needing to be designed around human crew limitations, it also has the disadvantage of needing mechanisms to replace human driving judgment. Thus, salient uncertainty surrounds how design requirements for UGV mobility platforms can be integrated with perception technologies to provide the capability to avoid obstacles, both positive and negative, that the platform is not hardened to overcome. The overall risk associated with building different mobility platforms will be less than that of developing sensors and software capable of successful mobility.

TABLE ES-3 Capability Gaps in Autonomous Behavior Technologies

Degree of Difficulty/Risk

Low
Medium
High

Technology Areas	Capability Gaps			
	Searcher	Donkey	Wingman	Hunter-Killer
Perception				
A-to-B mobility on-road		Algorithms and processing fast enough to support 40 km/h (road-following, avoidance of moving and static obstacles).	Algorithms and processing fast enough to support 100 km/h (road-following, avoidance of moving and static obstacles).	Algorithms and processing fast enough to support 120 km/h (road following, avoidance of moving and static obstacles).
			Sensors with long range.	Sensors with long range.
A-to-B mobility off-road	Algorithms for real-time two-dimensional mapping and localization.	Detect and avoid static obstacles (positive and negative) at 40 km/h day or night.	Sensors and strategies for fine positioning in bushes.	Algorithms for multiple sensor and data fusion.
	Miniature hardened range sensors.	Classify terrain (traversable at speed, in low visibility).	Detect and avoid obstacles at 100 km/h.	Detect and avoid static obstacles at 120 km/h.
	All-weather sensors.	Classify vegetation as "push through" or not, detect water, mud, and slopes.	Classify terrain and adapt speed, control regime.	Classify terrain and adapt speed, control regime.
		Algorithms for GPS mapping and corrections.	Continually assess terrain for potential cover and concealment.	Continually assess terrain for cover and concealment.
			Multiple sensor fusion.	
Situation awareness		Algorithms for detecting humans (even lying down, versus other obstacles).	Track manned "leader" vehicle.	Algorithms and sensors to recognize movement and identify source.
		Sensors and algorithms for detecting threats.	Select suitable OP (provides LOS cover and concealment).	Select suitable OP (provides LOS cover and concealment).
			Detect, track, and avoid other vehicles or people.	Detect, track, and avoid other vehicles or people.
			Distinguish friendly and enemy combat vehicles.	Distinguish friendly and enemy combat vehicles.
			Detect unanticipated movement or activities.	Detect unanticipated movement or activities.
			Acoustic, tactile sensors for recognition.	Detect potential human attackers in close proximity.
				Sensors while concealed (indirect vision).

continues

TABLE ES-3 Continued

Technology Areas	Capability Gaps			
	Searcher	Donkey	Wingman	Hunter-Killer
				Localization to coordinate multirobots.
				Identify noncombatants.
Navigation		Relative navigation utilizing communications and GPS.	Integration of GPS, digitized maps, and local sensors.	Error detection and correction.
Planning				
Path		Use DTED maps; 1-km replanning for obstacle avoidance.	Plan relative to leader; reason about overlapping views.	Tactical formation planning.
		Electronic "breadcrumbs."	Plan to rejoin or avoid team; use features other than terrain.	Adjust route based on external sensor inputs.
		Decision template for alternative routing.	Reasoning algorithms to identify and use concealment.	Plan to optimize observation points, target kill arrays, and communication links.
				Multiobject and pursuit-evasion path planning for multiple UGVs.
Mission			Mimic leader actions.	Plan for complex missions including combat survival.
			Independent actions.	Plan for team and marsupial operations.
				Independent actions.
Behaviors and skills				
Tactical skills	Basic nonlethal self-protection if touched or compromised.	Avoid enemy observation.	Hooks for specialized mission functions (e.g., RSTA, indirect fire).	Independent operations; fail-safe controls for lethal missions.
		"Flee and hide."	Self-protection.	Self-preservation and defensive maneuvers.
			Complex military operational behaviors.	Complex military operational behaviors.
Cooperative robots			Formation controls of multiple UGVs.	Formation controls of multiple UGVs and UAVs.
			Cooperation for such tasks as hiding in bushes.	
Learning/adaptation		Basic learning for survivability.	Advanced terrain classification.	Advanced fusion of multiple sensor and data inputs.
			Basic machine learning augmentation of behaviors.	Advanced machine learning augmentation of behaviors.

TABLE ES-4 Capability Gaps in Supporting Technology Areas

Degree of Difficulty/Risk

Low
Medium
High

Technology Areas	Capability Gaps			
	Searcher	Donkey	Wingman	Hunter-Killer
Human–robot interaction (HRI)	Telesystem HRI algorithms that support 1 operator per robot.	Semiautonomous HRI algorithms that support 1 operator per 5 homogeneous robots.	Natural user interfaces.	Natural user interfaces.
	Multimodal interfaces (nlp, gesture).	Natural user interfaces.	Diagrammatic and multimodality interfaces.	Methods for interacting and intervention under stress.
			Semiautonomous HRI algorithms that support multiple operators.	Near-autonomous HRI algorithms that support multiple operators and robots.
Mobility	Ability to right itself in restrictive passages/areas.	Platform capable of handling 40 km/h on smooth terrain with sensitive payload.	Platform capable of handling 100 km/h on smooth terrain with sensitive payload.	Heterogeneous marsupialism to transport specialized robots and sensors.
		Platform capable of handling 40 km/h on rough terrain with sensitive payload.	Platform capable of handling 100 km/h on rough terrain with sensitive payload.	Platform capable of handling 120 km/h.
Communications	High bandwidth for secure video; local to group.	Low bandwidth for "breadcrumbs," local to group.	Medium bandwidth for mobile network, local to group.	High bandwidth for secure and reliable network-centric communications.
	Wireless backup for line-of-sight communications.			Large amounts of distributed information.
Power/energy	High energy density rechargeable battery.	High energy density rechargeable battery.	Highly efficient stealth energy system.	Long standby (30 days).
	Small, hybrid energy system.			Highly efficient stealth energy system.
				High-speed mobility enablers.
Health maintenance	High physical reliability, low maintenance.	High physical reliability, low maintenance.	Design for combat survivability.	Self-repair by reconfiguring components.
		Cooperative diagnostics for remote operator.	Algorithms for self-diagnosis.	Self-repair by self-reprogramming.
		Ability to know when to call for help.		

Communications

Shortcomings in communications technologies that apply to manned systems also apply to UGVs. But communications are much more crucial to UGV system performance. Near-term wireless solutions, for example, are problematic. Network connectivity could easily be lost due to non–line-of-sight interference caused by terrain or other obstacles. Security attacks on dispersed unmanned systems could include denial of service, compromising of classified, high-value tactical information, corruption of information, and, in the extreme, usurpation of the system.

Communications for mobility management, of increasing importance to network-centric operations, must ensure that there will always be network participants on station to provide relay when needed. Disparate efforts in communications for both manned and unmanned systems must be based on a common vision and technical architecture and conform to common interface standards.

Power/Energy

Until one specifies a mission time requirement in kilowatt-hours (kWh), power/energy technologies may not be of concern. Short-duration, low-energy mission requirements can be met now. However, there are problems with providing extended-duration communications, such as streaming video, for small UGVs. For high-energy missions, the following issues must be addressed: catalysts for reforming fuel, thermal rejection processes, stealth, and energy storage and replenishment.

Health Maintenance

Tool sets consisting of sensors, diagnostics, and recovery methods must be based on operational requirements. Coming up with a calculus of diagnosis and recovery, similar to the triple-redundancy development systems and design logic used for NASA systems, will be a major challenge. Like human–robot interaction technologies, health maintenance developments will require extensive experimentation and testing.

TECHNOLOGY INTEGRATION AND ROADMAPS TO THE FUTURE

The Army's UGV technology development program is not organized so as to enable the acceleration of system-level UGV developments. Development and insertion of individual "entry-level" UGV technologies is possible, but necessary technology components for HRI, mobility, power/energy, communications, and health maintenance all depend heavily on system-level requirements that are currently unknown. Efforts on multiple technology development fronts now cover several different operational applications. Performance levels for the Army and DARPA FCS efforts should be synchronized to facilitate the definition of valid technical parameters. The Demo III project demonstrated many UGV system components, but it did not test a UGV system. Standards must be developed for measuring important system characteristics, such as perception for autonomous mobility. While relevant technologies will be enabled, the lack of user pull is a major impediment to achieving timely integration of capable UGV systems.

To accelerate the pace of technology development, the current UGV program should be upgraded to emphasize the collection of data under a variety of stressing conditions. The data will enable predictive models of system performance and support the systems engineering process essential for integration of relevant technologies. The models will also support the development of realistic simulations to use in developing concepts of operation and in exploring the utility of different classes of UGVs.

The committee was able to estimate milestones from a purely engineering perspective for a UGV system development program leading to the production of the four systems postulated by the study. Figure ES-2 estimates milestone dates for development of the four systems based on the TRLs in relevant technology areas. The milestones assume that the system developments are interdependent, each building on R&D accomplished to achieve capabilities needed for a preceding system. It was also assumed that the advanced capabilities needed by the Hunter-Killer would be identified as goals from the outset and that all capability gaps identified by the committee would be filled in a timely fashion. It should be emphasized that the milestones represent optimistic estimates for UGV systems of unquestioned utility on the battlefield and not for "entry-level" systems or prototypes.

RECOMMENDATIONS

The following recommendations address technical content, time lines and milestones of the UGV efforts for FCS. First, the Army should focus S&T efforts on the perception technology area, with other priority areas dependent on capability class. Second, the Army should adopt a "skunkworks" approach to develop the essential perception technologies to enable autonomous A-to-B mobility capabilities that can be fielded with multiple UGV systems as experimental prototypes for possible insertion in the FCS program. Third, the Army should begin immediately to fill a void in systems engineering by defining system requirements, planning for life-cycle support, establishing milestones for development of assessment methods and metrics for UGV systems, and taking advantage of modeling and simulation tools. Finally, the Army should designate a high-level advocate to accelerate S&T time lines and take the lead in integrating UGV technologies into prototypical systems.

The committee reasoned that only a "skunk-works" approach would bring together the necessary resources, focus,

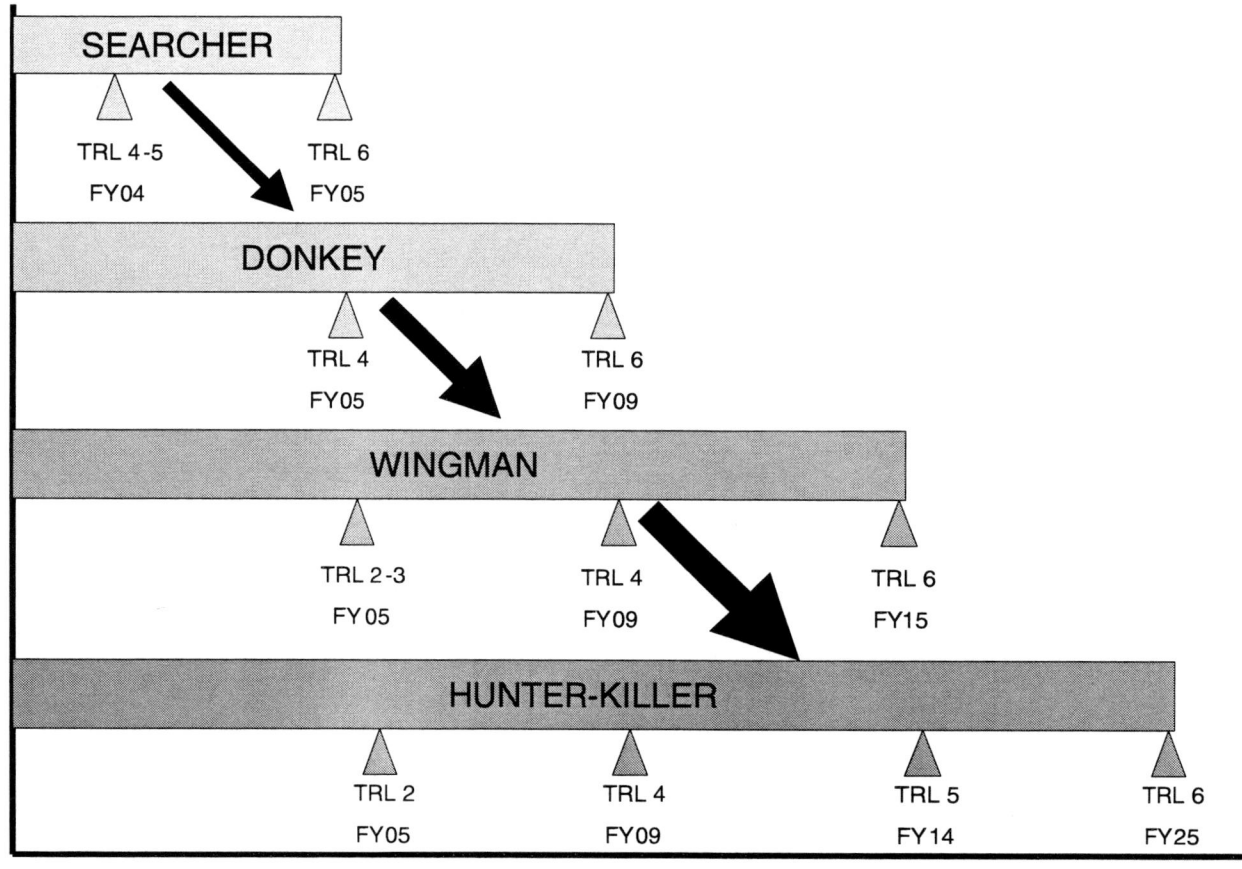

FIGURE ES-2 Time lines for development of example UGV systems, assuming progressive capability developments.

and leadership to ensure successful development of UGV systems on an accelerated basis. Such an approach would enable metrics to evolve as system prototypes are developed on a sound system-engineering basis and are made available for rigorous testing by the user. Above all, the approach relies on a strong advocate to guide the process and make sure that parallel efforts are mutually supportive.

Development Priorities

Clearly, the highest priority for the Army should continue to be to develop perception technologies for autonomous mobility. On-road and off-road mobility is fundamental to the acceptability of three of the four systems postulated by the study. The maturity of on-road capabilities, on both structured and unstructured roads, must be emphasized. The current level of perception capability cannot support an autonomous cross-country traverse of tactical significance, at tactical speeds, under combat conditions.

Perception technologies, including sensors, algorithms (particularly for data fusion and for "active vision" in multiple modalities), and processing capabilities, are essential. The Army can improve individual sensor capabilities and algorithms, but a big problem that has been largely unacknowledged is optimizing the perception system hardware and software architecture: sensors, embedded processors, coded algorithms, and communication buses. The perception subsystem thus presents a very complicated system-engineering problem that is exacerbated by having work carried out by separate organizations in separate programs. There is currently no way to know how perception performance is reduced by suboptimized architecture or where improvements might be made.

Other priorities for earliest attention vary with capability class. For teleoperated ground vehicles (TGVs), human–robot interaction, health maintenance, communications, and power/energy technologies assume major prominence. Current robots rely on teleoperation using PC user interfaces, such as keyboards and touch screens that are demanding to operate and have not been validated by human factors experts. Techniques to augment external navigation controls with algorithms for real-time mapping and localization for outdoor missions would reduce the stress on operators and enable a single operator per robot. Current TGV systems also require many more technicians for repair and preventive maintenance than do manned systems. Future TGVs (and UGVs in all classes) must be able to self-monitor and to provide information to remote locations for diagnosis and possible recovery.

Such vehicles should be designed with behaviors and characteristics that facilitate their own survivability.

For the semiautonomous preceder/follower (SAP/F-UGV) class of vehicles, mobility, navigation, tactical behaviors, and health maintenance technologies are all high priorities. Successful integration of navigation technologies with an all-terrain mobility platform could enable preceder/follower UGVs to serve not only as logistics carriers but also as lead elements for small-unit patrols or for soldier-portable vehicles for use on security outposts. These UGVs must be operationally reliable to a degree many times greater than manned vehicles. Depending on application, basic tactical behaviors will be required to ensure that SAP/F-UGVs can perform missions without becoming a burden on the battlefield.

Priority technologies for platform-centric autonomous ground vehicles (PC-AGVs) include those for the SAP/F-UGV class (mobility, navigation, tactical behaviors, and health maintenance) plus learning/adaptation technologies. To be useful for extended-duration missions, PC-AGVs must be capable of adapting embedded tactical behaviors to changing situations without requiring reprogramming in the field. Ideally, lessons learned would be cumulative and could be transferred to other PC-AGV systems.

Communications, including mobile self-configuring networks and distributed knowledge bases, become all-important for the network-centric AGV class. To respond to multiple demands, NC-AGVs must be tightly networked with other FCS elements and information systems on the battlefield. Other priority technologies include mobility (to provide versatile, multifunction platforms), human–robot interaction (to ensure proper task allocation between soldier–robot and robot–robot), and learning/adaptation (to expand the range of autonomous behaviors).

Recommendation 1. The Army should give top priority to the development of perception technologies to achieve autonomous mobility. In addition, it should focus on specific technologies depending on unmanned ground vehicle capability class.

Focus on Compelling Army Applications

The capabilities of UGVs should complement what humans can do better, with the aim of maximizing the additional benefits that can be gained by introducing UGVs. Each UGV class should be specified and designed to do what robots can do better (or at lower risk) than humans, rather than trying to imitate what humans already do very well.

Until requirements are validated in the Army user community, there can be no commitment to UGV systems and applications. The existing statements of requirements are insufficient to guide and stimulate technological evolution. This has forced UGV development into a technology-push mode, rather than a balance between technology-push and requirements-pull modes. For UGV systems to be a major factor in the Objective Force, a process of spiral development involving the user will be necessary, including successive iterations of application and capability refinements. Technologies that merit special development attention can then be identified and developed using a "skunk-works" approach that achieves the focus and centralized leadership necessary to reach goals set by the user community.

Such a "skunk-works" approach would consolidate and focus the development of technologies essential to FCS UGVs under a single manager, eliminate duplication of effort, and provide the basis for standardized research platforms to be used in the spiral development of UGV systems. Prototypes resulting from targeted technology development and integration can be used for higher-level developments or for experiments involving particular mission-package applications by the user. Such technology-integrating experiments will help users determine which concepts have the most value and facilitate the development and acceptance of UGV systems.

For the above reasons, the committee believes that the Army UGV program is best served by developing a small number of experimental vehicle types capable of applications with compelling value for FCS and the Objective Force. The aim would be to develop and integrate the technologies required for several classes of vehicle capability. The development process will be best served by systematic and extensive testing and refinement under severe operating conditions. Army mission needs and operational requirements for UGV systems can evolve in a spiral development process as a technology integration program advances, provided the program is focused on maturing the underlying technologies and achieving system integration of those technologies at several useful levels of vehicle capability.

Focusing on a few specific applications for the experimental prototypes, some of which may be simulated, is essential to maturing the needed technologies and resolving the significant issues of system integration. The focus on applications organizes the capabilities development effort into manageable components, each with a clear operational outcome to be achieved. While capabilities may mature at different rates, the program as a whole would address technical challenges of all applications concurrently. The application prototypes should be selected to develop capabilities needed for FCS and the Objective Force. The roadmaps developed as part of the study were built around four such applications, but they illustrate only one of many possible combinations for evolutionary development.

Recommendation 2. The Army should adopt a "skunk-works" approach to develop technologies necessary for autonomous A-to-B mobility, so that such capabilities can be fielded with a small number of unmanned ground vehicle (UGV) classes, each of which is an experimental prototype for a compelling military application. TRADOC and the research and development community should commit to a spiral development process for refining and evolving concept-based requirements for

UGVs, depending on what is learned from these technology-integrating prototypes.

Systems Engineering Challenge

Even when all underlying technologies for a UGV application have reached TRL 6, a great deal of work will be required for integrating specific technologies into one or more UGV systems capable of accomplishing FCS missions. In fact, the committee concluded that the greatest technical challenge for fielding UGVs of significant value to FCS and the Objective Force is likely to be technology integration and systemization. Adequate time must be allowed for the technologies that are developed to be combined and tested in the field in ways that give feedback to the developer and the user communities on how to improve a given concept.

The user and developer communities must work together to provide direction for the technology integration to implement vehicle experiments. These directions should feed into the spiral development process from experimental prototypes to requirements-based systems following the established development process. For example, application parameters must be formulated to address the integration of the mission-package technologies, mobility technologies, and communications technologies that are necessary for each experimental prototype.

Lack of system engineering will hinder development of acceptable UGV systems. Performance metrics and other assessment methodologies must be established that provide objective feedback to developers and users on how well an application-oriented experimental prototype is performing as an integrated system. Supporting technologies for HRI, mobility, power, communications, and vehicle health maintenance, while not part of the autonomous behavior architecture, will nonetheless be critical to UGV system developments. Software quality must also be regarded as an important issue, requiring extensive software engineering, re-implementation, and performance assessments to field a given system. Technology development in all areas will be heavily dependent on system prototypes.

Recommendation 3. The Army should begin planning for unmanned ground vehicle (UGV) system development now. Systems-engineering processes should be used to inform and guide the development of UGV operational concepts and technology.

Advocate for UGV Development

The Army's UGV program must cut across existing program "stovepipes" and increase dedicated resources. If the objective is to field a network-centric autonomous ground vehicle for the Objective Force, then the Army must dedicate resources now to basic and advanced research and development with a common focus on achieving this end. A strong central advocate is needed.

Experience has shown that the Army responds well to challenges that are represented by high-level positions or organizations dedicated to a single purpose. The Army Digitization Office, established in the 1990s by the Army Chief of Staff, provides a good example of how such focus can be used to move a project forward that might otherwise become lost in the bureaucracy. Similarly, special Army selection boards exist to select highly qualified personnel for designation as program managers (PMs) for technology and system developments of high-level importance.

Although UGV system concepts and requirements are not sufficiently advanced to merit the same approach at this time, extraordinary measures analogous to the Digitization Office initiative should be considered as the UGV program matures beyond the S&T stage. In the interim, a board-selected PM for UGV technology and system developments would be able both to serve as an advocate for autonomous systems and to focus development effort on achieving A-to-B mobility capabilities and developing experimental prototypes, thereby advancing the experimentation for and acceptance of UGV systems. This new position would contrast with the present program manager positions (for FCS and the Objective Force), which are focused on objectives that can be achieved with or without a dollar of investment in underlying UGV technology. The new position would not duplicate the functions of the DOD UGV PM position, which is focused on integrating UGV systems using existing technologies in response to specific, DOD-endorsed requirements.

Recommendation 4a. The Army should designate a Program Manager for Unmanned Ground Vehicles (PM-UGV) to coordinate research, development, and acquisition of Army UGV systems. The PM-UGV would act for the Assistant Secretary of the Army (Acquisition, Logistics, and Technology) to manage Army UGV technology developments, approve technology base planning, provide acquisition guidance, and oversee resource allocation. The PM would be the Army's principal advocate for unmanned ground systems and single point of contact for UGV developments with the Joint Program Office, the Defense Advanced Research Projects Agency, and other agencies.

Recommendation 4b. As the unmanned ground vehicle (UGV) program matures beyond the S&T stage, the Army should consider additional extraordinary measures, analogous to the successful Army digitization initiatives, to ensure sufficient focus on developing and fielding UGV systems for the Future Combat Systems and the Objective Force.

1

Introduction

Robots, including unmanned ground vehicles (UGVs), have many valuable attributes that will aid and complement soldiers on the battlefield. They are well suited to perform routine and boring tasks. They are fearless and tireless. They do repetitive tasks with speed and precision. They can be designed to avoid or withstand enemy armaments and to perform specific military functions. Most importantly, robots can reduce casualties by increasing the combat effectiveness of soldiers on the battlefield. The scenario in Box 1-1 illustrates many of these advantages.

The Army has recognized the potential of robotics for well over 20 years. Capitalizing on early work by the Defense Advanced Research Projects Agency (DARPA), the Army has sponsored basic and advanced research in intelligent systems and led developments in crew automation technology and in tactical unmanned air vehicles (UAVs). At the same time, it has successfully adapted commercial teleoperated ground vehicles for specialized military uses such as mine clearing, and it has made initial progress toward developing semiautonomous ground systems for combat. Appendices C and D provide detailed descriptions of much of this early work.

Since the Gulf War, an urgent need has surfaced to transform the Army from one characterized by heavy armor and firepower into a lighter, more responsive force that is at once more lethal and survivable. Concepts for the Army's Future Combat Systems (FCS) include unmanned systems, both ground and air, and will be required for fielding with other elements of the FCS as early as 2010. This report documents a study to assess the readiness of UGV technologies to support the development of the Army's Future Combat Systems. This first chapter provides background information, including the statement of task, study approach, and report organization.

BACKGROUND

The impetus for the study came from increasing awareness that shortcomings in robotic research and development have a potential to disrupt the ambitious schedule for design and acquisition of the FCS. The FCS program, now in early conceptual design phase, will play a central role in providing the combat systems that enable a "transformation" of the present-day Army into a future Objective Force. Figure 1-1 illustrates the three-pronged thrust of the transformation campaign and how the Objective Force is dependent upon timely research and development of new system concepts.

The Army desires that UGVs be utilized as part of the FCS (TRADOC, 2001a). Early FCS concepts considered UGVs that would serve in such roles as logistics carriers, remote weapons platforms, soldier companions, or surrogates for reconnaissance, surveillance, and target acquisition (RSTA). Recognizing their potential impact, Congress has mandated that at least one of every three future Army combat systems be unmanned (Congress, 2000). So, UGVs will definitely be part of the future Army, but the question is "When will requisite UGV technologies be in place to support the development of UGV systems?"

Statement of Task

The Assistant Secretary of the Army (Acquisition, Logistics, and Technology) asked the National Research Council to conduct a study to examine the overall Army program for unmanned ground vehicle (UGV) technology, with attention to the following tasks:

1. Review Army operational requirements for UGVs, including the Army Future Combat Systems (FCS)

> **BOX 1-1**
> **A Glimpse of the Future**
>
> The United States and its allies have enjoyed notable success in the long campaign against worldwide terrorism. A significant contributor to this success has been the evolution of manned ground and airborne forces into integrated teams of unmanned ground and airborne vehicles and highly trained manned forces. As a result unmanned ground vehicles have evolved from logistics support, rear guard activities, and simple scout missions to full membership in integrated combat teams. These teams live, train, and deploy together and the combination of manned and unmanned ground and airborne elements has significantly improved the survivability of manned forces. A key to success in achieving these goals was a definitive set of Army requirements and mission needs statements that were driven and guided by the Training and Doctrine Command (TRADOC) concept guidelines.
>
> In 2025 these integrated forces, largely made up of globally deployable Hunter-Killer teams, are supported by a superb sensor, information processing, and communications infrastructure. Unmanned and airborne and satellite assets provide birth-to-death tracking of adversaries for tactical forces anywhere on the globe. This capability combined with global maps achieving 1-meter resolution is used to pinpoint geographic locations of adversary forces, logistics, and equipment. Information from integrated, multiband sensor suites aboard airborne and some ground vehicles is processed and fused into a common operating picture of the battlefield. The capability to support this information sharing/processing is provided by a wireless network of unmanned airborne and ground nodes using LPI communications and redundant communications links. Individual combatants, both human and machine, are linked to each other, and to their commanders by this information system. Ground commanders are able to direct unmanned airborne and ground weapons and sensor systems in real time in support of their operations.
>
> The underlying technology to support this integrated force structure has been based on a system architecture designed to accommodate interoperation and evolution through rapid prototyping and subsequent field testing. A critical element in achieving this goal is the superior mobility and survivability of the unmanned ground vehicles. This was made possible through advanced combinations of wheeled/tracked technology and direct drive electric technology and such highly efficient electric power generation systems as fuel cells and energy recovery systems. The heretofore missing element of cooperative behavior has been enabled by significant improvement in cognitive systems ranging from simple self diagnosis and repair to planning/adaptation and recognition/understanding of targets and the surrounding environment and subsequently to reasoning and decision capabilities. This cognitive capability has also been used to implement the supporting information networks that now can not only carry out fault management as in earlier times but also repair and reconfigure the network based on policy-driven feedback. Such systems are also capable of incorporating new behaviors and learning from peer elements.
>
> Modeling and simulation have played a key role in the evolution of the integrated force concept largely by exploiting training, simulation and entertainment industry technologies. This has enabled realistic training of soldier and machine and led the evolution of tactics needed for advanced combat systems. Onboard computing has allowed the training and refinements to extend beyond garrison and field exercises to actual deployments to remote locations around the globe. The resultant feedback to system designers and planners has accelerated the rate of improvements.
>
> In June 2025 intelligence networks determined that a major initiative by hostile forces would take place in Central Asia aimed at toppling a government friendly to U.S. interests. In response to this threat a clandestine joint Hunter-Killer team was sent on a mission to block infiltration by a terrorist group through remote highland desert. As part of this mission, a squad of 10 networked robotic units was sent to guard a strategic mountain pass 150 km from the base camp and to prepare an ambush for any forces trying to penetrate through this pass. The ordnance available for the mission ranged from long-range rockets to machine guns and armor-piercing projectiles. The units had been on station for four days before the small advance scout units, which were carried to the ambush site by the large units, signaled the approach of a lightly armored camouflaged force with significant infantry in attendance. By means of a sophisticated communications scheme humans and robots made the decision to engage. Unmanned air vehicles were launched from the squad at the beginning of hostilities to provide targeting information and battle damage assessment. The resultant battle totally decimated the attacking force and prevented infiltration by a large force into the territory of a U.S. ally. Four of the team members sustained minor damage from small-arms fire before completely neutralizing the infiltrating force. In the process Unit 10, a rocket platform, was damaged, rendering it unable to return to the staging area. A fellow team member took the damaged unit in tow and successfully returned it to friendly territory for repair. Increasingly, robotic team members are "taking the point" and keeping soldiers out of harm's way.

baseline program, the Army Research Laboratory (ARL) UGV science and technology objective (STO), and other UGV requirements.

2. Review the current Army UGV efforts at ARL and the Army Tank-Automotive Research, Development and Engineering Center (TARDEC) and respond to the following questions:

2.a. Will the Follower UGV Advanced Technology Demonstration (ATD) lead to a capability that will meet stated Army operational requirements in time to be integrated into the baseline FCS development program?

2.b. Will the ARL STO program result in significant advances in UGV autonomy beyond that achieved

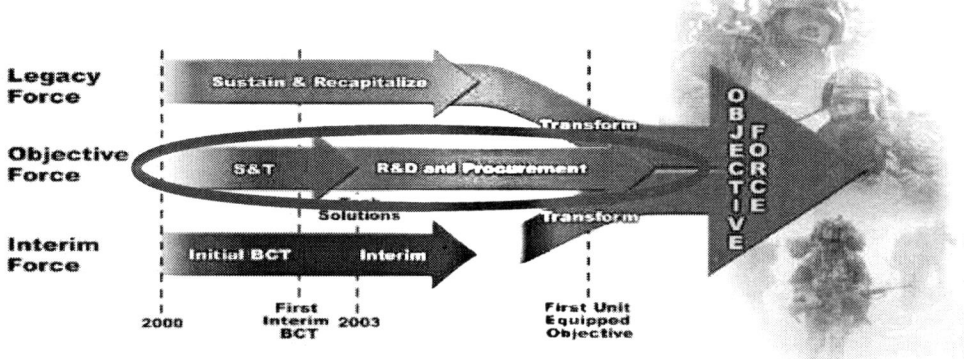

- Legacy Force—Today's Army
- Interim Force—Our bridge to the future
- The Objective Force—The Army's full-spectrum, decisive ground combat force

FIGURE 1-1 Army transformation to the Objective Force. SOURCE: Andrews (2001).

in the Follower UGV ATD (advanced technology demonstration), and what operational requirements would the resulting capability be able to address?

2.c. How do the Army UGV efforts interrelate with other government ground robotics initiatives (e.g., National Aeronautics and Space Administration [NASA] rovers, Department of Energy [DOE] programs, Defense Advanced Research Projects Agency [DARPA])?

3. Review the state of the art in unmanned vehicle technologies applicable to UGV systems (e.g., "intelligent" perception and control, adaptive tactical behaviors, human–system interfaces) and respond to the following questions:

    3.a. What technologies should next be pursued, and in what priority, to achieve a UGV capability exceeding that envisioned in the ARL STO?

    3.b. Are all the necessary technical components of a UGV technology program identified and in place, or if not, what is missing?

    3.c. Are there foreign UGV technology applications that are significantly more developed than those of the U.S. that, if acquired by the U.S. government or industry through cooperative venture, license, or sale, could positively affect the development process or schedule for Army UGV systems?

    3.d. What technology areas merit further investigation by the Army in the application of UGV technologies in 2015 or beyond?

4. Identify issues relating to technical risks and the feasibility of implementing applicable UGV technologies within the FCS baseline program time frame and respond to the following questions:

    4.a. What are the salient uncertainties in the "intelligent" perception and control components of the UGV technology program, and are the uncertainties technical, schedule related or bound by resource limitations as a result of the technical nature of the task, to the extent it is possible to enunciate them?

    4.b. What are the salient uncertainties for the other main technology components of the UGV technology program (e.g., adaptive tactical behaviors, human–system interfaces, mobility, communications)?

    4.c. Do the present efforts provide a sound technical foundation for a UGV program that could meet Army operational requirements as presently defined?

5. Document the results of the examination in a study report that will be provided to the Army. The report will contain a recommended roadmap for the development of UGV technology and systems, including topics that could be the subjects of investigations of longer-term (2015 and beyond) UGV technology applications. It will also respond to the following questions:

    5.a. From an engineering perspective, what are reasonable milestone dates for a UGV system development program leading to production? For example, does the current FCS program have a coherent plan and roadmap to build UGVs for FCS and the Objective Force?

    5.b. What can be recommended on the technical content, timelines and milestones based on these assessments?

    5.c. Are there implications for Army support infrastructure for a UGV system? For example, will other technologies need to be developed in parallel to support a UGV system, and are those likely to pose significant barriers to eventual success in demonstrating the UGV concept or in fielding a viable system?

Answers to the questions in the latter four tasks are highlighted in text boxes at appropriate points in the report, and a list of these text boxes with page numbers may be found in the front pages. The study was not requested to focus on unmanned aerial vehicles (UAVs), but these systems are very likely to be included in the FCS. While focusing on UGVs, the study does evaluate technology areas that are common to both air and ground autonomous systems.

Study Approach

The starting point for the study was a meeting of the NRC Board on Army Science and Technology (BAST) convened at the request of the Assistant Secretary of the Army (Acquisitions, Logistics, and Technology) (ASA [ALT]) in June 2001. This meeting discussed the statement of task for the study and determined the expertise that the study committee would need.

A primary source document for the study was the Army Science and Technology Master Plan (ASA [ALT], 2001a), which describes the Army science and technology program, and the UGV technology roadmap developed by the Office of the ASA (ALT) (ASA [ALT], 2001b). The latter document includes an evaluation of UGV technologies based on capabilities for UGVs as determined by the Army's Training and Doctrine Command (TRADOC) in 2000, and it provided a baseline for the committee's technology assessments. Another notable source was an Army Science Board (ASB) study that summarized representative force capabilities appropriate for unmanned ground vehicles (ASB, 2001).

Lacking an approved requirements document for the FCS, or for any specific UGV systems that might be included in the Objective Force, the committee hypothesized four compelling examples of likely UGV systems (described in Chapter 2) consistent with the FCS mission-needs statement (TRADOC, 2001b) and the statement of required operational capabilities for the Objective Force (TRADOC, 2001a). These examples were essential to provide points of reference for the application of standard Army criteria for technology readiness levels.

REPORT ORGANIZATION

This report documents the findings and recommendations of the study. It provides specific technology development objectives and a science and technology (S&T) roadmap to guide the Army as it includes relevant UGV technologies in its Objective Force systems. The report is organized in chapters as follows.

Chapter 1 (Introduction) provides background information and the statement of task for the study. Chapter 2 (Operational and Technical Requirements) summarizes documented and undocumented requirements for UGV technologies and provides examples. Chapter 3 (Review of Current UGV Efforts) describes the various programs and demonstrations that comprise the Army's overall UGV development program and other important UGV development activities. Chapter 4 (Autonomous Behavior Technologies) summarizes the state of the art in core autonomous behavior technology areas and provides technology readiness level criteria and estimates for each area. Chapter 5 (Supporting Technologies) summarizes the state of the art in supporting technology areas essential to UGV systems with technology readiness level (TRL) estimates. Chapter 6 (Technology Integration) describes system-level considerations important to successful UGV implementation in the FCS. Chapter 7 (Roadmaps to the Future) summarizes capability gaps for example systems and provides technology roadmaps based on the TRL estimates, both to achieve the technologies required for systems similar to the examples and to provide a basis for future Army S&T planning. Finally, Chapter 8 (Findings and Recommendations) explains the study findings and recommendations.

2

Operational and Technical Requirements

Historically, the acquisition of military systems has been driven by two activities: development of new technologies that apply to military missions (technology push); and definition of new battlefield requirements by operational users (requirements pull). For today's transformational Future Combat Systems (FCS), both technology push and requirements pull are being synergized by the Army, which is considering unmanned vehicles as a key enabler of increases in force effectiveness, protection, and economy.

In the case of unmanned ground systems in particular, there has been significant technology push over the past several decades. As early as World War II, unmanned platforms were experimentally evaluated for such missions as minefield breaching. By the early 1980s, robotic systems to support space exploration and unmanned aerial vehicles were also being developed. By the 1990s several programs within the Defense Advanced Research Projects Agency (DARPA) and the Unmanned Ground Vehicle (UGV) Systems Joint Program Office were developing technologies to support unmanned ground vehicles.

In planning military operations the warfighter considers hardware capabilities integrated with doctrine, force structure, and training. The Training and Doctrine Command (TRADOC) centers and battle laboratories study the battlefield impact of new systems and concepts and translate user insights into operational architectures and requirements. Because many operations performed by unmanned systems are the same as those provided by existing forces, the transformational nature of these systems has often been ignored. Although a number of unmanned system concepts have been explored, there have been few requirements defined or doctrinal issues evaluated. Much needs to be done by the warfighter before a reasonable understanding of the impact of unmanned systems on both FCS and the Objective Force can be assessed.

Because the committee could find no specific operational requirements for UGVs in support of FCS, it became enormously difficult to establish a logical technology evaluation. A virtually unlimited number of concepts exist for unmanned systems, ranging from mechanical insects to robotic trains, and practically all have at least some military potential. The committee reasoned that UGVs for FCS would have to be ground mobile and have other compelling capabilities to be desired as part of the FCS, and decided to define four capability classes (teleoperated, semiautonomous, platform-centric autonomous, and network-centric autonomous). It then defined a set of notional capabilities to use as compelling examples of UGV systems. These postulated examples were essential to guide the evaluation of UGV technologies and to focus technical issues. By articulating the details on specific required capabilities, the examples provided "marks on the wall" that enabled the committee to assess applicable technology readiness levels (TRLs).

OPERATIONAL REQUIREMENTS

The Army requirements determination system has evolved over the last several years with the increased influence of joint operations, the publishing of a Joint Vision (Joint Vision 2020) by the CJCS and establishment of the Joint Forces Command (JFCOM), responsible for developing joint concepts and recommending joint requirements. The TRADOC commander is responsible for developing and publishing the Army Capstone Concept (TRADOC Pam 525-5), which becomes the guide for all other concept development activities. Integrating and supporting concepts are developed by TRADOC centers and schools and are published in a series of TRADOC pamphlets. The Future Operational Capabilities (FOC) document published in TRADOC Pamphlet 525-66 is a statement of the required operational capability needed by the Army and is intended to help the Army Science and Technology activities as well as industry research and development initiatives. A review of this document shows few references to unmanned ground systems, and those are mostly confined to intelligence, logistics, explosive ordnance detection, and mine clearing.

The FOC document includes the "autonomous unmanned capability to achieve total situational awareness (on the ground or in the air), evaluate data received, develop courses of action consistent with the commander's intent, and employ combat power (lethal and non-lethal "smart" munitions) to achieve the commander's objectives. This "economy of force" element will control terrain, reduce the risk to soldiers in certain areas, and complement and maintain maneuver dominance at the strategic, operational, and tactical levels. Additionally, this capability will substantially enhance peacemaking and peacekeeping operations" (TRADOC, 1997).

On November 21, 2001, a formal request for proposal was released for a lead system integrator (LSI) for FCS, which included a Draft Mission Needs Statement and a Statement of Required Capabilities (SORC). These documents have provided some clarity to the definition of capabilities required for UGVs as part of the FCS. During a two-day seminar on the Objective Force and FCS in November 2001, the Army defined threshold-level capabilities for the FCS (TRADOC, 2001a,b) to include

- Manned and unmanned ground, air, and space means to extend vision beyond line of sight to gain timely combat information through passive and aggressive RSTA (reconnaissance, surveillance, and target acquisition) networked into an integrated common operational picture (COP) for unprecedented situational awareness and understanding.
- Integrated synergistic use of joint and Army manned and unmanned, air and ground RSTA to gain and maintain contact with enemy elements and to provide high-resolution combat information on terrain and weather.
- Robots to perform manpower intensive, high-risk functions, such as RSTA missions in urban operations (inside buildings and the subterranean dimension) and reconnaissance/reduction of minefields.
- Revolutionary means of transporting and sustaining people and materiel to leverage new ground and aerial concepts for delivery, including manned and unmanned systems.
- Mule-like robotic capability to perform a variety of sustainment/replenishment functions on a highly agile, light but survivable platform to include:
  —carrying dismounted soldier loads
  —operating in terrain requiring dismounted operations
  —performing non-standard Casualty Evacuation and other services, such as battery recharging
  —delivering classes of supply from battalion through company to the soldier to include resupply of ammunition
  —performing combat tasks such as reconnaissance of high-risk areas.

The use of unmanned aerial vehicles in the current war in Afghanistan is a prime example of user confidence in unmanned systems. The use of UAVs in military operations has been studied and tested for over 40 years. Over the last 20 years the use of UAVs in niche roles (predominantly reconnaissance) has driven the development of systems as well as requirements. The recent successful demonstration of UAVs in a lethal role in a combat situation has given the user the confidence necessary to push future development of the technology. This same level of confidence must be developed in the ability of unmanned ground vehicles in order for a leap-ahead to occur.

Assessment of FCS Operational Requirements

The Army is at a critical stage in the development of the FCS. The LSI has been selected. Over the next 3 years, the designs for the threshold version of the FCS will be determined, with requisite technologies brought to the prototype demonstration level. The prototype FCS demonstrator is intended to be capable of performing all desired functional requirements described in the FCS mission needs statement (MNS).

An Objective Force Task Force, reporting to senior Army leadership, has been created to accelerate the acquisition and deployment of FCS and other Objective Force systems. Through coordination and assessments the Objective Force Task Force has been tasked to expedite FCS-related efforts in the concepts, requirements, S&T, and acquisition communities. The task force is responsible for FCS design and preparation for the technology readiness decisions in FY2003. In FY2004, the task force will focus on achieving success of the FCS demonstrator phase and support a transition to SDD in FY2006. Key tasks are to develop and maintain the FCS campaign plan, synchronizing the plan with the Army transformation campaign plan and ensuring FCS integration into the Objective Force.

The LSI for FCS is tasked to assist the Training and Doctrine Command in developing the requirements documentation. The LSI should work closely with the TRADOC battle laboratories to develop operational architectures based on the technology readiness of the many technologies included in FCS. Missions currently envisioned to be performed by unmanned ground vehicles include reconnaissance and surveillance, rescue, eavesdropping, and mapping. Unmanned ground vehicles for lethal missions, including both direct and indirect fires, have been discussed, but that prospect clearly gives some military officials cause for concern. However, to be truly transformational, unmanned ground vehicles need to be at the forefront of the FCS program.

At the time of the study, neither the Objective Force Task Force nor the TRADOC had been successful in laying out a viable plan for integrating evolving UGV technologies and capabilities into the FCS roadmap. In the absence of definitive requirements, the acquisition community has struggled to identify where to place the emphasis for technology development. In April 2002, as one of its first actions the LSI for FCS requested industry proposals for 43 different technologies for FCS, including three UGV systems:

1. Soldier UGV—a small soldier-portable reconnaissance and surveillance robot
2. Mule UGV—a 1-ton vehicle suitable for an RSTA or transport/supply mission
3. Armed Reconnaissance Vehicle (ARV) UGV—a 6-ton vehicle to perform the RSTA mission, as well as a fire mission, carrying a turret with missile and gun systems.

While the FCS program may lack system requirements, there is general agreement that autonomous mobility from point A to point B, known as "A-to-B mobility," is a key if not *the* key enabler for future UGV systems. Requirements for autonomous mobility have consisted of describing speed of maneuver over differing classes of terrain. They do not include descriptions of which tactical behaviors may be required to perform mission functions in combat, including such actions as:

- Terrain reasoning—The ability to use information about natural terrain features (elevation, vegetation, rocks, water), manmade features (roads, buildings, bridges), obstacles (mines, barriers), and weather
- Military maneuver—Using terrain reasoning, mission, friendly and enemy locations to determine the best maneuver and selection of positions for stealth and to support mission package needs (e.g., hull down for direct fire, clear of overhead obstructions for indirect fire)
- Agility—Using rapid, significant changes in speed and direction to reduce an enemy's ability to acquire and hit a UGV
- Self protection—Sensing threats (e.g., mines, weapon systems, humans) in sufficient time for the UGV to avoid them; using onboard weapons systems or command and control (C2) links to friendly weapons systems to neutralize an enemy.

For UGVs to act as part of a system of systems, dynamic, extended range, redundant, and networked communications are essential. The FCS statement of required capabilities describes communications that are

- Highly integrated, self-organizing, ubiquitous, distributed, extendable, and capable of increased yet scalable data rates
- Open, multilayered with multiple paths that provide redundancy for assured communications, with voice and data routing around inoperative nodes without interruption
- Using platforms as integrated nodes that do not rely on stationary attended nodes.

TECHNICAL REQUIREMENTS FOR UGV CAPABILITIES

The category of ground-traversing vehicles without a human operator onboard covers a broad range of mission capabilities and degrees of autonomy with respect to command and tasking functions, terrain reasoning, military maneuvering, and mobility design. For this reason and to facilitate its analysis the committee characterized four generic classifications of UGV capabilities based on relevance to potential Army missions and level of autonomy and the challenge required to implement.

The classes are described, with a specific example system for each, in the sections that follow. Table 2-1 lists the UGV capability classes with potential mission function applications.

Each of the four classes (teleoperated, semiautonomous, platform-centric autonomous, and network-centric autonomous) varies in its need for different UGV technologies. For example, the dependence on technology in the communications area varies as follows:

- Teleoperated ground vehicle (TGV): high requirement at all times
- Semiautonomous preceder/follower (SAP/F-UGV): mostly moderate requirement (placing "breadcrumbs"), except when it moves off course or when a crisis situation (e.g., minefield, enemy attack) arises
- Platform-centric autonomous ground vehicle (PC-AGV): little need for human control, minimal connectivity requirements while executing its mission
- Network-centric autonomous ground vehicle (NC-AGV): little need for human control, high need for network connectivity.

Table 2-2 summarizes the committee's assessment of the relative dependence of relevant technology areas to each of the defined UGV classes. As can be seen, differences also exist in the other applicable technology areas including perception, navigation, planning, behaviors and skills, learning/adaptation, human–robot interaction, mobility, power/energy, and health maintenance.

UGV CONFIGURATIONS

The four capability classes categorize UGVs in order of increasingly complex military applications. For each class the committee developed an example military application to provide a "mark on the wall" against which to measure tech-

TABLE 2-1 UGV Capability Classes, Example Systems, and Potential Mission Function Applications

Example System	Capability Class	Other Possible Applications
Small robotic building and tunnel searcher ("Searcher")	Teleoperated ground vehicle	Mine detection, mine clearing, engineer construction, EOD/UXO, materials handling, soldier-portable reconnaissance/surveillance
Small-unit logistics mover ("Donkey")	Semiautonomous preceder/follower	Supply convoy, medical evacuation, smoke laying, indirect fire, reconnaissance/surveillance, physical security
Unmanned wingman ground vehicle ("Wingman")	Platform-centric autonomous ground vehicle	Remote sensor, counter-sniper, counter-reconnaissance/infiltration, indirect fire, single outpost/scout, chemical/biological agent detection, battle damage assessment
Autonomous hunter-killer team ("Hunter-Killer")	Network-centric autonomous ground vehicle	Deep RSTA, combined arms (lethal direct fire/reconnaissance/indirect fire for small unit defense or offense), static area defense, MOUT reconnaissance

EOD/UXO = explosive ordnance disposal/unexploded ordnance; RSTA = reconnaissance, surveillance, and target acquisition; MOUT = military operations in urban terrain.

nology maturity levels in each of the relevant technology areas. It is emphasized that these examples are not and should not be interpreted as recommended Army operational requirements. They are intended to illustrate the interplay of the applicable UGV technologies, as well as to show when the levels of technology could be developed to achieve reasonable military capabilities.

In the following sections each example application is described in terms of an overarching concept, operational approach, basic capabilities, and UGV-human interface. These descriptions will be the basis for subsequent analysis to determine when the various technologies will be sufficiently robust to support a system development (i.e., reach Technology Readiness Level 6 [TRL 6]). See definitions for Technology Readiness Levels in Table 4-1 of Chapter 4.

Teleoperated Ground Vehicles

In teleoperation a human operator controls a robotic vehicle from a distance. The connotation of teleoperation is that the distance is or can be great enough that the operator cannot see directly what the vehicle is doing. Therefore, the operator's information about the vehicle's environment and its motion in that environment depends critically on sensors that acquire information about the remote location, the display technology for allowing the operator to visualize the vehicle's environment, and the communication link between the vehicle and the operator. The operator controls the actions of the vehicle through a control interface (Murphy, 2000).

For the purposes of this report, the operator of a teleoperated ground vehicle (TGV) is assumed to be responsible for the majority of the command and tasking functions for the vehicle and its mission package. Control is similar to piloting a UAV. A TGV has no onboard terrain reasoning or military maneuvering capability, nor does it access this information from any other source. The operator conducts all cognitive processes. The sensors onboard the vehicle and the communications link allow the operator to visualize the

TABLE 2-2 Relative Dependence of Technology Areas for Each UGV Class

Technology Area	Need/Relevance			
	TGV	SAP/F-UGV	PC-AGV	NC-AGV
Perception				
For A-to-B mobility	2	3	5	4
For situation awareness	$0(2^a)$	$0(3^{a,b})$	5	5
Navigation	3	5	5	5
Planning				
For path	$0(2^b)$	3	5	5
For mission	1	1	4	5
Behaviors and skills				
Tactical skills	$1(2^b)$	$1(2^b)$	4	5
Cooperative robots	1	3	5	5
Learning/adaptation	$1(2^b)$	3	3	3
Human–robot interaction	5	2	4	4
Mobility	5	5	5	5
Communications	5	3	3	5
Power/energy	5	5	5	5
Health maintenance	1	3	5	5

TGV = teleoperated ground vehicle, SAP/F-UGV = semiautonomous preceder/follower ground vehicle, PC-AGV = platform-centric autonomous ground vehicle, NC-AGV = network-centric autonomous ground vehicle.
[a]Needed during crisis.
[b]Needed during lack of communication with operator.

Key to Ratings

0 = no need   3 = average need
1 = low need   4 = above average need
2 = below average need   5 = high need

UGVs location and movement within its environment through information displays, which typically include:

- Screen display(s) of the TGV's location using a geolocation system
- Images of the vehicle's environment based on data transmitted through the communications link from sensors onboard the TGV.

The operator may also have direct line-of-sight observation of the TGV, either with the unaided eye or with optical devices. Missions appropriate for TGV capabilities include minefield breaching, mine and ordnance clearing, tunnel reconnaissance, and some military operations in urban terrain (MOUT). TGVs come in all sizes.

Example 1: Ground Vehicle Building and Tunnel Searcher ("Searcher")

Overarching Concept. The world is becoming increasingly urbanized. The Army will find more and more situations where enemy forces are attacking U.S. forces from urban or rural buildings, tunnels, culverts, caves, and similar confined areas. Dismounted troops will have to clear areas such as these to regain control of the terrain. These close, cramped areas favor the defender and can cause extremely dangerous situations for soldiers tasked with the clearing mission. A teleoperated Ground Vehicle Building and Tunnel Searcher ("Searcher") could be of significant assistance in accomplishing these tasks. The Searcher would be a small ground vehicle that would carry high-resolution sensors and other lightweight payloads. The soldier-operator would be equipped with visual display and control devices such as a joystick and touch pad. Communications between the soldier and the ground vehicle would be by radio or wire as the situation dictated.

Operational Approach. The Searcher would be small and light so that the ground vehicle and all associated payloads could be carried by a single dismounted soldier. The self-contained power supply would be sufficient for the Searcher to climb stairs and search all hallways and rooms in a typical 10-story urban building. It would also be able to enter and search tunnel, culvert, or cave complexes (out to 1 km and back) that are capable of being traversed by a small adult human. It would be capable of automatically righting itself in the event of a rollover. The Searcher would travel at variable speeds on all surfaces, up to the speed of a running soldier on flat surfaces. The Searcher would be weather resistant, so as to be able to operate in locations exposed to the weather or in buildings in which fire-fighting sprinkler systems have been activated.

Basic Capabilities. At the most basic level, the teleoperated Searcher's every move would be controlled by an operator.

The basic payload would be a package consisting of any mix of infrared (IR), visual, acoustic, or other sensors. The sensor input would be transmitted to a display held by the operator. The display would provide high-level resolution for the operator to quickly identify humans, weapons, booby traps, supplies, and obstacles. At a minimum the Searcher would have an arm and manipulator that would allow it to open unlocked doors, safely detonate booby traps, and move small objects. Additionally, the Searcher would be able to mark areas or rooms that have been searched and deemed clear at the time of the search. The Searcher would be capable of carrying non-lethal payloads that could be detonated by the operator as necessary. The Searcher, upon being directed to do so, would be able to project limited synthetic voice commands either in English or in the appropriate foreign language. If communications were lost with the operator the Searcher would go into a fail-safe mode.

UGV-Human Interface. The Searcher's "level of initiative" is that it would normally wait until it is told what to do by its operator.[1] The amount of control required from the operator would be continuous. Wire or radio frequency (RF) would allow real-time communications between the operator and Searcher. The Searcher would require one trained dedicated operator during operations. The Searcher would be reliable enough to be maintained by the operator and not more than one additional technician.

Table 2-3 summarizes the basic capabilities of the Searcher UGV.

Semiautonomous Preceder/Follower UGVs

Like the TGV, semiautonomous preceder/follower (SAP/F) UGVs can come in all shapes and sizes. They are characterized by limits on the scope of autonomous mobility. Follower UGVs are the focus of current Army development and demonstrations. Preceder UGVs are follower UGVs with advanced navigation capability to minimize the need for operator interaction to achieve A-B mobility.

For the purpose of description in this report a SAP/F UGV would traverse its environment by following a trail of markers (often called "breadcrumbs") left by a "leader," which could be a dismounted human, a manned vehicle, or an autonomous vehicle. It would use some cognitive processes to select the best route from marker to marker. For example, the onboard processing could determine the heading to the next breadcrumb using geolocation information and simple terrain reasoning. The terrain reasoning might include identifying a road and its edges, traversable paths across open terrain, obstacles to be avoided or negotiated,

---

[1]Covey et al., 1994 defines six levels of initiative: 1. Wait until told; 2. Ask; 3. Recommend; 4. Act and report immediately; 5. Act and report periodically; and 6. Act on own. These are used to describe each of the example systems.

TABLE 2-3 Searcher: Basic Capabilities for an Example of a Small, Teleoperated UGV

Function	Basic Capabilities
Mobility	• Day and night • Climbs stairs and searches rooms in urban buildings at least up to 10 stories • Searches tunnels, culverts, caves out to 1 km and back • Variable speed based on situation but up to speed of a running soldier on flat surfaces
Mission packages	• Sensors with sufficient resolution to allow an operator to quickly identify humans, obstacles, and other information while searching buildings, tunnels, or other enclosed areas • Manipulators for opening doors and moving small objects • Marking system to indicate to follow-on soldiers which areas were cleared and deemed safe at the time of search • Synthetic voice projection • Non-lethal self-protection
Communications	• Wire and/or RF from UGV to an operator's hand-held display
Human control	• Control by joystick, touch screen, or similar type input device • Continuous for planning and navigation
Automated UGV control and decision making	• Climb stairs
Other	• Weather proof • Self-righting
Human support	• Maximum of one operator and one maintenance technician per Searcher

RF = radio frequency; UGV = unmanned ground vehicle.

and other path-planning elements. Because the follower UGV would move through a "known environment" that has been successfully traversed by its leader, the leader is assumed to possess the majority of cognitive processes and makes decisions for military maneuvering. The leader would also have some degree of local situational awareness of its following UGVs through sensing modalities similar to those available to an operator of a TGV. The sensor suite is typically more complex than that on a TGV and may include:

- A geolocation sensor
- Daytime and nighttime viewing cameras (for leader override)
- Laser detection and ranging (LADAR) and multispectral sensors (providing digital representations of terrain and obstacles near the UGV)
- Foliage-penetrating sensors (to assess trafficability through grass and other light vegetation).

Missions appropriate for follower UGV capabilities include (1) a soldier's "mule" to carry weapons, ammunition, and other items cross-country behind dismounted soldiers; (2) logistics resupply vehicles to follow a leader vehicle in road-traversing convoy mode (sometimes called close following); and (3) logistics resupply cross-country (including poor roads and paths) following a leader vehicle by an interval of minutes to hours.

A semiautonomous preceder UGV represents a step up in mobility autonomy from follower capabilities. The preceder must have sufficient autonomy to move in advance of its controller, which could be a dismounted soldier, manned vehicle, or autonomous vehicle. It would have sufficient cognitive processes onboard to select the best route to traverse an objective designated by the controller without following a breadcrumb trail (traversing "new ground"). The controller could provide course correction updates to the preceder, either at the controller's option or upon request from the UGV. However, the preceder should be able to achieve its traverse objective for its normal mission/environment envelope with few calls to the controller for help.

The cognitive processes of a preceder would be focused on: (1) determining heading using general instructions from the control vehicle; (2) geolocation information about itself, other UGVs, and the controller; and (3) complex terrain reasoning (but not as complex as the controller's capability). It would also have limited capability to support the command and tasking needs of its mission package. The controller would continue to have the majority of cognitive processes associated with the mission: friend, foe, or neutral (FFN) locations; higher-level terrain reasoning functions (such as identifying "no go" terrain); and military maneuvering. The controller would also have local situational awareness of the environment and could intervene and override movement decisions. The sensor suite of the preceder would be essentially the same as that of a follower.

Missions appropriate for SAP/F-UGV capabilities could include (1) RSTA missions 1–5 km in advance of the controller's position; and (2) forward fires or supply prepositioning up to several kilometers in advance of the controller. A preceder could possibly lead one or more followers by dropping the breadcrumbs for its followers and perhaps maintain some geolocation awareness of them. Local situational awareness for each of the followers, with the capability for command intervention and override, would remain with the controller.

Example 2: Unmanned Small Unit Logistics Mover ("Donkey")

Overarching Concept. Since the time of Caesar, it has been pointed out that the battlefield load carried by the dismounted soldier in general and by the dismounted infantryman in particular is too heavy. A heavy load can drain a soldier of energy and critically slow down the reactions of a soldier placed in a dangerous situation. Soldiers are reluctant to carry lighter loads because the fog of war and chaos of the battlefield invariably foil the best plans to effectively conduct timely small unit resupply. For want of another box of ammunition or antitank weapon or night vision goggles soldiers may have perished. This situation can be improved if there is near certitude that needed weapons, ammunition, food, clothing, equipment, and other items will be delivered in a timely manner to precisely the right place.

The Donkey[2] Unmanned Small Unit Logistics Mover would be a medium-sized, semiautonomous preceder/follower UGV that could lighten the soldier's load and allow an instantaneous reaction to the fight. The Donkey would be capable of automatically following a path through urban and rural terrain from a start point to a release point. It would be able to carry a load of at least 1,000 pounds and be able to operate on a nonlinear battlefield where small troop units would constantly be moving and operating out of small, dispersed observation/fighting positions.

Operational Approach. The Donkey would be capable of operating day and night under all but the most extreme weather conditions. It would be capable of crossing terrain at least as well as that negotiated by current state-of-the-art, commercial all-terrain vehicles (ATV). The Donkey would be able to carry supplies, rucksacks, and other equipment forward. Although not its primary purpose, it could carry soldiers. On return trips it could, for example, carry used food containers, items needing higher-level maintenance, batteries for recharging, and in the extreme, casualties. The round-trip distance for automatic movement along an electronic path would be at least 50 km. It would be able to automatically cross small streams up to axle level in depth. The Donkey's speed would be adjustable based on its sensing of the terrain and/or programmed instructions for a specific electronic path. At a minimum a human would mark the initial electronic breadcrumb paths for Donkeys to follow later. On one hand the Donkey could cautiously pick its way along the electronic path through unexpected rubble or battlefield debris. On the other it could maintain speeds at least 40 km/h on electronic paths that follow roads or are in open terrain. It would be able to automatically adjust its path by up to 1 km at each problem area to get around unexpected debris or obstacles and return to the correct electronic path. The Donkey would have rudimentary sensors and range finders that could detect other vehicles or humans. Based on sensor input, it would be programmed to slow down (for safety near friendly forces) or to take predetermined avoidance actions in areas where friendly troops are not anticipated. It would have a very high probability of successfully moving from start to release point. As a resupply vehicle, the Donkey's technological emphasis would be on simplicity, low cost, and very high maintenance reliability. In very demanding situations (e.g., during early entry operations when friendly forces are initially heavily outnumbered) one Donkey would normally be sufficient to support at least a 30-person dismounted unit. In less demanding situations (e.g., in a more mature theater when friendly forces have a significant numerical advantage) one Donkey could support at least a 200-person dismounted unit.

Basic Capabilities. The basic Donkey would have a minimum requirement for higher-technology payloads. Since the probability of the Donkey's accomplishing its mission would have to be very high, the Donkey would need a secure communications package. The package would allow the Donkey to interact with humans located near the start and release points of the electronic path. During movement the communications capability would allow operators to keep track of the exact location of each Donkey. Additionally, operators would be informed of and could take action if there was a maintenance breakdown, unexpected large area that could not be traversed, or unauthorized tampering. The communications package could allow operators to direct the Donkey, while enroute, to move from one electronic path to another or to immediately return to the start point if a change in situation or priorities occurred. To preserve bandwidth and minimize signal clutter the Donkey and base station would communicate only as needed.

UGV–Human Interface. The Donkey's level of initiative would be that it would act automatically on assigned portions of its task as it follows its electronic path about the battlefield. However, it would be capable of asking for human guidance when unprogrammed or unanticipated situations occurred. The Donkey would require some human interaction. Through use of a small keyboard, touch screen, or other input device a human would program start times, pro-

---

[2]The name "Donkey" was selected to distinguish the example from the FCS "Mule" application.

gram paths to be followed and provide decision template information. Other human interaction would be to maintain and fuel Donkeys, as well as loading supplies on the carrying decks. As the situation dictated, Donkeys could be operated as part of a small (10–12 Donkeys) team. The number of humans on each team would be no more than one supervisor and two maintenance technicians.

Table 2-4 lists basic capabilities for the Donkey.

Platform-Centric Autonomous Ground Vehicles

The desired endpoint of unmanned vehicle evolution has commonly been described as an *autonomous vehicle*. The closest dictionary definition for "autonomous" is "independent in mind or judgment," but this definition only shows that "autonomous" is a metaphor when applied to a robotic vehicle. The implication appears to be that an autonomous vehicle can be assigned a complex task or mission and will then execute it, perhaps acquiring information from other sources as it goes but without further guidance on what to do. In developing a systems architecture for the next-generation remote-controlled vehicle, DARPA (2001) defined "autonomous" as

> A mode of control of a UGV wherein the UGV is self-sufficient. The UGV is given its global mission by the human, having been programmed to learn from and respond to its environment, and operates without further human intervention.

The committee spent considerable time pondering where to draw an operationally useful and technologically meaningful distinction between *supervised* vehicles that require less and less intervention from their commander to accomplish a complex mission (the incremental evolution of supervised control) and autonomous vehicles that can receive simply expressed orders for complex missions and accomplish them without needing to be told how. A further military subtlety is the characteristic of "responsible" autonomy, since UGVs capable of lethal weapons require fail-safe interrupt or override mechanisms.

For the kinds of robotic vehicle applications Army planners and developers have discussed, the committee decided

TABLE 2-4 Donkey: Basic Capabilities for an Example of a Medium-Sized, Preceder/Follower UGV

Function	Basic Capabilities
Mobility	• Day and night under all but the most extreme weather conditions • Negotiates terrain as well as a current state-of-the-art ATV • Crosses water obstacles up to axle depth • Follows electronic "bread crumb" paths for a round-trip distance of at least 50 km • Variable speed but at least 40 km/h on roads or open terrain
Mission packages	• Cargo bed with minimum of 1,000 pounds load capacity to carry logistical supplies between a base station and dismounted troop locations • Sensors and range finders that can identify and locate other vehicles and humans
Communications	• Secure package allowing communications between Donkey and base station • Bandwidth and signal clutter minimized; communication occurs only when programmed, the UGV is queried, or when human guidance is needed
Human control	• Mark electronic paths for Donkey to subsequently follow • Load and unload cargo • Program the electronic path to be followed, start points and stop points; change paths or start/stop points while en route as situation dictates • Monitor communications from Donkey that unanticipated situation has occurred at a certain time and location; take appropriate action
Automated UGV self-control and decision making	• Adjusts path up to 1 km to skirt unexpected obstacles or debris • Identifies humans; slows for safety reasons when near known friendly locations; speeds away in locations where humans are identified but no friendly troops anticipated
Other	• Very high probability of successfully moving from start point to stop point and back • Very high levels of maintenance reliability • Emphasis on simplicity and low cost
Human support	• A maximum of one supervisor and two maintenance technicians will be able to operated a small (10–12) Donkey UGV team

that two conditions, at least, must be met for a robotic vehicle to be considered autonomous. First, the vehicle must have A-to-B autonomy. A vehicle has A-to-B autonomy when it is at point A, can be given a direction to go to point B, and can get to B with no help along the way from a human operator. For Army applications the benchmark for distances between A and B is from a few kilometers to a hundred (Eicker, 2001) and should include most of the terrain conditions where the Army would operate with manned vehicles of similar mobility characteristics.

The second condition that the committee set for an autonomous Army UGV is that the vehicle must be able to carry out its assigned mission in a hostile environment. As with negotiating difficult terrain, the benchmark here is that the UGV should have survivability and self-defense roughly equivalent to a similar manned vehicle sent on the same mission. For example, on a forward-scouting mission it may not need to survive a direct hit by a rocket or anti-armor round, but it should not be incapacitated by an unarmed human running up and throwing a blanket over its sensors. This self-survival capability adds another set of technological requirements.

The vehicle must have local-area RSTA capabilities, beyond obstacle detection and identification for navigating, to detect the presence of potential threats and either take evasive action, stealthy maneuver, or offensive self-defense.

- The vehicle must be capable of identifying friends, foes, and noncombatants (or "neutrals") (IFFN).
- It must have adequate lethal or non-lethal weapons for self-defense; in the event of hostile actions it should be able to survive.
- It must be able to refuel itself by means likely to be available, such as prepositioned or air-dropped fuel supplies or rendezvous with a fuel supply vehicle (manned or unmanned).
- It must have sufficient reliability and robustness to absorb and overcome the common mishaps of cross-country maneuver.
- It must have the higher-level cognitive processing needed to support tactical maneuvers and self-protection/self-defense behaviors.

Once the committee recognized that this second requisite was essential for "autonomy" in Army applications, it became obvious that an autonomous UGV could easily have multimission capability. Relatively "dumb" mission-specific modules could be attached to the basic vehicle. Specific instructions for how to work the equipment, equivalent to a soldier's training on a weapon or other specialized equipment, could be in software loaded into the UGV's main computing capability when the mission package is attached. The intelligence to know when and how to employ the weapon system or equipment appropriately (software equivalent of doctrine and rules of engagement) would already be present in the behaviors that give the vehicle A-to-B autonomy in hostile environments.

Example 3: Unmanned Wingman Ground Vehicle ("Wingman")

Overarching Concept. Small mechanized infantry and armor combat units normally operate and fight as teams. In mounted operations the smallest team consists of two vehicles in which a section leader (with crew) in one combat vehicle is accompanied by a subordinate wingman (with crew) in another vehicle. The section leader, following broad guidance from his superiors, usually determines routes to be followed, positions to be occupied, formations to be used during movement, and actions to be taken during various situations. In combat, when two or more enemy targets are encountered, the section leader will give fire commands to determine which enemy targets he will attack and which will be attacked by the Wingman. The Wingman is constantly observing his designated sector and reports any changes or dangers to the leader. The Wingman's observations and communications often result in the section leader changing his initial plans based on more complete information. A Wingman could perform section missions normally assigned to a manned Wingman vehicle. The unmanned Wingman would be capable of functioning at the highest states of alert indefinitely. It would be able to routinely conduct missions that would normally be considered extremely risky for a manned system. The Wingman would be able to automatically move about the battlefield—remaining within assigned areas or on designated routes—as directed by the section leader. Through its sophisticated sensor package the Wingman would provide eyes and ears to the section leader who could give early warning of danger and increase the survival rate for the manned vehicle. The Wingman could provide 360-degree observation or focus solely on a specific sector as directed by the section leader. It would provide an RSTA capability that would allow it to recognize natural and manmade features as well as nearby people, vehicles, obstacles, and other information. The Wingman would transmit observed information to the section leader. This intelligence would allow the section leader to continue with the current plan, adjust movement, maneuver to a more advantageous position, directly attack an enemy target, request other assistance to accomplish the task, or take any number of actions as necessary. The Wingman would be able to perform continuous local security while soldiers slept or were otherwise occupied.

Operational Approach. The unmanned Wingman, a medium to large platform-centric autonomous ground vehicle, would be capable of operating day and night under all weather conditions. It would be capable of moving at variable speeds, based on conditions encountered, up to at least 100 km/h on roads and in open terrain. Both the Wingman and the section leader's vehicle would be able to traverse urban and rural terrain and swim across ponds, lakes, or other slow moving bodies of water with equal ability. It would automatically move in relation to the section leader (precede, follow, or on a flank), as directed. The Wingman UGV would be the approximate size of the section leader's vehicle to preclude the enemy from more easily identifying and firing at the section leader. Stand-off distances that the Wingman would achieve relative to the section leader would be based on doctrine (tactics, techniques, and procedures), local terrain, higher headquarters guidance, and the section leader's order. The Wingman would be able to sense the section leader's location, the location of other nearby manned or unmanned vehicles, perceive the local terrain, and automatically adjust movement direction and speed, as necessary. The section leader's human–machine interface capability would allow him to very easily order the Wingman to move to a designated area or a designated point to conduct RSTA tasks. Depending on the situation, the Wingman would be given more or less latitude in determining routes to move to the observation position. Similarly, it could occupy and adjust its precise location in stationary positions to take best advantage of cover, concealment, and lines of sight. Additionally, it could be able to automatically tie in with manned/unmanned systems in its vicinity to achieve overlapping fields of observation and fires. It would have a continuous local security capability that would immediately signal the section leader when unanticipated movement or activities occur. The Wingman would achieve high levels of maintenance reliability. It would be able to self-diagnose and store anticipated noncritical maintenance problems for later downloading and correction. More critical maintenance or other problems that would impact on its mission would be reported immediately to the section leader.

Basic Capabilities. The Wingman would be electronically tethered to the section leader through a secure local area network.

- The Wingman would carry a sophisticated sensor and range-finding RSTA/BDA (RSTA battle damage assessment) package. Its sensors may be any combination of seismic, acoustic, magnetic, visual, IR, or other capabilities.
- The Wingman would have access to a sophisticated automatic target recognition (ATR) capability that as a minimum, can distinguish between friendly and enemy combat vehicles.
- The Wingman's computational capability, tied to its sensors and ATR, would allow for the rapid creation and transmission of recommended direct and indirect fire commands to the section leader.
- The Wingman would provide for its own survival by sensing when it is under attack from direct or indirect fire and by taking immediate programmed action, such as rapidly changing location. It would sense danger from approaching humans and be able to launch close-in non-lethal effects.

UGV–Human Interface. The Wingman's level of initiative would be to do assigned tasks and report the results of its acts. It would also alert and recommend courses of action, as programmed, for certain dangerous, complex, or ambiguous situations, but wait for guidance from a human before acting in these cases. Even though the Wingman would be capable of considerable automatic actions there would still be a need for a close human interface. The objective would be to allow the section leader to keep focused on the battle and not be distracted by the Wingman UGV. Therefore, emphasis would be on minimizing human interaction with the Wingman while maximizing the Wingman's automatic capabilities. The assistant section leader would have the ability to teleoperate the Wingman electronically in situations where human override was necessary due to safety or highly complex situations; however, this would not be the preferred method of control and would be used infrequently.

Table 2-5 lists the basic capabilities of the Wingman UGV.

Network-Centric Autonomous Ground Vehicles

The committee decided that a third condition not essential for vehicle autonomy was likely to be essential for autonomous UGVs to be elements of FCS. Without diminishing their A-to-B autonomy in hostile environments, the vehicles must also be competent as independent nodes in a network-centric warfare model. They must be able to receive information from the communications network and incorporate it in their mission execution. They must respond to appropriate information requests and action commands received from the network, including resolution of conflicts when requests or commands interfere with each other or with the original mission assigned to the UGV. Again the rough benchmark for operational success as an independent network-centric node is an equivalent manned system similarly tasked.

Example 4: Autonomous Hunter-Killer Team ("Hunter-Killer")

Overarching Concept. Potential enemy forces, ranging from sophisticated mechanized units to Third World guerillas, need to move about the countryside, through villages, and in cities to accomplish their missions. Ambushes conducted against enemy forces can have a powerful effect. Ambushes can disrupt and slow the pace of logistics and combat operations. After experiencing initial losses to ambushes an enemy is often forced to stop movement on well-defined roads and trails that could be the most likely sites of fatal

TABLE 2-5 Wingman: Basic Capabilities for an Example of a Medium-Sized to Large Platform-Centric UGV

Function	Basic Capabilities
Mobility	• Day and night under all weather conditions • Crosses urban and rural terrain with same ability as section leader's manned vehicle • Swims water obstacles without additional preparation • Variable speeds depending on situation but up to 100 km/h on roads or in open terrain
Mission packages	• Sophisticated sensors and range finders that allow Wingman to become the "eyes and ears" of its manned section leader • Sophisticated RSTA/BDA package; ATR capable of differentiating between friendly and enemy combat vehicles • Non-lethal self-protection package
Communications	• Secure communication package forms basis for "electronic tether" control/information sharing between Wingman UGV and human section leader • Near real-time transfer of sensor and other information to section leader
Human control	• Human very easily directs Wingman to new locations and describes tasks to be performed by UGV while en route and upon arrival at new location • Human monitors sensor and other input from Wingman • Actively makes go or no go decisions on all Wingman recommended calls for direct or indirect fire • Electronically directs/overrides Wingman movements into very confined, dangerous, or complex locations
Automated UGV self-control and decision making	• Automatically moves in relation (precedes, follows, or on a flank) to manned vehicle as initially directed; adjusts speed and movement direction based on terrain, vegetation, nearby vehicles, or other objects • Occupies and adjusts its precise location in stationary positions; ties in observations and fields of fire with adjacent manned or unmanned systems • Automatically calls for recommended direct or indirect fire missions when sensing an enemy • Senses when under attack from direct or indirect fire and takes appropriate action • Recognizes commands to change allegiance to a different human section leader, as necessary
Other	• Wingman UGV is about same size as section leader's manned vehicle • High levels of maintenance reliability • Self-diagnosis and storing of anticipated noncritical maintenance problems; immediate reporting of critical maintenance issues to section leader
Human support	• No more than one assistant section leader (Wingman controller) and one maintenance technician both of whom ride in the manned section leader's vehicle

ambushes. This can reduce enemy movement speed to that of dismounted soldiers. With the growing sophistication of U.S. air-delivered smart weapons and night vision capability, a thinking enemy may be forced to take advantage of urban terrain where there is better protection from U.S. observation and attack. Ambushes can be especially effective in urban areas where streets channel movement and buildings offer excellent cover and concealment for ambushing units. Additionally, an ambush can be one of the most doctrinally straightforward (albeit among the most dangerous and certainly not simple) missions conducted by a small unit. The ambush site is selected in advance on terrain that favors the friendly unit and puts the enemy at a disadvantage. Outposts can be established to provide advance warning of the enemy approach. Detailed fire plans can be made that include prearranged calls for indirect fires. Withdrawal routes and assembly areas can be planned and reconnoitered in detail. An Unmanned Autonomous Hunter-Killer Team (in an "ambush unit" scenario) can give the U.S. Army the ability to foil enemy movement on the ground, inflict heavy enemy casualties, and minimize friendly casualties in the process. The Hunter-Killer team would consist of several unmanned vehicles that would be capable of moving to an ambush site. Upon arrival they would arrange themselves in an effective observation/kill array, would be able to transmit relevant information to a human base station, and would be able to monitor the environment to know when the enemy is approaching and when he is in the kill zone. The team would kill enemy forces in the kill zone with overwhelming lethal force. Subsequently, it would take programmed actions to ensure its survival, such as moving to a new location.

Operational Approach. The Hunter-Killer team would consist of at least 10 medium-sized "killer" unmanned network-centric autonomous ground vehicles. Each killer vehicle would carry internally (in a "marsupial" manner) up to five small network-centric autonomous "hunter/observer" ground vehicles; as the situation dictated the ground hunters/observers could be replaced or augmented with aerial hunters/observers. All UGV would be capable of operating day and night under all weather conditions. The Hunter-Killer team would be able to operate in rural and urban terrain and swim across slowly flowing rivers. Initially, the Hunter-Killer team would be programmed by humans at a base station. For intra-Hunter-Killer team communications purposes, the UGVs would be tied together through a local wireless network. Information gathered by each unmanned vehicle could be passed to others, including updating and modifying the human-provided input as long as those modifications remain within programmed hard decision rules. The ambush vehicles would have sophisticated sensors that detect humans or other vehicles and would be allowed to take evasive actions, as necessary. The Hunter-Killer team would be capable of conducting a round-trip distance between start point and ambush site up to at least 300 km. Speed of movement would be variable depending on the terrain but would be capable of reaching at least 120 km/h on roads or in open terrain. At the ambush site the medium-sized killers would launch small hunter/observer UGVs to provide area security and detailed information on approaching people and vehicles. Once the stealthy hunter/observers were in position the killers would go on power standby. The Hunter-Killer team would be able to remain on site in standby mode without human interaction for at least 30 days. When hunters/observers sensed that humans or vehicles were approaching they would automatically activate higher-power sensors and ATRs to confirm the identity of an enemy force. This information would be sent to the killers, who would move from power standby to full alert. Based on data from the hunters/observers, the killers would be able to calculate the size of the enemy force. If the enemy force were too large to ambush it would be allowed to pass, otherwise the Hunter-Killer team would risk being overpowered by the enemy and being destroyed. The killers would also determine which actions needed to be taken by an appropriately sized enemy to trigger the ambush. They would notify the human base station of actions about to happen, as programmed. Depending on the rules of engagement a human in the base station may have to signal approval. When the ambush was triggered the killers would launch an overwhelming and precisely targeted lethal response using onboard targeting systems: The general rule is "one shot, one kill." In addition to onboard weapons, the killers would be able to automatically send fire commands to the base station or directly into the appropriate battlefield C2 networks, resulting in massive indirect fires being provided. Reactions to counterattack would be dependent on the situation but would likely initially include calling for additional indirect fire and maneuvering one or more killers to achieve a tactical and firepower advantage over the enemy. The killers would also be provided with decision criteria that would allow them to immediately move from the ambush site to a remote assembly area to await further instructions from the base station. Even if not attacked by the enemy, moving out of the ambush site quickly would be required. If time permitted, the hunters/observers would rejoin and board the killers for transportation to the next mission. Otherwise, the hunters/observers would go on standby mode or self-destruct, as programmed. Maintenance reliability would be very high. In addition to being able to self-diagnose maintenance problems the Hunter-Killer team will have rudimentary self-repair.

Basic Capabilities. The unmanned Hunter-Killer team would need a sophisticated local and global terrain-sensing capability. The team's sensors could read local terrain, vegetation, obstacles, and other information in great detail, and store this information for downloading to allow updating the global database. It would need sensors, range finders, and ATR that could accurately identify enemy forces and distinguish friend from foe from noncombatant. It would need highly sophisticated communications packages and local area networks to pass information among all necessary UGVs in the unit. It would also need to be part of a secure wide area network to communicate with its base station and other unmanned and manned systems, as necessary. It would need longer-range communications and relays to allow it to contact a distant base station or directly into C2 networks to receive human approval to take certain actions, as required, and to request additional support. Examples include passing on the arrival of a new ally on the battlefield that has similar vehicles to the enemy, the anticipation of noncombatant movements in the area, or the signing of a cease-fire.

UGV–Human Interface. There would be very little human interface with the Hunter-Killer team once launched on its mission. The Hunter-Killer team would essentially be autonomous except when programmed human intervention and communications must occur.

The Hunter-Killer team would also be a step beyond where each robot has self-diagnostic capabilities, because each would also do some self-repair. The hunters/observers' marsupial UGV, for example, could actually do field repairs on the Hunters-Killers or themselves or vice versa. Another concept is that maintenance could be conducted between missions, in safe areas, by other highly specialized robots. Because of this, it is anticipated that a small team of mechanics and technicians (as few as 10) could maintain up to 5 Hunter-Killer teams.

Table 2-6 lists basic capabilities for the Hunter-Killer team.

TABLE 2-6 Hunter-Killer Team: Basic Capabilities for a Small- and Medium-Sized Marsupial Network-Centric UGV Team

Function	Basic Capabilities
Mobility	• Operate day and night under all weather conditions • Operate in all rural and urban terrains • Swim across slowly flowing rivers and other bodies of water without additional preparation • Move and perform missions without active human intervention over a round-trip distance of at least 300 km • Move at variable speeds depending on the situation but up to at least 120 km/h on roads • Remain in position for at least 30 days without human intervention
Mission packages	• Local and global terrain, vegetation, obstacle sensing • Highly sophisticated sensors and range finders • Highly sophisticated ATR that can discriminate among vehicles (combat and commercial) and humans (friend, foe, and noncombatant) • Sensors able to read detailed terrain, vegetation, obstacle, and other data that can be downloaded upon command to update global databases • Stealth capabilities that make enemy detection of any UGV very difficult • Precisely targetable, highly lethal kill systems; "one shot, one-kill" • Lethal self-protection package
Communications	• Secure local area network allows all UGVs to pass information among themselves • Secure wide area network allows team to call for backup support and to communicate with base station as well as other networked systems, both manned and unmanned.
Human control	• Program various movement, communications, intelligence, rules of engagement, decision making, and other initial inputs • Monitor communications from UGV team for programmed reports or situations requiring human guidance • Override in case of changes in situation
Automated UGV self-control and decision making	• Automatic "intelligent" decision making based on programmed human instructions augmented or modified with real-time UGV sensing of the situation • Fully automated movement; capable of moving as a team or infiltrating separately • Killers able to launch hunters to gather intelligence on terrain, vegetation, obstacles, or human activity • Upon arrival at a mission location all UGVs able to close down all energy dependent systems except for the most energy efficient; capable of "waking up" other systems as the situation warrants • Only attacks enemy forces that are within its ability to devastatingly destroy; otherwise, follows programmed decision rules • Understands enemy tactics and reacts to enemy actions with coordinated UGV tactics, as necessary
Other	• Minimum size of one team is 10 medium-sized "killer" UGVs that each internally (in a marsupial fashion) carries up to 5 small hunter/observer ground or aerial UGVs • Very high level of maintenance reliability; self-diagnosis and repair of maintenance problems
Human support	• Control is by on-duty staff officer at appropriate headquarters • Maintenance beyond scope of UGV and other programming requirements performed by a small team of humans (no more that 10) to support up to five Hunter-Killer teams (approximately 50 killers and 250 hunters)

Although the committee aligned the Searcher, Donkey, Wingman, and Hunter-Killer examples with the defined TGV, SAP/F, PC-AGV, and NC-AGV capability classes, respectively, many of the example applications could be performed by UGVs in more than one class, depending upon the Army operational requirements. For example, it would be possible to develop a teleoperated Donkey or a platform-centric Donkey providing more or less operational potential than the semiautonomous preceder/follower Donkey. It would also be possible to develop an autonomous Searcher with cooperative robot capabilities.

3

Review of Current UGV Efforts

This chapter discusses efforts to achieve unmanned ground vehicle (UGV) capabilities that may be applicable to the Future Combat Systems (FCS), including Army Science and Technology Objectives (STOs) and initiatives of the Department of Defense (DOD), Defense Advanced Research Projects Agency (DARPA), other government agencies/services, as well as foreign UGV activities.

ARMY SCIENCE AND TECHNOLOGY PROGRAM

The Army Science and Technology program is administered through a well established process for initiating and managing a series of initiatives described in terms of Science and Technology Objectives. In this section relevant Science and Technology Objectives are described briefly and analyzed in terms of capabilities and impact on the Future Combat Systems program. The Army FCS program as presently planned is in a pre-systems acquisition phase of ongoing activities in development of user needs, science and technology, and concept development. Critical decisions regarding which technologies will be available to be integrated into the baseline FCS system design and development will be made in fiscal years 2003 and 2006.

As described in the Army Science and Technology Master Plan (ASTMP), there are two primary STOs in support of UGV developments: (1) the Tank-Automotive Research, Development, and Engineering Center UGV follower Advanced Technology Demonstration and (2) the Army Research Laboratory Semiautonomous Robotics for the Future Combat Systems.

Robotic Follower Advanced Technology Demonstration (STO: III.GC.2000.04)

The goal of the robotic follower Advanced Technology Demonstration (ATD) is to develop, integrate, and demonstrate in a relevant environment an unmanned follower vehicular system capability for future land combat vehicles. The technology will support a wide variety of applications, such as transporting ruck sacks and logistics supplies and providing security for army rear areas, leading to a lethal capability for beyond line-of-sight fire. As with the FCS program, the ATD has been accelerated to meet FCS Block I and Block II objectives. The follower ATD and a sister Crew Integration and Automation Testbed ATD (see later section) have been combined under a Vetronics Technology contract to provide two systems as follows:

1. A follower UGV system capable of following a manned lead vehicle or dismounted soldier, both on-road and off-road
2. A crew integration and automated testbed (CAT) vehicle system consisting of a two-soldier crew, with multimission capabilities allowing the crew to accomplish "shoot," "scout," and "carrier" missions while coordinating and controlling unmanned ground and air systems, including the follower vehicle.

The program is structured to provide, in close consonance with the Army Research Laboratory Semiautonomous Robotics STO (see following section), the robotic technology to support the performance requirements for the Mule UGV and the Armed Reconnaissance Vehicle UGV, as described by the FCS lead system integrator team.

To support the Mule mission the ATD program is planned in FY03 to demonstrate leader-follower capability with up to 1-km separation, including limited semiautonomy to avoid obstacles. Capability for teleoperation will be included as a backup. For the armed reconnaissance vehicle (ARV) mission, two robotic followers (a Stryker and an updated experimental unmanned ground vehicle [XUV]) will be included in the demonstration. In FY06 the ATD program is scheduled to demonstrate multilane on-road operation as well as off-road operation with up to 24-hour separation, with limited semiautonomy.

This description of the robotic follower ATD program, coupled with the CAT ATD program, provides the basis for the committee's answer to Task Statement Question 2.a in Box 3-1.

Semiautonomous Robotics for Future Combat Systems (STO: IV.GC.2001.03)

The focus of this STO is to develop semiautonomous mobility technology critical to achieving the transformation envisioned for the FCS. A number of on-road and cross-country mobility experiments have successfully demonstrated initial capabilities in controlled environments (e.g., known terrain, daylight, and favorable weather). Continuing technical efforts are focused on perception and sensor technologies, intelligent vehicle control, tactical environment behaviors, and supervision of unmanned ground systems. Key technologies include obstacle avoidance, terrain characterization and classification, and fusion of data from multiple classifiers. Several technologies developed under this STO were integrated with the experimental unmanned vehicles (XUV) used for the DEMO II and DEMO III UGV demonstrations (see DOD Joint Robotics program).

A restructured program has been proposed to demonstrate during FY03 terrain-dependent semiautonomous navigation at ranges over 1 to 3 km and limited mission capability for such survivability functions as obscurant dispersal and remote sensing capabilities for self-protection missions. The continuing program has the goal of demonstrating limited scout capabilities by FY2006, based on improved LADAR (laser detection and ranging) and stereo vision sensing, and on advances in perception and intelligent vehicle control. An extensive set of field experiments in a variety of settings is an essential part of this effort, but additional funding will be required to support development and instrumentation of test-bed vehicles to support the field tests. This description of the Army Research Laboratory STO program provides the basis for the answer to Task Statement Question 2.b in Box 3-2.

Other Army R&D Activities Contributing to FCS

The robotic follower STO and the semiautonomous robotics for FCS STO are the principal science and technology efforts for developing unmanned capabilities for the FCS. In addition to these primary TARDEC and ARL STOs, a review of other Army STOs indicates that 19 additional STOs use the term "robotics" or "unmanned" in the description of the STO activity. Of these, the following seven are of principal interest to the further development of UGV technology and offer potential support and risk mitigation for achieving the overall goals of the FCS program.

Future Combat Systems (STO: III.GC.2000.03)

A single STO covers the collaborative DARPA/Army FCS program described in Chapter 2. It is directed toward lightweight, lethal, deployable, self-sustaining, and survivable combat systems, collectively known as the Future Combat Systems (FCS), for the 2008–2015 time frame. UGVs are considered to be integral to the FCS. The DARPA efforts under this STO are described in a separate section below.

BOX 3-1
Task Statement Question 2.a

Question: Will the follower UGV Advanced Technology Demonstration (ATD) lead to a capability that will meet stated Army operational requirements in time to be integrated into the baseline FCS development program?

Answer: While it is likely that the follower UGV ATD objectives will be achieved, the program does not consider relevant supporting technologies or system-level requirements. Consequently, it is unlikely that all of the technologies needed for a follower UGV system will reach TRL 6 in the 2003–2006 time frame that is required for the baseline FCS development program.

The committee is aware that the follower ATD is being restructured (in consonance with the Army Research Laboratory STO and the Crew Integration and Automation Testbed STO at the Tank-Automotive Research and Development and Engineering Center (TARDEC)) to focus on Mule and Armed Reconnaissance Vehicle capabilities, consistent with the concepts requested of industry by the FCS lead system integrator. To the extent that this assists in defining system requirements, it is clearly a step in the right direction. It remains to be seen, however, whether this dual focus will accelerate development of UGV systems for FCS.

BOX 3-2
Task Statement Question 2.b

Question: Will the Army Research Laboratory (ARL) STO program result in significant advances in UGV autonomy beyond that achieved in the follower UGV ATD, and what operational requirements would the resulting capability be able to address?

Answer: The ARL STO is being restructured in consonance with the follower ATD to focus on Mule and Armed Reconnaissance Vehicle capabilities, as described by the FCS lead system integrator. In light of this the ARL STO will not significantly advance UGV autonomy beyond that likely to be attained in the follower UGV ATD for those particular applications.

The original ARL STO addressed capabilities that would support autonomous requirements not addressed in the follower ATD, including planning, navigation, and human–robot interaction.

Crew Integration and Automation Testbed ATD (STO: III.GC.1999.02)

As described previously, this STO is contributory to and has been combined with the robotic follower ATD because as the technology is directed toward the reduction of the workload on the human crew and therefore is directly applicable to unmanned crew functions. Specific technologies include intelligent driving decision aids, the application of semiautonomous driving technology, and automated route planning, all of which are pertinent to both manned and unmanned vehicles. The complete development of driving technologies and decision aids is planned for FY03, including demonstration on a vehicle test bed. This STO will contribute technologies to the Mule and ARV UGV missions as defined by the FCS lead system integrator.

Obstacle Marking and Vehicle Guidance (STO: IV.EN.2000.02)

The focus of this effort is to dispense "smart" markers to transmit and receive critical navigation information through and around obstacles or minefields. Successful development of this technology could be applicable not only to manned vehicles and dismounted forces but also to the path planning and path following of unmanned vehicles. It is planned by FY03 to have a complete smart marker system that will be timely for evaluation with the robotic follower ATD and the Semiautonomous Robotic Vehicle.

Mobility Support for Objective Force Maneuver (COE-ERDC STO COE.2002.04)

The results of this research will provide the capabilities and algorithms to quantify mobility and physical agility parameters that are essential for characterizing unmanned systems. The work addresses current deficiencies in modeling the breaching and crossing of complex obstacles with lighter vehicles than are now in the inventory. By FY06 the current program plan projects that the technologies to quantify mobility in urban environments, assess and negotiate obstacles, and model reliable driving behaviors will be available. The products of this program will include vehicle performance assessment tools and measures of performance of physical agility.

Advanced Robotic Simulation (STRICOM STR-03)

The objective of this activity is to develop intelligent behaviors for robotic systems within an environment that will provide complex mission and coordination behaviors for real robots. The basic premise is to leverage the demonstrated abilities to create complex military behaviors in semiautomated forces simulation and extend the expertise to create and experiment with military behaviors in live robots. Development of tactics and procedures for employment of new robotics systems in the battlespace are limited at present. This advanced robotic simulation STO provides the capability to train and control unmanned forces in collaborative simulation environments. It will provide personnel with the capability to train with robotic systems and also provide for development and experimentation with manned and unmanned interfaces, for instance, to determine how many unmanned systems an FCS crew can control under a variety of scenarios.

Sensors for the Objective Force ATD (STO: III.IS.2001.02)

Unmanned networked sensors can provide remote monitoring and advanced warning for robotic system operation in and beyond friendly force lines. The goal is to develop sensor packages for UGVs using advanced sensor technologies integrated with a robust network architecture. A component of this program involves the use of virtual simulation and live experiments in operational environments to establish baseline architectures, address operational integration issues, and investigate new operational concepts. If successful, this technology will be available in FY05 and will be well positioned for technology insertion in the FY06 FCS system block upgrade. This generic type of low-cost, distributed sensor system could provide advanced self-protection for robotic components of the FCS.

Airborne Manned/Unmanned System Technology Demonstration (STO: III.AV.1999.01)

This effort will demonstrate through simulation and flight test the control, tactics, and procedures for the operation of manned and unmanned air vehicles. Technical barriers associated with manned and unmanned teaming will also be addressed. The resultant software products for manned and unmanned teaming will be available in FY03 and may provide valuable information to UGV developers by means of lessons learned.

OTHER INITIATIVES

Joint Robotics Program

The DOD is actively developing special-purpose robotic UGVs for such applications as range and mine clearance, force protection, breaching, neutralization of ordnance and explosives devices, and reconnaissance, among others (DOD, 2001). These special-purpose robotic vehicles represent a first step toward achieving functional capabilities for ground systems. The earliest DOD robotic systems employed teleoperation, making use of an operator-in-the-loop, but this is clearly not enough. According to the DOD Joint Robotics

Program Coordinator, "Some users can accept teleoperation initially, but all users want autonomy ultimately" (Toscano, 2001a).

There are a variety of robotic UGVs being developed under the auspices of the DOD Joint Robotics program (Toscano, 2001a,b,c). The mission of the Joint Robotics Program (JRP) is to develop and field a family of affordable and effective mobile ground systems, develop and transition technologies necessary to meet evolving user requirements, and serve as a catalyst for insertion of robotic systems and technologies into the force structure. The JRP is structured to field first-generation robotic systems, mature promising technologies, and upgrade these capabilities by means of an evolutionary strategy. In the near term the acquisition programs emphasize teleoperation, operation on diverse terrain, more autonomous functioning for structured environments, and extensive opportunities for users to operate UGVs.

The JRP oversees efforts for specific robotics programs in all services. Its work on UGVs is managed primarily by the Unmanned Ground Vehicles/Systems Joint Project Office, Redstone Arsenal, Alabama; the Air Force Research Laboratory, Tyndall Air Force Base, Florida; and the Army Tank-Automotive Research, Development and Engineering Center in Warren, Michigan.

Several programs managed under the Joint Robotics Program umbrella include technologies with a potential for FCS UGVs:

- The *Standardized Robotics System (SRS)* is a kit with components that can be used to provide tele-operations to various fielded systems. The Vehicle Teleoperation occupational requirements document (ORD) was approved in 1997. Operational employment for the SRS includes obstacle/minefield breaching and route and area clearing. The vehicles currently planned include D7G, T3, and Deployable Universal Combat Earthmover (DUECE) bulldozers, and an upgrade to the M1 Abrams chassis of the M-60 Panther currently being employed by U.S. forces in Bosnia.
- The *Robotics Combat Support System (RCSS)* is a short-range (300 meters) line-of-sight remote control vehicle that uses a set of interchangeable attachments to perform a variety of engineering missions including landmine neutralization, wire breaching, dispensing of obscurants, and demolition emplacement.
- The *Man-Portable Robotic System (MPRS)* will provide lightweight, man-portable UGVs to support light forces and special operations missions, focusing on reconnaissance during military operations in urban terrain (MOUT). The Joint Robotics Program Office has developed and matured the necessary technology through the conduct of concept demonstrations, fielding Matilda (Man Portable Robot) UGVs to the National Guard and the active Army for special purposes, such as cave clearance operations, and monitoring the DARPA Tactical Mobile Robotics program.
- The *Basic UXO Gathering System (BUGS)* employs a semiautonomous reconnaissance platform that controls several small man-portable, expendable UGVs to clear submunitions and other unexploded ordnance (UXO) from the battlefield. After the reconnaissance platform locates the UXO and downloads pertinent location information to smaller robots, the smaller robots pick up and remove, or conduct "blow in place" operations. BUGS is in the technology demonstration and evaluation phase.
- The *Remote Ordnance Neutralization System (RONS)* was started by the JRP, transitioned to production, and then upgraded by the Services. RONS has been fielded to a number of explosive ordnance disposal (EOD) units and additional systems are currently being procured.
- The *Robotics for Agile Combat Support (RACS) program* consists of several robotic systems that can perform different types of missions. The All-purpose Remote Transport System—Force Protection (ARTS-FP) allows an operator to investigate and disable suspicious packages and vehicles. The All-purpose Remote Transport System—Range Clearance (ARTS-RC) includes a frangible blade assembly that provides an initial shock to UXO so EOD personnel can safely move through a cleared path. The ARTS-FP/RC systems are currently being procured and fielded. The Automated Ordnance Excavator (AOE) uses a commercial excavator with an extended reach capability that can precisely locate itself and dig up and remotely grasp subsurface UXO. The Remote Crash Rescue Vehicle (RCRV) will provide an autonomous crash/fire rescue platform that can respond in aircraft accidents.
- The *Mobile Detection Assessment Response System—Interior (MDARS-I)* is a mobile robotic security platform that can conduct random patrols inside warehouses and storage areas, as well as conduct electronic inventories. MDARS-I is currently in development. The Mobile Detection Assessment Response System—Exterior (MDARS-E) is the exterior version of MDARS-I. MDARS-E can conduct robotic security functions at large fixed installations, such as warehouses and ammunition storage sites. It can also conduct such physical security tasks as intruder detection, assessment, lock/barrier checks, and alarm response. This latter system is being designed for a single operator to simultaneously control up to 32 robots.

- The *Mobility Enhancement Program (MEP)* is aimed at improving the mobility of small, man-portable unmanned systems in support of urban warfare, combat engineering, physical security and force protection missions. The program has two main thrusts: the Omni-Directional Inspection System (ODIS) and the T3 High Mobility Platform. ODIS is a small high-agility platform that can read license plates and be driven under and survey the underside of suspect vehicles. The T3 High Mobility Platform is a novel 6 × 6 platform that is being used to explore UGV mobility over rugged terrain.

Demo III Program

The *Demo III program* is a technology base effort begun under the DOD Joint Robotics Program being conducted by the Army Research Laboratory and its government and industry partners. The program has been focused on developing and demonstrating technology that can provide supervised autonomous mobility in an unstructured environment. The JRP transferred responsibility for the Demo III program to the Army at the beginning of FY2001. A successor effort to the Demo III program is the ARL *Collaborative Technology Alliance (CTA) in Robotics* involving many of the same industry participants.

The committee was invited to view the formal Demo III demonstration at Fort Indiantown Gap in November 2001. Key technologies demonstrated included: perception algorithms for the fusion of information from multiple mobility sensors (daylight video, FLIR, LADAR, radar), object classification, and active vision; semiautonomous controls; dynamic planning/replanning; tactical behaviors; and soldier-robot interactions (Shoemaker, 2001). Appendix C includes an analysis of the Demo III contribution to semiautonomous A-B mobility development.

DARPA Unmanned Vehicle Programs

In addition to working directly with the Army on the FCS conceptual design, DARPA has undertaken four advanced research programs with high potential to benefit the development of future Army UGV systems. These include the Unmanned Ground Combat Vehicle (UGCV), PerceptOR (Perception off-road) Tactical Mobile Robotics (TMR), and Organic Air Vehicle (OAV) programs.

The *Unmanned Ground Combat Vehicle (UGCV)* program has as an objective the development of prototypes to demonstrate advanced vehicle design and interaction to achieve new levels of mobility, endurance, and payload capacity. Two payload classes of vehicles are being developed, a 150-kg payload version and a 1,500-kg payload version. Missions for both classes of vehicles are expected to evolve with the overall concept of the Army's FCS.

Performance benefits are expected to be harnessed in the UGCV program as a result of being unrestrained by conventional design parameters associated with accommodating onboard human crew. These parameters include:

- Shock constraints (collision, rollover, blast)
- Vibration constraints (absorbed power)
- Life support in nuclear, biological, and chemical environments
- Man–machine interface constraints (e.g., pedals, seats, steering wheels)
- Human comfort constraints (temperature, humidity, cockpit dimensions)
- Time constants associated with human reactions
- Risk constraints (loss of life versus loss of vehicle)
- Survivability constraints (signature, shot lines)
- Geometric constraints (e.g., volume for humans, window sizes) and
- Human vulnerability issues (e.g., gun gas, energy leaks).

Because the resulting vehicles are not required to accommodate crew-associated constraints, innovative methods of design as well as operational use can be considered.

Operational endurance in terms of the range of travel of the vehicle and the time duration between refueling is also a key parameter of the UGCV program. The UGCV is expected to execute missions over much longer resupply periods than its manned counterparts. When resupply is needed, it is expected to be limited to fuel drops that the UGCV can self-serve quickly and resume its mission. Fourteen-day duration and ranges of at least 450 km are considered objectives of the program.

Because the UGCV can be expected to operate with imperfect knowledge of the environment, it may occasionally crash or roll over. The program is seeking designs that can recover from impacts with trees, walls, and rocks and have the ability of self-recovery from rollover or operation in inverted mode.

Designs must consider operations in forested or urban environments (complex terrain). In general, narrow vehicle concepts (or novel mobility concepts) will have higher maneuverability in these confined environments. In earlier programs vehicle width proved to be a limiting factor in attempts to move through forested areas. Higher speed operations in open but broken terrain show that a large wheelbase and/or low center of gravity have obvious advantages for stability. Narrow vehicles (e.g., mono-tracks, motorcycles) are possible options for this confined access, but issues associated with lateral stability (which may be low if practical self recovery is addressed) continue to need to be addressed.

The *PerceptOR* program seeks to quantify and develop ground robot perception for off-road and urban mobility under a variety of terrain and environmental conditions. There is a strong emphasis on experiments in real-world conditions. Inexpensive surrogate vehicles are being used. Experiments are to be conducted with various changes in spectral, thermal, and material compositions changes to allow true all-weather, day-and-night operations.

The DARPA *Tactical Mobile Robotics (TMR)* program investigated ways to penetrate denied areas and project operational influence in ways that humans cannot by using reliable semiautonomous robotic platforms. Its approach was to integrate sensors, locomotion, power, and communications with limited autonomy on a compact, man-portable platform capable of penetrating into denied areas and serving as an extension of the soldier.

Program technical challenges included close-to-the-ground mobility in cluttered and complex terrain, perception for obstacle negotiation, and autonomous fault recovery. Prospective users were special operations and early-entry forces of all Services. The program has been successful in developing and evaluating robust and agile small-robot platform prototypes and in achieving advances in onboard sensor integration and data processing, permitting increased autonomous behavior, specifically in route selection and navigation, obstacle detection, classification and avoidance, and fault recovery from communications failure and platform destabilization.

Four different types of TMR research platforms were provided with volunteers from the TMR contractor companies and academic institutions to assist the search-and-rescue response at the site of the destroyed World Trade Center in New York City. During the final year of the program the research results were integrated onto a semiautonomous Packbot platform prototype with selected mission package options. The capabilities developed are similar to the Searcher example system postulated by the committee and could provide the foundation for developing a soldier-portable robot as envisioned by the FCS LSI.

The DARPA *Organic All-Weather Targeting Air Vehicle (OAV)* program is a key FCS enabling technology program sponsored by DARPA and the Army. The OAV program merges technologies for small, vertical takeoff and landing (VTOL) unmanned air vehicles (UAVs) with autonomous capabilities in order to enhance the situational awareness and effectiveness of soldiers in a network-centric battlefield.

The OAV concept employs ducted fan configurations with duct diameters ranging from 6 to 36 inches with both hover and cruise flight capability. VTOL eliminates the need for a separate launcher or airfield from which to operate. The vehicle can land autonomously to provide continuous surveillance from the ground (or a building ledge) using sensor packages currently available or in development. Operators can remotely order the OAV to "perch and stare" or to move to other locations or to return to base, adapting the capability to changing battlefield conditions.

Such a relocatable sensor capability for FCS could be linked with network-centric autonomous ground vehicles, such as the Hunter-Killer, and provide continuing updates of intelligence for situational awareness in ground operations.

Air Force Initiatives

While UAVs are a primary development focus, the Air Force is also concerned with UGVs for special purposes.

Unmanned Combat Air Vehicles

The Unmanned Combat Air Vehicle (UCAV) program is a joint DARPA/Air Force ATD program that will demonstrate the technical feasibility of UCAV systems that can effectively and affordably prosecute twenty-first century SEAD/strike missions within the emerging global command and control architecture. The objective of the UCAV ATD, also called the UCAV Demonstration System (UDS), is to design, develop, integrate, and demonstrate critical and key technologies, processes, and system attributes pertaining to an operational UCAV system. The critical technology areas are adaptive autonomous control; advanced cognitive aids integration; secure robust command, control, and communication; and compatibility with integrated battlespace. Through its UDS the Air Force seeks to validate assertions from its prior studies that a future UCAV Operating System (UOS) is both effective and affordable.

The UCAV program, along with other Air Force programs to improve the early Predator and Global Hawk UAVs, will rely on many of the same technology developments in perception, planning, human–robot interaction, and communications that are needed by UGV systems.

Air Force Unmanned Ground Systems

Air Force UGV efforts, including those managed under the Joint Robotics Program, are centered at the Air Force Research Laboratory at Tyndall Air Force Base, Florida. The laboratory group has a mission to "conduct research and development of advanced robotic technologies and systems to protect, support and augment the war fighter in the accomplishment of dirty, dull, dangerous and impossible missions" (AFRL, 2002). To this end they are developing robotic systems that will provide a spectrum of devices for agile combat support (ACS) and have developed an impressive array of vehicles based on commercial off-the-shelf (COTS) piece parts employing the Joint Architecture for Unmanned Ground Systems (JAUGS) common architecture, focusing on modularity and interoperability. These units are teleoperated (rf and tethered fiber optic) and cover such tasks as explosive ordnance clearing from training ranges and other areas where unexploded ordnance is present. The units are robotized Caterpillar D-8 bulldozers, Caterpillar 325L excavators, and an All-Purpose Remote Transport System (ARTS) that can employ a number of bolt-on attachments (laser ordnance neutralization system, articulated remote manipulator system II, water-cutting head, charge-setting device, flail, and firefighting nozzle) to process the ordnance uncovered by the larger units. The evolutionary path for this vehicle is to provide such additional capabilities as infrared imaging system, GPS, and inertial navigation, and increased autonomy through path planning and obstacle avoidance software.

The Advanced Robotic Modules and Systems project and the Autonomous Vehicles Technologies project are conducted with industry partners and the University of Florida. These programs currently address vehicle positioning sensors, path planning, vehicle control, and modular architecture development and integration. Coupled to the autonomous vehicle technologies are mission packages that will allow validation of common architecture and modularity of the robotic system. Applications envisioned are RSTA (reconnaissance, surveillance and target acquisition), mule, security, medical evacuation, marsupial missions, and multiple vehicle control and vehicle–vehicle interactions. The initial focus is on the development of low-cost IMU/GPS (inertial measurement unit/global positioning system) units for position sensing, obstacle avoidance through innovative data fusion from a host of sensors, and path planning both from terrain models and vision based sensors. The test bed approach is to integrate the technology into such COTs machines as the caterpillar and ARTS units through a common architecture bolt-on package (Malhiot, 2002).

Navy and Marine Corps Initiatives

The Navy is involved in developing unmanned systems for air, sea, and ground. It is developing a separate UCAV for carrier-based operations. The Marine Corps developed the lethal UGV for ground combat known as "Gladiator" as part of the JRP. The Naval Research Laboratory conducts research in artificial intelligence and other technology areas relevant to autonomous systems on all platforms.

UCAV Development

The Navy UCAV would provide a versatile, multipurpose vehicle capable of performing surveillance/reconnaissance and strike missions. The Navy's UCAV effort leverages the Air Force development efforts discussed above, but there are differences in the operational environments that must be considered. Naval technological development addresses additional constraints, such as design compatibility with catapult takeoffs, carrier approaches, and arrested recoveries, among others. The Navy considers that the UCAV must be well orchestrated into the mix of aircraft and that the coordination of manned and unmanned aircraft must be seamless. As in the case of the Air Force UCAV program, it is important that DOD ensure a close coordination between the Air Force, Navy, and Army UAV programs, as well as between the UGV programs.

Artificial Intelligence Research

The Navy Center for Applied Research in Artificial Intelligence at the Naval Research Laboratory (NRL) conducts research in adaptive and autonomous systems, intelligent decision aids, and other technology areas with direct applicability to future autonomous systems. While research in voice and hand gestures for human control of robots is in its infancy, the Samuel intelligent testing system designed for software-based systems is now ready to support Army systems development (Meyrowitz and Schultz, 2002).

NASA Initiatives

The National Aeronautics and Space Administration has conducted research and development and has deployed UGVs (such as the Lunar Rover) for use in the exploration of celestial bodies. The program uses technologies directly applicable also to Earth-bound UGVs, including perception and sensors, path planning, navigation, mobility, communications, and power/energy.

Much of the NASA research and development is accomplished at the Jet Propulsion Laboratory (JPL) in Pasadena, California, and JPL has also been involved in defense programs in these technology areas. As indicated below, there is strong research cross-fertilization among these technologies due to the number of agencies and contractors working on the various NASA and military UGV programs, most of whom are involved with more than one program.

National Institute of Standards and Technology Initiative

The National Institute of Standards and Technology (NIST) developed a "4-D/RCS" reference model architecture for behavior generation, world modeling, sensory processing, and value judgment processes used in the Demo III program (Albus and Meystel, 2001). NIST is also teamed with industry to develop a common software architecture that can be applied to unmanned ground systems of all services. This effort is called the Joint Architecture for Unmanned Ground Systems (JAUGS). The objectives of JAUGS are to define and model the unmanned systems domain and then standardize the interfaces and behaviors among software components. JAUGS will be applied to all UGVs developed under the JRP.

Department of Transportation Initiatives

The U.S. Department of Transportation (USDOT) from time to time has sponsored research relevant to UGV issues since the mid-1960s. The DOT-sponsored research has been applied to vehicles intended to operate on roads or special guideways, and has not been applied to off-road operations. Hence, its applicability to Army UGVs is associated primarily with on-road operations. However, some of the vehicle positioning and sensing research can be more broadly applicable. Projects have been sponsored by:

- USDOT
- Federal Highway Administration (FHWA)
- National Highway Traffic Safety Administration (NHTSA)
- Federal Transit Administration (FTA)
- Intelligent Transportation Systems Joint Program Office (ITS-JPO)
- California Department of Transportation (Caltrans) and
- Minnesota Department of Transportation (Minn-DOT).

These projects have been aimed at supporting the development of a variety of ground transportation services:

- Collision warning systems
- Collision avoidance systems
- Automated guideway transit systems
- Automated highway systems and
- Bus rapid transit systems.

Note that none of the vehicles that would use these systems are intended to be "unmanned," but some of them are intended to be driven under completely automatic control. Others use sensors and user interfaces to assist drivers, but the same sensors could also be applicable to UGVs.

The customers for these systems are typically private vehicle purchasers (individuals and corporate fleets) and local public agencies (cities, counties, and transit districts), rather than the federal or state departments of transportation. These systems are developed by private vendors, and the major share of development costs is assumed by these developers rather than by the USDOT. Thus, the USDOT funding should be considered only the "seed funding" to advance the technology to the feasibility demonstration stage, and does not approach the total investments being made on these systems as they approach deployability.

Systems that are developed for commercial trucks or private passenger vehicles are intended to be mass produced in large quantities at low unit costs. These offer excellent opportunities for gaining production economies of scale that could benefit UGV systems using similar or related technologies.

The key technologies from these road transportation applications that could be applicable to UGVs are

- Sensing of vehicle position relative to roadway lanes
- Sensing of absolute vehicle position (global positioning system [GPS] and inertial navigation system [INS] in combination with others)
- Sensing of proximity to other vehicles (ranges up to 150 m)
- Identification and hazard assessment of targets within sensor range
- Vehicle–vehicle data communication
- Vehicle–roadside data communication
- Vehicle dynamics control
- Road condition sensing and estimation
- Automatic steering control and
- Vehicle-following speed and spacing control.

Department of Energy Initiatives

The DOE has mobile robotics programs at several of its national laboratories as well as within the DOE University Research Program in Robotics (URPR). DOE applications include security, environmental remediation, materials storage and monitoring, and response and cleanup of accidents involving nuclear materials.

At Sandia National Laboratories the Intelligent Systems and Robotics Center has developed several mobile robotic platforms ranging in size from smaller than 1 cubic inch to as large as a military wheeled vehicle (HMMWV) (SNL, 2001). Extremely small platforms were developed to establish the current limits in autonomous micromechanical systems that can be applied to covert surveillance missions. The larger platform Accident Response Mobile Manipulator System was developed for DOE accident response to nuclear weapons or other hazardous materials. Some of the other platforms such as Fire Ant, Dixie, SARGE (Surveillance and Reconnaissance Ground Equipment), Gemini, and Sand-Dragon were developed for military agencies. For example, SARGE was developed as a production prototype for the DOD Joint Program Office for Unmanned Ground Vehicles/Systems, Gemini was developed for the Special Operations Command, and Sand Dragon was developed for the Marine Corps Warfighting Laboratory. Other mobile platforms such as the RATLER vehicles and the hopping robots have been used in such DARPA programs as TMR, Distributed Robotics, Software for Distributed Robotics (SDR), Mobile Autonomous Robots Software (MARS), and Self-Healing Minefield. In addition to designing, building, and testing robotic platforms, Sandia has considerable experience in developing distributed cooperative controls for mobile robotics. They have developed and demonstrated decentralized control algorithms for formation following, perimeter surveillance, facility surround, building search, minefield reconfiguration and healing, and chemical plume localization missions.

At Oak Ridge National Laboratory the Center for Engineering Science Advanced Research is developing autonomous multirobot learning algorithms for inherently cooperative tasks. They have been "studying autonomous multi-robot learning for inherently cooperative tasks and have developed two new approaches to learning in the domain of cooperative multi-robot observation of multiple moving targets. These new techniques now allow robots to build up memories of their experiences in the environment, evaluate the utility of alternative cooperative actions, and then select actions to take that increase the likelihood that the desired global team goals will be achieved through the individual robot decisions. These multi-robot learning techniques are the first in the field that enable robot teams to automatically learn new inherently cooperative control tasks, rather than having to be programmed explicitly. These capabilities facilitate the solution to a wide variety of applications, including environmental cleanup, space exploration, military applications, and industrial operations" (ORNL, 2002). Oak Ridge has also participated in such DARPA mobile robotic programs as TMR and MARS.

At Idaho National Engineering and Environmental Laboratory the Remote, Robotics, and Automated Systems group is developing large and small robotics and automated systems to simplify efforts in the protection of DOE workers and the environment (INEEL, 2002). They are working primarily on robotics for mixed waste operations, deactivation and decommissioning of underground storage tanks, chemical analysis automation, and cooperative telerobotic retrieval. A small group of staff has also been involved with such DARPA mobile robotic programs as SDR and MARS.

Within the DOE URPR the University of Michigan has worked on mobile robot navigation and radiation mapping (<http://www.urpr.org>). Their work includes innovative mobile robot design, obstacle avoidance, and advanced mobile robot positioning (UMICH, 2001). The principal investigator, Johann Borenstein, has also been involved in DARPA's TMR program and developed obstacle avoidance technology that allows mobile robots to navigate cluttered indoor environments filled with dense smoke.

Collaboration Among UGV Programs

Collaboration is achieved between UGV efforts because there are relatively few defense projects involving robotics technologies in general, and the field of unmanned systems is limited to a relatively small number of universities and companies. The small circle of robotics experts from academia and industry participating in multiple programs facilitates a common awareness of UGV advances and requirements.

Over the last 10 years DOD has initiated many Multidisciplinary University Research Initiatives (MURIs). Some of these have involved frameworks, models, algorithms, and software in the areas of perception, navigation, learning, and decision making in uncertain environments. There is also a thriving, worldwide university community in robotics, and progress has been made in multiple areas applicable to unmanned systems, including intelligent controls, robotic vision, computational geometry for intelligent systems, soft computing for intelligence augmentation, terrain modeling, and communication and control.

Because of their diversity, many of these crucial investments in research may not be fully utilized in the Army UGV program as it now exists. Further, pathways for rapidly transferring basic research knowledge to advanced technology test beds do not exist unless directly related to the Army through the particular MURI program. The mechanisms for technology transfer are inconsistent at best, with technology exchange meetings involving the university groups tending to be very perfunctory. This gap in the R&D continuum must be bridged to concentrate the spectrum of efforts that will develop UGV technologies and systems.

Many of the same industry teams that have participated in the Joint Robotics, Demo III, and PerceptOR programs, for example, are also part of the recently established Robotics Collaborative Technology Alliance (CTA). The JPL has supported NASA, DARPA, and Army robotics programs, so collaboration is high. Similarly, NIST participated with the Army in the Demo III program.

The committee sensed the competition among prospective government and industry participants as they vied for the PerceptOR down-select. There are a very limited number of UGV contracts. An unwillingness to share proprietary information in particular technology areas could easily offset the collaborative advantages of having a small playing field.

While having the same individuals participate in multiple programs may have advantages, it could prove a significant weakness if the demand for expertise increased dramatically or if the same few players monopolized and wittingly or unwittingly discouraged new participants. This assessment of interrelationships among the principal government agencies involved with UGV development provides basis for the answer to Task Statement Question 2.c in Box 3-3.

Technological collaboration between multiple programs will become more important, even essential, if the Army decides to focus its energy on developing a UGV for the FCS. The committee believes that it will take a designated advocate to do this, and a principal function of such an advocate will be to leverage UGV developments and promote collaboration toward explicit goals.

Automotive Industry Developments

Considerable effort is being invested within the automotive industry and related transportation organizations to develop systems that will enhance driving safety, assist drivers in controlling their vehicles, and eventually automate the driving as well. These activities, which are truly international in scope, offer the potential for technology spin-offs that could benefit the Army's UGVs by lowering costs and accelerating the availability of components and subsystems. These could include sensors, actuators, and possibly software, control, and communication systems as well.

> **BOX 3-3**
> **Task Statement Question 2.c**
>
> **Question:** How do the Army UGV efforts interrelate with other government ground robotics initiatives (e.g., National Aeronautics and Space Administration [NASA] rovers, Department of Energy [DOE] programs, Defense Advanced Research Projects Agency [DARPA])?
>
> **Answer:** With the exception of DARPA (the Army funds several of the DARPA robotics programs) interrelationships of other UGV efforts with those of the Army are informal and unstructured. The small size of the robotics industry and the small number of robotics experts tend to encourage technical collaborations. The Jet Propulsion Laboratory, for example, has supported NASA, DARPA, and Army programs, so collaboration is high. Similarly, NIST was part of the Demo III program. Collaboration in particular technology areas may be inhibited by intense competition for a limited number of UGV-related contracts.

Automotive Night Vision Sensors

General Motors introduced the first automotive night vision system on its Cadillac deVille in the 2000 model year, and has sold them as fast as its supplier Raytheon has been able to make them (6,000 per year) (Scientific American, 2001a). This is a passive infrared system that detects infrared (IR) emissions from objects in front of the vehicle and projects the IR image on a head-up display. The system is being sold for $2,250 retail, and considering typical automotive mark-ups, this implies that the system is being supplied from Raytheon to GM for a little more than $1,000.

More recently, DaimlerChrysler in Germany has announced that it is developing an active IR night vision system that depends on active IR illumination of the driving scene by the vehicle (Scientific American, 2001b). This is claimed to offer the ability to "see" more objects that are the same temperature as the background and at a greater distance.

Ultrasonic Proximity Sensors

Various manufacturers in the United States, Japan, and Europe are offering parking assistance systems that warn drivers when they are approaching too close to obstacles at very short range (up to 1.5 meters). These systems typically use arrays of up to eight ultrasonic ranging sensors surrounding the vehicle at bumper height, and retail for up to $2,000. To be detected the sensing requires active ultrasonic emissions; it is intended for very short ranges and is therefore only applicable for very low-speed maneuvering.

Intermediate-Range Ultra-Wideband Radar Sensors

Several automotive companies are developing ultra-wideband impulse radar sensors for use at intermediate sensing range (perhaps 15 meters) to detect other vehicles that could represent hazards. These are not yet on the market and do not yet have FCC approval, but if the sensors come to market they could represent an inexpensive way of achieving wide-angle detection of hazards surrounding a vehicle at intermediate range.

"Long-Range" Automotive Radar Sensors

The automotive industry is beginning to provide its customers with adaptive cruise control systems (cruise control that can adjust speed to follow another vehicle at a suitable distance) and forward collision warning systems. These are based on use of forward-looking IR laser or millimeter wave (24 or 77 GHz) radar sensors, which typically have a narrow field of view (perhaps 12 degrees) and a range of 100 to 150 meters. The signal processing of the sensor systems is designed to discriminate "other vehicle" targets in a road environment with considerable clutter (e.g., bridges, signs, roadside lighting fixtures, and vegetation). Considerable adaptation would be required for use in an off-road battlefield environment; however, they have the advantages of multiple suppliers and probably rapidly decreasing costs. Current fully integrated systems, with the interfaces to throttle and brake and the HMI, retail for up to $3,000 on luxury cars and heavy trucks, but the prices are expected to decrease significantly in the next few years.

Lane-Tracking Vision Sensors

Because road departure crashes are a serious cause of death and injury on our highways, there is considerable interest in developing systems to warn drivers of imminent road departures. The most common technology for detecting this is machine vision, based on use of a small charge-coupled device (CCD) camera capturing the image of the road ahead of the vehicle and identifying the lane markings. The technology could be applied to Army UGVs that are intended to follow roads or well-marked trails for supply missions but would not be transferable to the more general off-road environment. Two American companies, AssistWare and Iteris, are marketing systems for use on heavy trucks at prices in the $2,000 to $2,500 range, and other systems have been made available on high-end passenger cars in Japan.

Drive-by-Wire Actuation Systems

Automotive vehicle designers are gradually adopting drive-by-wire technology for actuator systems for reasons unrelated to vehicle automation. Throttle by wire is available on some high-end passenger cars, although the analogous capability is already standard on most modern heavy-duty trucks. Electronically assisted power steering has been used on the Acura NSX sports car for several years and will soon be available on less expensive cars. Brake-by-wire are being introduced on new Mercedes-Benz automobiles, and other manufacturers are likely to follow. The motivations for these introductions have been associated with providing a higher degree of control, simplifying the installation of the systems in the vehicles, and saving energy. However, ancillary benefits can be gained by making the vehicles more amenable to automation. As the actuation systems are proven for automotive use, durability and robustness will become well established and prices will decline with volume production. This could make them promising candidates for use on Army UGVs.

Vehicle-to-Vehicle Wireless Communications

Interest is growing in the automotive world in the possibilities for wireless communication of data among vehicles to enhance driving safety and to provide new traveler information services. Recent work on standardization of vehicle–roadway communication (dedicated short-range communications [DSRC], at 5.9 GHz) has opened the door to inclusion of vehicle–vehicle communications within the same spectrum allocation and devices. The IEEE 802.11a R/A wireless standard is being adopted for vehicular use. This means that within the next few years the equipment for vehicle–vehicle communication could become very inexpensive and widely available. The DSRC standard will be applied to several different ranges of operation, perhaps extending as far as 1,000 meters, and with a sufficiently general geographic (i.e., GPS-based) addressing scheme it could be applied to any off-road battlefield environment for communications among UGVs and even between UGVs and nearby UAVs.

Automatic Vehicle-Following in Convoys

Several automotive research projects have developed approaches for enabling vehicles to follow each other automatically at close separations on highways. This could be applicable to Army supply convoys operating on-road or possibly even off-road. In the on-road applications it is necessary for the vehicles to operate very close together in order to produce the primary benefits (fuel savings and better utilization of highway infrastructure capacity) and to prevent nonequipped vehicles from cutting in front of the follower vehicles. In a battlefield environment it may be desirable to operate the vehicles much further apart, which introduces a contrasting set of system design requirements. The primary research on automatic vehicle-following for convoys of road vehicles has been performed by DaimlerChrysler in the CHAUFFEUR project in Germany (Schulze, 1997) and by the University of California PATH program (Rajamani and Shladover, 2001).

Foreign Government UGV Activity

A number of foreign governments sponsor research and development in robotics, including unmanned ground vehicle systems. The level of military interest ranges from high in some cases to no active interest. The military development paths include dedicated military research and development (R&D) programs as well as the application and exploitation of commercial robotic technologies. For many years commercial industries have developed robotic devices to relieve workers of repetitive and labor-intensive tasks; Japan is a notable example of commercial applications. Examples include such tasks as welding in difficult locations, spray painting, and repetitive assembly processes.

Potential military applications include a similar objective of relieving personnel of such repetitive and labor-intensive tasks as ammunition handling and loading, but more importantly the replacement of soldiers in hazardous tasks such as mine clearing, obstacle breaching, and disposal of unexploded ordnance. As semiautonomous and autonomous unmanned ground vehicles are further developed, it will become possible to replace personnel as well in such noncombat tasks as guard duty and logistic vehicle driving. Most importantly, as robotics technologies continue to develop both in commercial and military programs, the prospect of enhancing individual soldier performance becomes a real possibility.

The range of military mission areas under investigation in various countries is significant:

- Reconnaissance and surveillance: France, Germany, Great Britain, Israel
- Mine clearing and ordnance disposal: France, Germany, Great Britain, Israel
- Logistics: Great Britain, France, Germany
- Targeting: France, Germany, Great Britain, Israel
- Unmanned weapons platforms: France, Germany, Great Britain
- Camouflage, concealment, and detection: France, Germany, Great Britain.

It should be noted that commercial efforts in foreign countries, particularly in automotive industries such as Japan's, continue to make significant advances that contribute to general progress in robotics technology. Presently, the United States has technology agreements with France, Canada, the United Kingdom, Israel, and Germany.

Examples of several foreign UGVs described below illustrate that the efforts are similar to those being investigated in the United States (Hutchinson, 2001).

There are a number of foreign teleoperated vehicles. The Israeli Pele is a tank-mounted mine-clearing and breeching vehicle. Several vehicles by Great Britain include the Bison and Groundhog ordnance disposal vehicles, the Armored Vehicle Royal Engineers (AVRE) mounted on a Chieftain chassis and the Combat Engineer Tractor (CET). The British MARDI (Mobile Advanced Robotics Defense Initiative) test bed is teleoperated with optical fiber, intended for possible applications in RSTA and smokescreen operations.

The German PRIMUS (Program of Intelligent Mobile Unmanned Systems) mounted on the 4-ton Wiesel vehicle is teleoperated with some semiautonomous capabilities. A cooperative program is being conducted with the U.S. Army Research Laboratory on the auto-navigation system.

The French SYRANO (Systeme Robotise d'Acquisition pour la Neutralisation D'Objectifs) is teleoperated with optical fiber and with some semiautonomous capabilities. Possible applications include RSTA.

The foreign UGV activities are focused principally on platform-oriented R&D and on advanced concept demonstrations. Teleoperated vehicles are the most developed and are expected to be the principal applications in the next 5–15 years, with an increased application of semiautonomous capabilities, based on continuing development, particularly in computational power and in command and control capabilities. It is anticipated that smaller vehicles with limited intelligence will be utilized. Beyond 15 years increased autonomy is anticipated.

This section provides the basis for the answer to Task Statement Question 3.c in Box 3-4.

BOX 3-4
Task Statement Question 3.c

Question: Are there foreign UGV technology applications that are significantly more developed than those of the United States that, if acquired by the U.S. government or industry through cooperative venture, license, or sale, could positively affect the development process or schedule for Army UGV systems?

Answer: No. Based on the information available to the committee, there are no foreign UGV technology applications that are significantly more advanced than those of the United States.

4

Autonomous Behavior Technologies

An unmanned ground vehicle (UGV) encompasses the broad technology areas depicted in Figure 4-1. The next two chapters review and evaluate the state of the art in each of these UGV technology areas. This chapter evaluates technologies needed for the autonomous behavior subsystems that are unique to unmanned systems: perception, navigation, planning, behaviors and skills, and learning/adaptation.

As part of the evaluation of each technology area the committee estimated technology readiness levels (TRL) relative to the development of specific UGV systems. Table 4-1 summarizes the basic criteria for TRL estimates.

Technology areas responsible for autonomous behavior are depicted in Figure 4-2. It is important to note that these technologies are software-based, except for sensors (needed for A-B mobility and situation awareness). The figure illustrates how the software subsystems depend upon each other and are linked together to provide "intelligence" for a UGV.

The Perception subsystem takes data from sensors and develops a representation of the world around the UGV, called a world map, sufficient for taking those actions necessary for the UGV to achieve its goals. It consists of a set of software modules that carry out lower-level image-processing functions to segment features in the scene using geometry, color, or other properties up to higher-level reasoning about the classification of objects in the scene. The Perception subsystem can control sensor parameters to optimize perception performance and can receive requests from the planner or from the behaviors and skills subsystem to focus on particular regions or aspects of the scene.

The Navigation subsystem keeps track of the UGV's current position and pose (roll, pitch, yaw) in absolute coordinates. It also provides the means to convert vehicle-centered sensor readings into an absolute frame of reference. It will generally use a variety of independent means such as an IMU (inertial measurement unit), GPS (global positioning system), and odometry with estimates from all combined by a Kalman filter or something similar. It may make use of visual landmarks if they can be provided by the Perception subsystem.

The Planning subsystem is a hierarchy of modules: the Mission Planner decides B is the destination; the Navigator does global A to B path planning based on an *a priori* map and other data; the Pilot does moment-to-moment trajectory planning. Using information from the Navigation subsystem and the world model, the planner can also plan sensor and sensor data-processing activities. For example, it can cue certain sensors to point in a particular direction or activate a specific feature detection algorithm.

Software for Behaviors and Skills combines inputs from Perception, Navigation, and Planning and translates them into motor commands for the UGV to move and accomplish work. This also includes software necessary for the robot to accomplish specific mission-functions, including those based on tactics, techniques, and procedures used in military operations.

Learning/Adaptation software is used to improve performance through experience. It offers a way for a system to become robust over time (i.e., to be able to handle variability not initially anticipated by the system's programmers). Learning is not implemented as a separate subsystem but is incorporated as part of Perception, Navigation, Planning, and Behaviors.

PERCEPTION

The perception technologies discussed in this section include the sensors, computers, and software modules essential for the fundamental UGV capabilities of A to B mobility and situation awareness. The section describes the current state of the art, estimates the levels of technology readiness, identifies capability gaps, and recommends areas of research and development needed. Additional details relating to perception for autonomous mobility are contained in Appendix C.

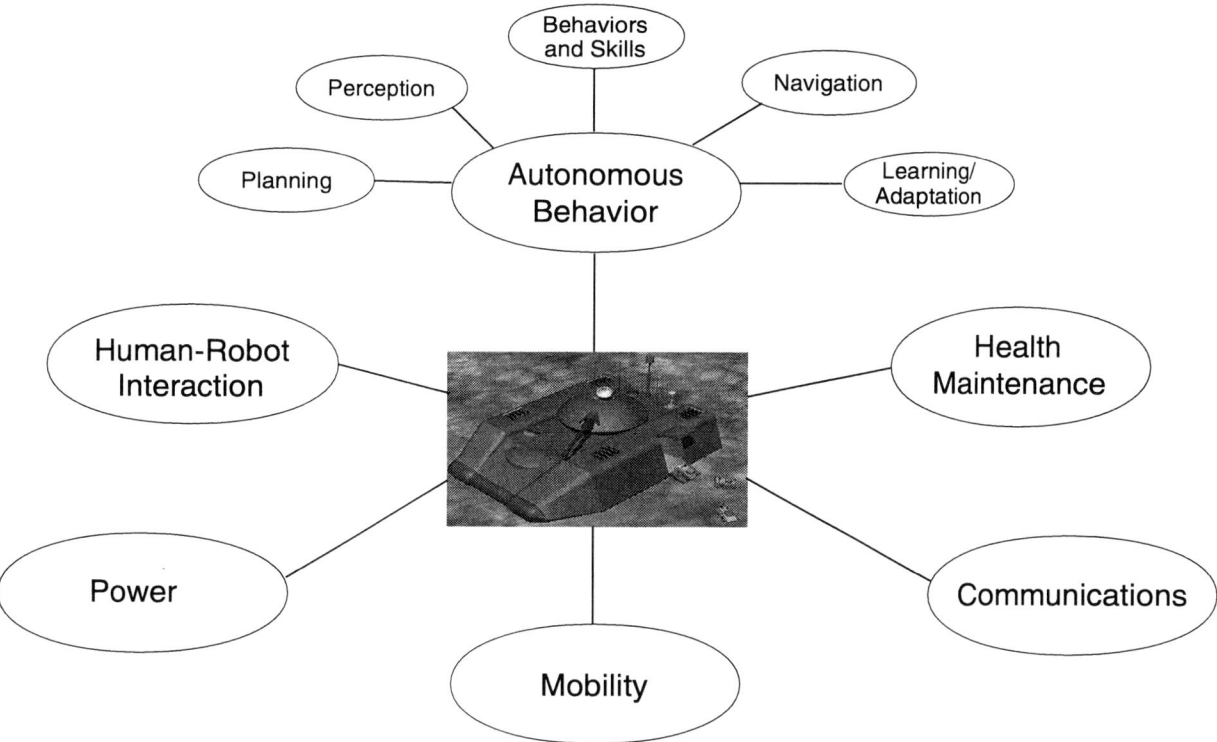

FIGURE 4-1 Areas of technology needed for UGVs.

A UGV's ability to perceive its surroundings is critical to the achievement of autonomous mobility. The environment is too dynamic and map data too inaccurate to rely solely on a single navigation means, such as the global positioning system (GPS). The vehicle must be able to use data from onboard sensors to plan and follow a path through its environment, detecting and avoiding obstacles as required.

The goal of perception technology is to relate features in the sensor data to those features of the real world that are sufficient, both for the moment-to-moment control of the vehicle and for planning and replanning. Humans are so good at perception, the brain does it so effortlessly, that we tend to underestimate its difficulty. It is difficult, both because the perception process is not well understood and because the algorithms that have been shown to be useful in perception are computationally demanding.

Technical Objectives and Challenges

The actions required by a UGV to move from A to B take place in a perceptually complex environment. An FCS UGV is likely to operate in any weather (rain, fog, snow), during day or night, in the presence of dust or other battlefield obscurants, and in conjunction with friendly forces opposed by an enemy. Perception system tasks are summarized in Table 4-2.

The UGV must be able to avoid positive obstacles such as rocks or trees (or indoors obstacles like furniture) and a negative obstacle such as a ditch. Water obstacles present special challenges; the UGV must avoid deep mud or swampy regions, where it could be immobilized, and must traverse slopes in a stable manner so that it will not turn over. The move from A to B can take place in different terrains and vegetation backgrounds (e.g., desert with rocks and cactus, woodland with varying canopy densities, scrub grassland, on a paved road with sharply defined edges, in an urban area), with different kinds and sizes of obstacles to avoid (rocks in the open, fallen trees masked by grass, collapsed masonry in a street), and in the presence of other features that have tactical significance (e.g., clumps of grass or bushes, tree lines, or ridge crests that could provide cover).

Each of these environments imposes its own set of demands on the perception system, modified additionally by such factors as level of illumination, visibility, and surrounding activity. In addition to obstacles it must detect such features as a road edge if the path is along a road, or features indicating a more easily traversed local trajectory if it is operating off-road. The perception system must be able to detect, classify, and locate a variety of natural and manmade features to confirm or refine the UGV's internal estimate of its location (recognize land marks); to validate assumptions made by the global path planner prior to initiation of the

TABLE 4-1 Criteria for Technology Readiness Levels

TRL Number	Description
1. Basic principles observed and reported	Lowest level of technology readiness. Scientific research begins to be translated into applied research and development. Examples might include paper studies of a technology's basic properties.
2. Technology concept and/or application formulated	Invention begins. Once basic principles are observed, practical applications can be invented. The application is speculative and there is no proof or detailed analysis to support the assumption. Examples are still limited to paper studies.
3. Analytical and experimental critical function and/or characteristic proof of concept	Active research and development is initiated. This includes analytical studies and laboratory studies to physically validate analytical predictions of separate elements of the technology. Examples include components that are not yet integrated or representative.
4. Component and/or breadboard validation in laboratory environment	Basic technology components are integrated to establish that the pieces will work together. This is relatively "low-fidelity" compared to the eventual system. Examples include integration of ad hoc hardware in a laboratory.
5. Component and/or breadboard validation in relevant environment	Fidelity of breadboard technology increases significantly. The basic technological components are integrated with reasonably realistic supporting elements so that the technology can be teased in a simulated environment. Examples include "high-fidelity" laboratory integration of components.
6. System/subsystem model or prototype demonstration in a relevant environment	Representative model or prototype system, which is well beyond the breadboard tested for TRL 5, is tested in a relevant environment. Represents a major step up in a technology's demonstrated readiness. Examples include testing a prototype in a high-fidelity laboratory environment or in simulated operational environment.
7. System prototype demonstration in an operational environment	Prototype near or at planned operational system. Represents a major step up from TRL 6, requiring the demonstration of an actual system prototype in an operational environment, such as in an aircraft, vehicle, or space. Examples include testing the prototype in a test-bed aircraft.
8. Actual system completed and "fight qualified" through test and demonstration	Technology has been proven to work in its final form and under expected conditions. In almost all cases, this TRL represents the end of true system development. Examples include developmental test and evaluation of the system in its intended weapon system to determine if it meets design specifications.
9. Actual system "fight proven" through successful mission operations	Actual application of the technology in its final form and under mission conditions, such as those encountered in operational, test and evaluation. In almost all cases this is the end of the last "bug-fixing" aspects of true system development. Examples include using the system under operational mission conditions.

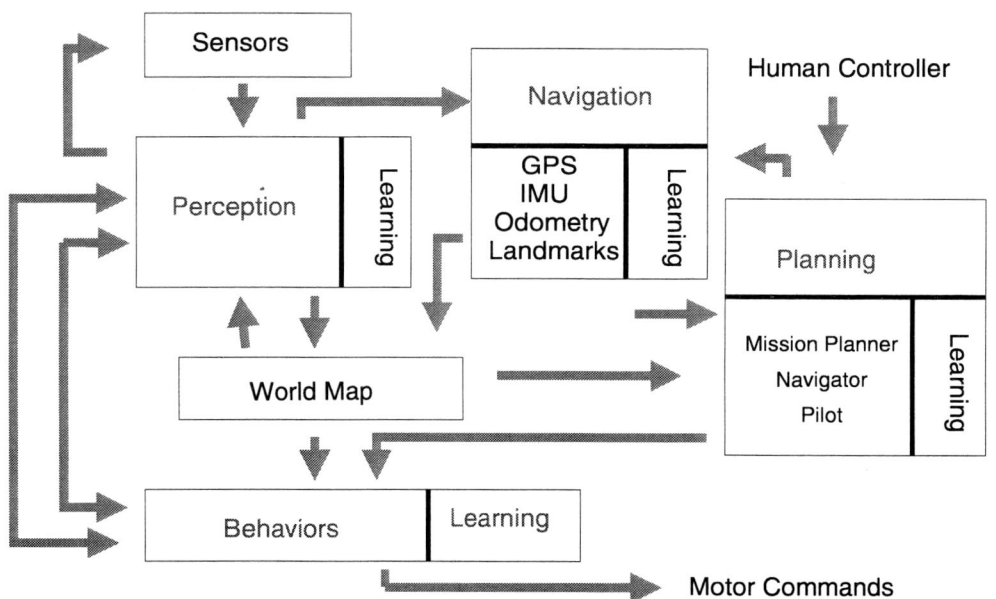

FIGURE 4-2 Autonomous behavior subsystems. Courtesy of Clint Kelley, SAIC.

TABLE 4-2 Perception System Tasks

On-Road	Off-Road
Find and follow the road	Follow a planned path subject to tactical constraints.
Detect and avoid obstacles	Find mobility corridors that enable the planned path or that support replanning.
Detect and track other vehicles	Detect and avoid obstacles.
Detect and identify landmarks	Identify features that provide cover, concealment, vantage points or as required by tactical behaviors. Detect and identify landmarks. Detect, identify, and track other vehicles in formation. Detect, identify, and track dismounted infantry in force.

traverse (e.g., whether the region of the planned path is traversable); to gather information essential for path replanning (e.g., identify potential mobility corridors) and for use by tactical behaviors[1] (e.g., "when you reach B, find and move to a suitable site for an observation post" or "move to cover"). The perception horizon begins at the front bumper and extends out to about 1,000 meters. Figure 4-3 illustrates the different demands that might be placed on a UGV perception system.

Specific objectives for A-to-B mobility are derived from the required vehicle speed and the characteristics of the assumed operating environment (e.g., obstacle density, visibility, illumination [day/night], and weather [affects visibility and illumination but may also alter feature appearance]). Table C-1 in Appendix C summarizes the full scope of environments, obstacles, and other perceptual challenges to autonomous mobility.

State of the Art

In the 18 years since the beginning of the Defense Advanced Research Projects Agency (DARPA) Autonomous Land Vehicle (ALV) program, there has been significant progress in the canonical areas of perception for UGVs: road-following, obstacle detection and avoidance (both on-road and off), and terrain classification and traversability analysis for off-road mobility. There has not been comparable progress at the system level in attaining the ability to go from A to B (on-road and off) with minimal intervention by a human operator. There are significant gaps in road-following capability and performance characterization particularly for the urban environment, for unstructured roads, and under all-weather conditions. Driving performance more broadly, even on structured roads, is well below that of a human operator. There is little evidence that perception technology is capable of supporting cross-country traverses of tactical significance, at tactical speeds, in unknown terrain, and in all weather, at night, or in the presence of obscurants. Essentially no perception capability exists (excluding limited UGV RSTA [reconnaissance, surveillance, and target acquisition] demonstrations) beyond 60 meters to 80 meters. Ability to detect tactical features or to carry out situation assessment in the region 100 meters to 1,000 meters is nonexistent as a practical matter.

The state of the art is based primarily on the DOD and Army Demo III project, the DARPA PerceptOR (Perception Off-Road) project, and research supported by the U.S. Department of Transportation, Intelligent Transportation Systems program. The foundation for much of the current research was provided by the DARPA ALV project, 1984–89, and the DARPA/Army/OSD Demo II project, 1992–98. Perception capabilities demonstrated by these and other projects are described in Appendix C and Appendix D.

On-Road

Army mission profiles show that a significant percentage of movement (70 percent to 85 percent) is planned for primary or secondary roads. Future robotic systems will presumably have similar mission profiles with significant on-road components. In all on-road environments the perception system must at a minimum detect and track a lane to provide an input for lateral or lane-steering control (road-following); detect and track other vehicles either in the lane or oncoming to control speed or lateral position; and detect static obstacles in time to stop or avoid them.[2] In the urban environment, in particular, a vehicle must also navigate intersections, detect pedestrians, and detect and recognize traffic signals and signage.

On-road mobility has been demonstrated in three environments: (1) open-road: highways and freeways; (2) urban "stop and go"; and (3) following dirt roads, jeep tracks, paths and trails in less structured environments from rural to undeveloped terrain. Unstructured roads pose a challenge because the appearance of the road is likely to be highly variable, generally with no markings, and edges may not be distinct.

---

[1]The tactical behaviors are assumed to also encompass the positioning of the UGV as required by the on-board mission packages (e.g., RSTA, obscurant generation, mine clearance, weapons). The mission packages may also have organic sensors and processing which will not be considered here.

[2]These behaviors are necessary but not sufficient for "driving" behavior, which requires many more skills.

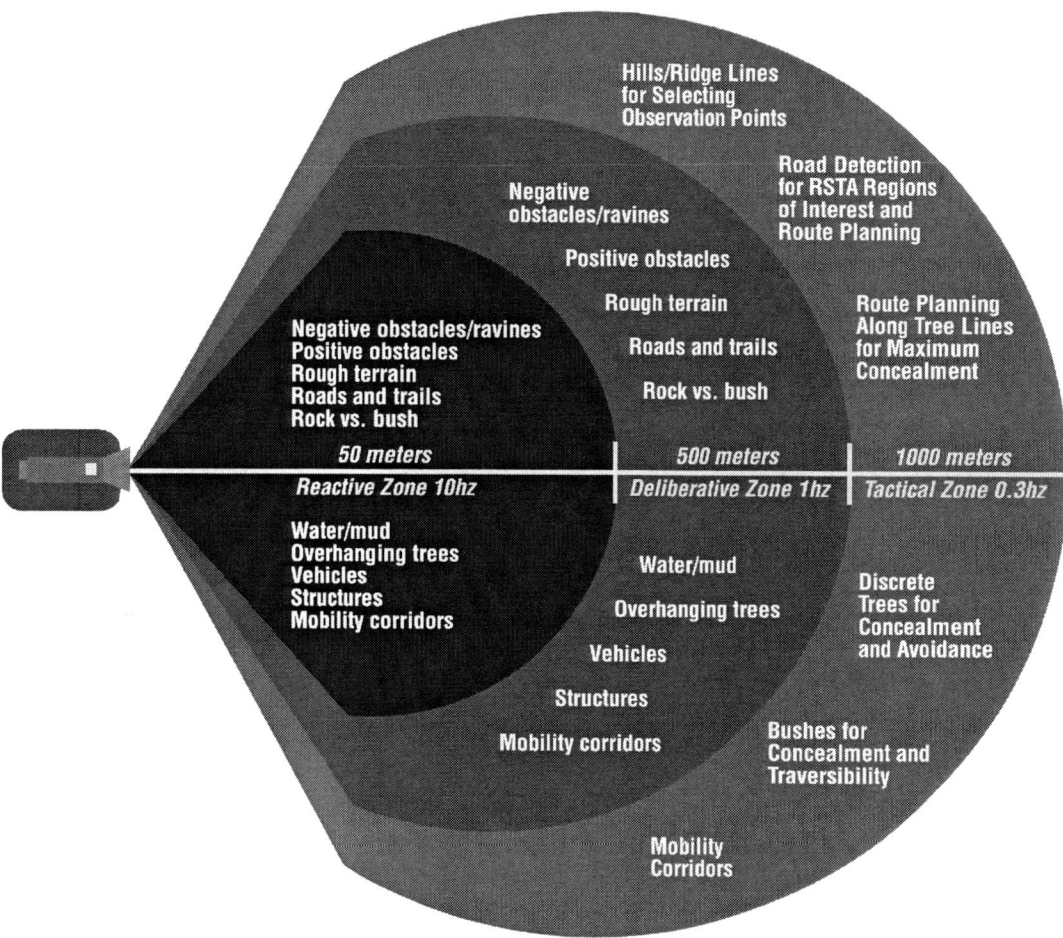

FIGURE 4-3 Perception zones for cross-country mobility. Courtesy of Benny Gothard, SAIC.

Perception for lane detection warning a driver of lane departures on structured, open roads are at the product stage. About 500,000 miles of lane detection and tracking operation has been demonstrated on highways and freeways. Lanes can be tracked at human levels of driving speed (e.g., 65 mph) or better under a range of visibility conditions (day, night, rain) and for a variety of structured roads (see Jochem, 2001). None of the systems can match the performance of an alert human driver using context and experience in addition to perception. Most systems are advisory and do not control the vehicle, although the capability exists to do so.

On-road mobility in an urban environment is very difficult. Road-following, intersection detection, and traffic avoidance cannot be done in any realistic situation. Signs and traffic signals can be segmented and read only if they conform to rigidly defined specifications and if they occupy a sufficiently large portion of the image. Pedestrian detection remains a problem. A high probability of detection is accompanied by a high rate of false positives. Because of the complexity of the urban environment, approaches must be data-driven, rather than model-driven. A variety of specialized classifiers or feature detectors are required to provide accurate and rapid feature detection and classification. Running all of these continuously requires considerable computing power. Research is required on controller strategies to determine which should be active at any time. Active camera control (active vision) is required for the urban environment because of the simultaneous need for wide fields of view and high resolution. Little research has been done on the use of active vision in an urban environment. (See Appendix C.)

Autonomous mobility on less structured roads has not received much emphasis despite its potential military importance. There is no experience comparable to the 500,000 miles or more of lane detection and tracking operation on highways and freeways; limitations are not as well understood and systems are not as robust. The limited experiments suggest UGVs can operate day and night at about human levels of driving speed (e.g., 10 mph to 40 mph) only on unstructured roads (secondary roads, dirt roads, jeep tracks, trails) where the road is dry, relatively flat with gentle slopes, no sharp curves and no water crossings or standing water. The road must be differentiable from background, using readily computed features. Current approaches may lose the road on sharp curves or classify steep slopes as obstacles.

Difficulty will be encountered when the "road" is defined more by texture and context. Performance on unstructured roads can be significantly affected by weather to the extent it reduces the saliency of the perceptual cues, and is likely to vary from road to road. Standing water can be detected but depth cannot be estimated. Mud can be detected in some cases, but not in others. Crown height can be measured but generally is not used. Texture could be used to provide warning of upcoming rough road segments, but computational issues remain.

On-road obstacle detection (other vehicles and static objects) is much less developed than lane tracking. Capability has been demonstrated using both active (LADAR, radar) and passive (stereo-video) sensors. For example, static objects 14 cm tall were detected at distances of 100 meters using stereo-video in daylight (Williamson, 1998). Vehicles were detected at up to 180 meters using radar (Langer and Thorpe, 1997). A very narrow sensor field of view is required to detect obstacles in time to stop or avoid them at road speeds, and sensors and processing must be actively controlled. Most demonstrations of obstacle detection were staged under conditions much less demanding than real-world operations and comprehensive performance evaluation has not been done. Capabilities developed for off-road obstacle detection are applicable to the on-road problem but have been demonstrated only at low speed. On-road obstacle detection has not generally been integrated with lane-tracking behavior for vehicle control.

Road-following assumes that the vehicle is on the road. A special case is detecting a road, particularly in a cross-country traverse, where part of the planned path may include a road segment. The level of performance on this task is essentially unknown.

Road-following capability can support leader-follower operation in militarily significant settings where the time interval between the preceder (route-proofing) vehicle and the follower is sufficiently short so that changes to the path are unlikely. Autonomous, unaccompanied driving behavior, particularly in traffic or in urban terrain with minimum operator intervention, is well beyond the state of the art. Consider the performance of the human driver relative to today's UGVs. The current road safety statistics for the United States reveal that the mean time between injury-causing crashes is tens of thousands of vehicle hours. By contrast, it would be a major challenge for a UGV to get through 0.1 hours of unassisted driving in moderate to heavy traffic, and it is doubtful that that could be accomplished consistently in a statistically valid series of experiments. Despite impressive demonstrations today's automated systems remain many orders of magnitude below human driving performance under a realistic range of challenging driving conditions.

Insufficient attention has been given on-road driving behavior in view of Army mission profiles, which call for vehicles to operate mostly on-road. Essentially no research has been done on the additional skills beyond road-following and obstacle avoidance required to enable driving behavior more generally.

Off-Road

Autonomous off-road navigation requires that the vehicle characterize the terrain as necessary to plan a safe path through it and detect and identify features that are required by tactical behaviors. Characterization of the terrain includes describing three-dimensional terrain geometry, terrain cover, and detecting and classifying features that may be obstacles including rough or muddy terrain, steep slopes, and standing water, as well as such features as rocks, trees, and ditches.

No quantitative standards, metrics, or procedures exist for assessing off-road UGV performance. It is difficult to know if progress is being made in off-road navigation and where deficiencies may exist. Unlike road-following, speed as a metric to gauge progress in off-road mobility is incomplete and may be misleading. No meaningful comparisons can be made without knowing the environmental conditions, the details of the terrain, and in particular, how much reliance was placed on prior knowledge to achieve demonstrated performance.

Published results and informal communications provide no evidence that UGVs can drive off-road at speeds equal to those of manned vehicles. Although UGV speeds up to 35 km/h have been reported, the higher speeds have generally been achieved in known benign terrain, and under conditions that did not challenge the perception system nor the planner. During the ALV and Demo II experiments in similar benign terrain, manned HMMWVs (high-mobility multi-purpose wheeled vehicles) were driven up to 60 km/h. In more challenging terrain the top speeds for all vehicles would be lower but the differential likely greater.

The ability to do all-weather or night operations or operations in the presence of battlefield obscurants has not been adequately demonstrated. In principle LADAR-based (laser detection and ranging) perception should be relatively indifferent to illumination and should operate essentially the same in daylight or at night. FLIR (forward looking infrared radar) also provides good nighttime performance. LADAR does not function well in the presence of obscurants. Radar or FLIR has potential depending on the specifics of the obscurant. There has not been UGV system-level testing in bad weather or with obscurants, although experiments have been carried out with individual sensors. Much more research and system-level testing under realistic field conditions are required to characterize performance.

The heavy almost exclusive dependence of DEMO III on LADAR may be in conflict with tactical needs. Strategies to automatically manage the use of active sensors must be developed. Depending on the tactical situation, it may be appropriate to use them extensively, only intermittently, or not at all.

RGB (red, green, blue, including near IR) video provides a good daytime baseline capability for macro terrain classification: green vegetation, dry vegetation, soil/rocks, and sky. Material properties can now be used with geometry to more accurately classify features as obstacles. This capability is not yet fully exploited. More detailed levels of classification during the day require multiband cameras (or a standard camera with filters), use of texture and other local features, and more sophisticated classifiers. Detailed characterization of experimental sites (ground truth) is required for progress. More research is required on FLIR and other means for detailed classification at night. Simple counts of LADAR range hits give a measure of vegetation density, once vegetation has been classified, and provide an indication of whether the vehicle can push through. Reliable detection of water remains a problem. Different approaches have been tried with varying degrees of success. Fusion may provide more reliable and consistent results.

Positive obstacles that subtend 10 or more pixels, that are not masked by vegetation or obscured for other reasons, and are on relatively level ground, can be reliably detected by stereo at speeds no greater than about 20 km/h, depending on the environment. LADAR probably requires 5 pixels and can reliably detect obstacles at somewhat higher speeds (e.g., 30km/h). LADAR, stereo-color, and stereo-FLIR all work well for obstacle detection. Day and night performance should be essentially equivalent, but more testing is required. Again, less is known about performance under bad weather or obscurants.

Little work has been done to explicitly measure the size of obstacles. This bears on the selection of a strategy by the Planner. No proven approach has been demonstrated for the detection of occluded obstacles. LADAR works for short ranges in low-density grass. There have been some promising experiments with fast algorithms for vegetation removal, which could extend LADAR detection range. Some experiments have been done with FOLPEN (foliage penetration) radar but the results are inconclusive. Radar works well on some classes of thin obstacles (e.g., wire fences). LADAR can also detect wire fences. Stereo and LADAR can detect other classes of thin obstacles (e.g., thin poles or trees). Radar may not detect nonmetallic objects, depending on moisture content. Much more research is required to characterize performance.

Detection of negative obstacles continues to be limited by geometry. While performance has improved because of gains in sensor technology (i.e., 10 pixels can be placed on the far edge at greater distances), sensor height establishes an upper bound on performance, and negative obstacles (depressions less than a meter wide) cannot be seen beyond about 20 meters. With the desire to reduce vehicle height to improve survivability the problem will become more difficult.

Little work has been done on detecting tactical features at ranges of interest. Tree lines and overhangs have been reliably detected but only at ranges less than 100 meters. Essentially no capability exists for feature detection or situation assessment for ranges from about 100 meters out to 1,000 meters.

Cross-country capability is very immature and limited. Demonstrations have been carried out in known, relatively benign environments; have seemingly been designed to highlight perception and other system strengths and potential military benefits; and have consequently done much less to advance the state of the art. Such demonstrations may potentially mislead observers as to the maturity of the state of the art.

Improvements in individual sensor capability, sensor data fusion, and in active vision are required to achieve autonomous A-to-B mobility. Improvements in LADAR range, frame rate, and instantaneous field of view (IFOV) are necessary and improvements in video resolution and dynamic range are desirable. Multi- or hyperspectral sensors could substantially improve the ability to do rapid terrain classification in daylight. Multiband thermal FLIR could potentially allow terrain classification at night. However, the conditions under which the UGVs must operate are so diverse that no single sensor modality will be adequate. Different operating conditions (missions, terrains, weather, day/night, obscurants) will pose different problems for each sensor modality, and complementary sensor systems with different vulnerabilities will be needed to provide system robustness through data fusion.

Much work will be required to translate existing research on sensor fusion into a capability for UGVs. Active vision must also be emphasized to address the trade-off between IFOV and required field of regard. Again, research exists but useful applications lag. Use of active vision could provide earlier obstacle detection and reduce the likelihood of the vehicle becoming trapped in a cul-de-sac. The development of appropriate algorithms for data fusion and active vision and their integration into the UGV perception system should be a high priority.

Technology Readiness

Except for the teleoperated Searcher UGV, the example systems defined in Chapter 2 presuppose a number of firm requirements for perception. The most fundamental are those to move autonomously from A to B either on roads or cross-country. Three maximum speeds were specified: 40 km/h, 100 km/h, and 120 km/h. Movement was to take place under day, night, or limited visibility conditions. Table 4-3 refines the TRL criteria used to estimate technology readiness for perception technologies. Tables 4-4 through 4-8 then provide TRL estimates as associated with particular sensor technologies for mobility, detection, and situation awareness. "Speed" in the tables corresponds with the example UGV systems as follows: 40 km/h (Donkey), 100 km/h (Wingman), and 120 km/h (Hunter-Killer). (No perception

TABLE 4-3 Technology Readiness Criteria Used for Perception Technologies

TRL	
TRL 6	Demonstrated in large numbers of representative environments. High probability of meeting performance objectives in any likely environment.
TRL 5	Components integrated and system tested in a few representative environments. Does not meet performance objectives. Relatively modest enhancements to algorithms required to meet objectives. More extensive testing required. 12–24 months to TRL 6.
TRL 4	Proof of concept demonstrated. Major components integrated. Additional components may have to be developed added. Moderate integration risk. Moderate length test-mobility-test cycle. Modifications to components likely after testing identifies shortfalls. 24–36 months to TRL 6.
TRL 3	Best approach identified. Some components exist and have been validated. Research base exists for developing additional components. Integration risk is high. Extensive test-mobility-test cycle. 36–60 months to TRL 6.
TRL 2	Uncertain as to best approach. Substantial research required to identify approach and develop algorithms. Very long test-modify-test cycle. 60–120 months to TRL 6.
TRL 1	Concepts available that can be implemented in software and limited evaluation initiated under controlled conditions. Uncertainty as to time to TRL 6, but no less than 10 years.

TRL = technology readiness level.

TABLE 4-4 TRL Estimates for Example UGV Applications: On-Road/Structured Roads

Speed (km/h)	Day	Night and Limited Visibility[a]
Lane-Following and Speed Adjustment (Collision Avoidance)		
40	TRL 4/[b]TRL 4 SV or MV+R	TRL 3/[c]TRL 3 SFLIR or MFLIR+R
100	TRL 4/TRL 3 SV or MV+R	TRL 3/TRL 2 SFLIR or MFLIR+R
120[d]	TRL 3/TRL 2 SV or MV+R	TRL 3/TRL 2 SFLIR or MFLIR+R
Obstacle Avoidance		
40	TRL 3/[e]TRL 3 SV or LADAR	TRL 3/[f]TRL 3 SFLIR, LADAR, radar
100[g]	TRL 3/TRL 2 SV or LADAR	TRL 3/TRL 2 SFLIR, LADAR, radar
120[h]	TRL 2/TRL 1	TRL 2/TRL 1

[a]Includes rain, snow, fog, and manmade obscurants.
[b]Demonstrated lane-following and speed adjustment.
[c]Demonstrated lane-following only, but speed adjustment with radar demonstrated in daylight should work equally well.
[d]Architecture must be optimized for real-time performance. The assumption for urban environments is that the vehicle will maneuver similarly to rescue vehicles or police in pursuit (i.e., as fast as circumstances permit but no faster than 120 km/h).
[e]Obstacle avoidance integrated with road-following.
[f]Will require data fusion (e.g., multiple IR bands, FLIR with radar).
[g]At about this speed or greater, active vision required.
[h]No obstacle detection capability demonstrated at 100 or 120 km/h.

Note: TRL = technology readiness level; SV = stereo video; MV+R = monocular video plus radar; SFLIR = stereo forward looking infrared; MFLIR+R = monocular forward looking infrared plus radar; LADAR = laser detection and ranging.

capabilities are required by the Searcher example.) The estimates are highly aggregated judgments of performance across a variety of situations:

- *On-road:* Includes performance on structured and unstructured roads from those designed to standards and are well marked to barely perceptible dirt tracks. Structured roads have known, constant geometries (e.g., lane width, radius of curvature) and clear lane and boundary markings. Unstructured roads may be of variable geometry, have abrupt changes in curvature, and may be difficult to distinguish from background (may be paved or unpaved). Environments range from open-road to urban stop-and-go to open road. Performance includes lane-following and speed adjustment to avoid vehicles in lane (moving obstacles). In an urban environment performance may also require intersection detection and navigation and traffic signal and signage recognition and understanding. Obstacle avoidance requires detection of stopped vehicles, pedestrians, and static objects. In a combat environment obstacles may include bomb craters, masonry piles, or other debris. Obstacle detection on unstructured roads, in particular, may be more difficult because curves or dips may limit opportunity to look far ahead.

- *Off-road:* Terrain types are highly variable (e.g., desert, mountains, swampy terrain, forests, tall-grass-covered plains); have positive and negative obstacles (e.g., ditches, gullies) some of which will be visible and others that will be hidden in cover. Performance requires segmenting the terrain into traversable and nontraversable regions using geometry (i.e., size of features and assessment of material properties [rock, soil, vegetation, including assessment of terrain roughness, fordable water, and trafficability of steep slopes, or muddy or swampy regions]).

- *Detection of tactical features:* Requires identifying natural and manmade features that could provide cover or concealment (e.g., tree lines or ridge crests, large rocks, buildings) or support mission packages (e.g., select a site for an observation post). Region: 100 meters to 1,000 meters.

TABLE 4-5 TRL Estimates for Example UGV Applications: On-Road/Unstructured Roads

Speed (km/h)	Day	Night and Limited Visibility
Lane-Following and Speed Adjustment		
40	TRL 3[a] SV or MV+R	TRL 3 SFLIR or MFLIR+R
100[b]	TRL 3	TRL 3
120	TRL 3	TRL 3
Obstacle Avoidance		
40	TRL 3[c]	TRL 3
100	TRL 3	TRL 2
120	TRL 2	TRL 2

[a]Need color or texture segmentation to cover all likely situations. ALVIN, RALPH, and Robin demonstrated road-following during day. Robin at night with FLIR. Not integrated with speed adjustment (e.g., radar) for unstructured roads nor with obstacle avoidance.

[b]Require active vision for lane-following at higher speeds due to possibility of abrupt curves.

[c]Obstacle avoidance demonstrated at 40 km/h but not integrated with road-following or speed adjustment on unstructured roads.

Note: TRL = technology readiness level; SV = stereo video; MV+R = monocular video plus radar; SFLIR = stereo forward looking infrared; MFLIR+R = monocular forward looking infrared plus radar.

TABLE 4-6 TRL Estimates for Example UGV Applications: Off-Road/Cross-Country Mobility

Speed (km/h)	Day	Night and Low Visibility
Terrain Classification		
40	TRL 4 Color video, multiband	TRL 2 Multiband FLIR
100[a]	TRL 2	TRL 2
120	TRL 1	TRL 1
Obstacle Avoidance[b]		
40	TRL 5 LADAR, SV, FOLPEN	TRL 3 LADAR, SFLIR, FOLPEN
100[c]	TRL 3	TRL 3
120	TRL 1	TRL 1

[a]Requires macro-texture analysis, terrain reasoning to predict terrain roughness.

[b]Uses geometry alone or applies geometric criteria to objects that pass through material classification sieve.

[c]Requires active vision.

Note: TRL = technology readiness level; LADAR = laser detection and ranging; SV = stereo video; FOLPEN = foliage penetration; SFLIR = stereo forward looking infrared.

TABLE 4-7 TRL Estimates for Example UGV Applications: Detection of Tactical Features

Example[a]	Day	Night
Donkey	TRL 4[b]	TRL 3
Wingman	TRL 3	TRL 3
Hunter-Killer	TRL 2	TRL 2

[a]Donkey: cover and concealment (natural and manmade); Wingman: cover and concealment; Hunter-Killer: cover and concealment, select observation post (OP), select ambush site and kill zone.

[b]Very limited, tree-lines and overhangs.

Note: TRL = technology readiness level.

TABLE 4-8 TRL Estimates for Example UGV Applications: Situation Assessment

Example[a]	Day	Night
Donkey	TRL 2	TRL 2
Wingman	TRL 2	TRL 2
Hunter-Killer	TRL 1	TRL 1

[a]Donkey: detect, track, and avoid other vehicles or people; Wingman: track manned "leader" vehicle, detect, track, and avoid other vehicles or people, distinguish among friendly and enemy combat vehicles, and detect unanticipated movement or activities; Hunter-Killer: detect, track, and avoid other vehicles or people, discriminate among friendly and enemy vehicles, detect unanticipated movement or activities, and detect potential human attackers in close proximity.

Note: The assumption is that the focus is on a region extending from 100 meters to 1,000 meters. RSTA is assumed to start at 1,000 meters. TRL = technology readiness level.

- *Situation assessment:* Requires identifying and locating friendly and enemy vehicles and dismounted personnel in a region extending from 100 meters to 1,000 meters.

In addition to the task-specific variables above, perception performance will be affected by weather, levels of illumination, and natural and manmade obscurants that affect visibility.

Salient Uncertainties

The success in detecting and tracking vehicles for traffic avoidance argues for the eventual success of on-road perception-based leader-follower operation.[3] Limited success in

---

[3]The leader-follower work carried out as part of the Demo II and Demo III programs was based on GPS, not perception.

detecting pedestrians suggests that off-road leader-follower, where the vehicle follows dismounted infantry, is also a long-term potential.

To be useful for any mission UGVs must be able to go from A to B with minimal intervention by a human operator; however, there are no quantitative standards, metrics, or procedures for evaluating UGV performance. There is uncertainty as to how much progress has been made and where deficiencies exist. For example: Is DEMO III performance improved over DEMO II? If so, by how much? For what capabilities and under what conditions? Because there is little statistically valid test data, particularly in environments designed to stress and break the system (e.g., unknown terrain, urban environments, night, and bad weather), there is considerable uncertainty as to how systems might perform in these environments. Similarly, there is no systematic process for benchmarking algorithms in a systems context and corresponding uncertainty as to where improvements are required.

The foregoing provides the basis for the answer to Task Statement Question 4.a in Box 4-1.

Recommended Research

As a high priority, the Army should develop predictive performance models and other tools for UGV autonomous behavior architecture system engineering and performance optimization. This work includes:

- Statistically valid data collection in unknown environments under stressing conditions leading to the development of predictive performance models, and
- Development of performance metrics and algorithm benchmarking.

An equally high priority should go to development and integration of real-time algorithms for data fusion and active vision. Other important areas include development and integration of real-time algorithms for terrain classification using texture analysis and multispectral data and development and integration of algorithms for sensor management, particularly active sensors.

NAVIGATION

Navigation for UGV is a large problem domain that includes such elements as current location (both absolute and relative); directions to desired location(s) such as final destination or intermediate waypoints; aiding in situational awareness (SA) including providing the location of friendly forces and targets over a large region; the mapping of immediate surroundings, how to navigate about the immediate surroundings and how to navigate to the next waypoint or final destination; and the detection of nearby hazards to mobility. Navigation overlaps and has interrelationships with several other key areas of this study, including perception, path planning, behaviors, human–machine interface, and communications. One of the major goals of the navigation module is to aid in providing enough information to allow near-autonomous mobility for the UGV.

State of the Art

Currently GPS/INS is often used for airborne and ground vehicles to determine current location and to provide directions to desired locations. GPS/INS is a proven technology that is currently used in many applications. For GPS/INS, the inertial navigation system (INS) provides accurate relative navigation with the normal drift of the INS corrected by the absolute position obtained by GPS. With selective availability turned off GPS provides accuracy of 10 to 20 meters. This accuracy is dependent upon the geometry of the satellites used to determine the position. Horizontal position accuracy is usually better than vertical position accuracy. Horizontal errors of only 3 to 5 meters are common. Accuracy of 1 meter or less can be obtained using differential GPS (DGPS). One relative navigation technique for a communication network is to determine the relative position of each member of the network by ranging on the network communication signals. By ranging on all or most of the communications signals of a network the topology of the members

---

BOX 4-1
Task Statement Question 4.a
Perception Component of "Intelligent"
Perception and Control

Question: What are the salient uncertainties in the "intelligent" perception and control components of the UGV technology program, and are the uncertainties technical, schedule related, or bound by resource limitations as a result of the technical nature of the task, to the extent it is possible to enunciate them?

Answer: The greatest uncertainties are in describing UGV performance and in determining the effect of perception (and other subsystems) on UGV system performance. No metrics have been developed and no statistically significant data have been collected in unknown environments under stressing conditions.

There are no procedures for benchmarking algorithms and hence considerable uncertainty if the algorithms are best-of-breed. In the absence of metrics and data there is little basis for system optimization and a corresponding uncertainty about performance losses due to system integration issues. There is no systematic way to determine where improvements are required and in what order.

The uncertainties exist because of a lack of resources in the Army's program.

of the network can be determined. To pin down the absolute location of this topology requires that the absolute location of some of the members of the network be determined by some other method. To provide situation awareness and information about geographical surroundings, the UGV's current position can be tied to geographical information system (GIS) databases, such as detailed terrain maps and current situation maps. These databases can be stored on board the UGV or very recent databases can be downloaded by means of communication links. Current position information integrated with GIS databases has been used in many commercial products. Other relevant SA information including non–line of sight (NLOS) and beyond line of sight (BLOS) targets can be provided to the UGV from other team members by the communication network. Onboard sensors (i.e., perception) can also be used to detect and locate potential line-of-sight (LOS) targets and nearby friendly units.

To illustrate the current state and future needs of UGV navigation, the navigation aspects of each of the four example military applications from Chapter 2 are described separately in the following paragraphs. The first application, the Searcher, is a teleoperated UGV used to search urban environments (e.g., buildings) or tunnels. Because this UGV is teleoperated, the range from the operator to the UGV is likely to be less than 1 km, and the Searcher may even be within sight of the operator. Therefore, all navigation decisions can be made by the operator, and there is little need for any sophisticated navigation sensors onboard the Searcher. Teleoperation is currently being used successfully in several military robotic programs, including Matilda and the Standardized Robotic System (SRS).

Another UGV example application is the Donkey, an unmanned small unit logistics server. The Donkey is envisioned as being in the semiautonomous preceder/follower UGV class. The Donkey will follow electronic paths (electronic "bread crumbs") through urban or rural terrain from a start point to a release point. Navigation along this electronic path (e.g., GPS waypoints, radio frequency tags, or defined points on an electronic map) is critical for successful performance of the Donkey. If the path were defined as GPS waypoints, latitude/longitude points, or other absolute position points then the Donkey would probably use GPS/INS (or another beacon navigation system integrated with a relative navigation system) as its main navigation system.

To move along the electronic path various techniques utilizing onboard sensors combined with navigation equipment will allow the Donkey to detect immediate hazards and to navigate around these hazards while still progressing along the path (see sections titled "Perception" and "Path Planning"). Navigation techniques for the Donkey must also consider threat capabilities. Since all navigation techniques have some vulnerabilities, multiple navigation techniques should be used in conjunction to reduce these vulnerabilities. For the Donkey, environmental conditions along the path may have changed since the path was defined. The Donkey may have to operate in areas of GPS denial (or denial of other navigation beacons), either intentional (jamming) or environmental/unintentional (urban canyon, indoors, heavy foliage). Also, communication networks may be jammed. Thus navigation may have to be performed without any outside aiding (at least for some period of time).

There is much current work being done to alleviate some of the vulnerabilities of GPS to jamming (including development of both new signals and frequencies); however, it must be assumed that GPS will always have some vulnerabilities. For some current applications the combination of GPS and INS is used to resolve this problem. If GPS were denied, navigation could be performed by "riding" the INS until GPS is restored. If the Donkey could recognize its environment (perception), it may be able to determine its position based upon comparison of external sensor data with onboard maps, utilizing its last known position. The Donkey must also be able to detect when its navigation solution is in error and exhibit the appropriate behavior when this occurs. For GPS, receiver autonomous integrity monitoring (RAIM) is one technique used to verify the validity of individual satellite signals. RAIM has requirements dictating how quickly errors must be detected and what probability of missed errors or false positives are allowable.

The third UGV example application is the Wingman, a platform-centric autonomous ground vehicle. The navigation requirements of the Wingman include the ability to navigate to designated areas without any path information supplied (drive from point A to point B) and to operate at predefined standoff positions relative to the section leader. Thus, the Wingman will have to determine its absolute position and its position relative to the section leader, and to navigate with little supervision. In some instances human interaction from the section leader may aid the Wingman in determining its navigation position. Again navigation is critical for the successful performance of this UGV. The Wingman will probably use GPS/INS (or another beacon navigation system supplemented by INS) as its main navigation system. If high-accuracy positions were needed (errors of less than 10 meters), DGPS might also be required. The Wingman's relative position compared to the section leader can be determined by communication between the section leader and the Wingman in which each tells the other its absolute position. It may be possible for the Wingman to range off of communications signals from the section leader to aid in determining its relative position compared to the section leader. Theoretically, near-autonomous mobility (point A to point B) can be obtained by various techniques utilizing onboard sensors combined with navigation equipment to allow the UGV to detect immediate hazards and to navigate around these hazards while still progressing towards the desired location (see sections on "Perception" and "Path Planning"). Tests to date have shown that all techniques have drawbacks and near-autonomous mobility has yet to be achieved. One technique includes vision detectors utilizing pattern recognition and neural networks to identify hazards (UGV perception of external environment). Utilization of detailed terrain maps combined with GPS/INS has also been attempted, but GPS/INS position errors along with the need to update terrain maps as rapidly fluctuating conditions warrant have been stumbling points. The vulnerabilities mentioned in the above section on the Donkey also apply to the Wingman. Because the Wingman may also have automatic, lethal, direct fire capabilities, it is even more imperative for this UGV to determine its absolute position, the relative positions of all friendly/innocent assets, and detect when its navigation position is in error.

The final UGV example application is the Hunter-Killer unit, a group of network-centric autonomous vehicles. These UGVs will be tied together through a local wireless network. Autonomy from human inputs will be greater for the Hunter-Killer than for the Wingman. The navigational criticality and capabilities for the Hunter-Killer will be very similar to the Wingman discussed above. The Hunter-Killer will also probably use GPS/INS (or another beacon navigation system combined with a relative navigation system) as its primary navigation system. Because of the communications network inherent in the Hunter-Killer, relative navigation/geolocation of individual units can be performed by ranging on these communications signals. This will help to overcome the vulnerability of GPS/INS, on which UGV depends for navigation/geolocation, because in areas of GPS denial (e.g., urban environments), the ultra wide band (UWB) network signal may remain viable. Individual units may be able to send paths to other units to aid the other units in navigating from point to point. The vulnerabilities mentioned in the above section on the Donkey also apply to the Hunter-Killer. Like the Wingman, the Hunter-Killer will also have automatic, lethal, direct fire capabilities. Therefore it is imperative for this UGV to determine its absolute position, the relative positions of all friendly/innocent assets, and detect when its navigation position is in error.

For all example applications UGV navigation will be highly dependent upon the level of autonomy required of the UGV (see section on "Perception"). The final navigation solution (for all but the teleoperated Searcher) will probably involve an integration of onboard sensor information (sensor fusion), GPS/INS (or another beacon absolute navigation system integrated with a relative navigation system), navigation integrated with communications signals, the sharing of navigation information by all relevant assets (satellites, pseudolites, UAVs, other UGVs), and some operator oversight. Navigation requirements will thus impact (and be impacted by) perception, path planning, behavior, and communication requirements.

Technology Readiness Estimates

GPS/INS is a proven navigation technology (TRL 9). Other mature forms of navigation include dead reckoning (e.g., INS) and other beacon techniques (e.g., LORAN [long-range navigation]). Relative navigation on communications signals is also mature but requirements on bandwidth and utilization of timing sources complicate the problem. Teleoperation has been shown to be viable (TRL 6) in various programs including Matilda and the SRS. The ability of various communications signals (e.g., UWB) to be viable for relative navigation when GPS is denied needs to be investigated; some work is being done in this area, and the method is probably at a TRL 4 or 5.

Perception and path-planning technologies are closely related to navigation technologies, and the fusion of navigation, perception, path planning, and communications is the key to autonomous A-to-B mobility. This fusion is also the most technically challenging area and the least technically mature. Assigning a TRL value to this fusion of navigation, perception, path planning, and communications is difficult, but a reasonable guess is TRL 1 or 2.

Salient Uncertainties

The major technology gap for beacon absolute navigation (e.g., GPS/INS) is the threat of denial of the signal. For GPS much current work is ongoing to improve GPS anti-jamming characteristics, corrections for multipath problems in urban environments, and other GPS signal-tracking improvements. Combining GPS (or any other beacon navigation) with relative navigation using communications signals will help in areas where GPS is sometimes precluded; this combination of GPS and geolocation on communications signals needs to be developed. Further integration between navigation and communications will help to create more robust positioning solutions. Perception can also be used to aid in navigation, for example, the ability to determine the angle to various landmarks for which the position is known a priori can be used to determine the current position. The integration of navigation and perception is another technology gap that needs to be filled. UGVs must be able to detect when they are lost and then exhibit the appropriate behavior. For GPS, RAIM helps to meet this requirement, but for other navigation systems this ability to detect navigation errors will have to be developed. To reach near or full autonomy, UGV navigation will require the integration of perception, path-planning, communications, and various navigation techniques. This integration of multiple systems is the largest technology gap in autonomous navigation.

One of the greatest risks for UGV navigation is the interrelationships of navigation, perception, communications, path planning, man–machine interface, behavior, and the level of autonomy of the UGV. Decreased performance in any one of these areas implies greater emphasis on one or all other areas. For example, an inability of perception to delineate obstacles requires that navigation may have to rely more upon GPS/INS and maps or upon the man–machine interface (teleoperation or semiautonomy), either of which imply a greater reliance upon communications. The interrelationship between these disparate areas must be recognized and planned from the start of the program.

Because no one navigation solution meets all conditions, several navigation techniques must be included in any UGV design. Probably both an absolute and a relative navigation technique will be necessary. The selection of which navigation techniques to utilize will be requirements driven. The four military applications presented in Chapter 2 are a good start toward defining the problem. Requirements that have to be pinned down include:

- Do we expect to be in an area of GPS denial?
- Do we expect operators to always be able to communicate to UGVs?
- Is the operator–UGV communication real-time?
- Is relative position to other assets as important as absolute position?
- Is relative position "good enough" for most assets?
- How long can the UGV be expected to operate without an updated absolute navigation position?
- What are the mission goals?
- What is the expected behavior of the UGV when it is lost?
- What navigation failure rate or position error is tolerable?

Note that these requirements can be modified as the program progresses but a first cut at these requirements is necessary to bound the navigation problem. If initial requirements are not defined, program risk grows greatly. The possible military applications presented in previous sections are a good first step in defining initial requirements.

Full autonomous navigation for all conditions is probably not feasible, especially in the near future. Therefore, requirements for the man–machine interface for operator aiding need to be included in all future programs. How the UGV recognizes that it is lost and what the behavior should be when the UGV knows it is lost needs to be defined. The FCS program should define semiautonomous navigation capabilities at different levels of operator control. As navigation improves operator control can be lessened but at each stage of development a viable product is produced. Note that the progression of the possible military applications from the Searcher through to the Hunter-Killer is an evolution from no autonomous navigation through semiautonomous navigation all the way to full autonomous navigation.

The foregoing provides the basis for the answer to Task Statement Question 4.a as it pertains to the navigation component of "intelligent" perception and control. See Box 4-2.

Recommended Areas of R&D

Currently there is much ongoing research to improve GPS anti-jamming characteristics, corrections for multipath

---

BOX 4-2
Task Statement Question 4.a
Navigation Component of "Intelligent" Perception and Control

Question: What are the salient uncertainties in the "intelligent" perception and control components of the UGV technology program, and are the uncertainties technical, schedule related, or bound by resource limitations as a result of the technical nature of the task to the extent it is possible to enunciate them?

Answer: Further integration between navigation and communication will help to create more robust positioning solutions. UGVs must be able to detect when they are lost and then react appropriately. The ability to detect navigation errors will have to be developed. To reach near or full autonomy, UGV navigation will require the integration of perception, path planning, communication, and various navigation techniques. This integration of multiple systems is the largest technology gap in autonomous navigation.

These uncertainties are bound by resource limitations and result from the technical nature of the task.

---

problems in urban environments, and other GPS signal-tracking improvements. UGV navigation should be able to "ride the coat-tails" of these efforts to obtain the best GPS navigation solutions. Research and development should be done in the following areas to yield the desired autonomous navigation solutions for all possible environmental conditions:

1. Relative navigation utilizing communications signals (especially ultra-wide band signals due to their ubiquitous characteristics) integrated with GPS.
2. Improved integration of GPS (or any absolute navigation position) with accurate digitized maps to aid in point-to-point mobility and perception.
3. Integration of perception with accurate maps to allow UGV to determine its position by comparison of sensor input and with map information.
4. Integration of absolute navigation system (i.e., GPS) with sensor information (perception) and map information to determine a more accurate position.
5. Utilization or development of improved active beacons that are viable in urban, heavy foliage environments, or jammed environments (e.g., pseudolites, beacon signals for indoor navigation).
6. Development of error detection techniques for any navigation system chosen for UGV navigation.
7. Development of UGV behavior when the UGV detects that it is lost in order for the UGV to recover its position.

8. Integration of absolute navigation techniques (e.g., GPS, LORAN), relative navigation techniques (INS, dead-reckoning, relative position estimates based upon ranging on communication signals), position estimates based upon perception, information received from other assets (including UAVs, pseudolites), and path-planning information. This is a large research area and probably the most important research area for UGV navigation. The absolute and relative navigation techniques chosen are interrelated with other system requirements, including perception, communications, power, stealth (i.e., is the UGV entirely passive), available computer power, and path planning. Various combinations of these navigation techniques and other position estimation methods should be integrated and evaluated. Note that the interrelationships of navigation, path planning, perception, and communication must be evaluated at a system level.

Impact on Logistics

The use of any beacon navigation system will require the installation of the navigation beacons. Even for GPS it may be necessary to set up pseudolites or utilize airborne pseudolites.

The electronic path will have to be determined and disseminated before a Donkey UGV can be utilized. Maps may have to be generated and disseminated to aid in navigation for the Donkey, Wingman, and Hunter-Killer. These maps will have to be as recent as possible and may have to contain much SA information.

Communications to support navigation inputs will have to be set up for the Hunter-Killer and possibly for the Wingman. These same resources may also be needed to support communications with other assets in the area of operations.

PLANNING

This section defines the scope of the planning technology area. It describes the mid- and far-term state of the art, and identifies the impact, if any, on Army operations or logistics. Planning for Army UGV systems encompasses software for path planning, which interacts with both perception and navigation, and mission planning.

Path Planning

Path planning is the process of generating a motion trajectory from a specified starting position to a goal position while avoiding obstacles in the environment. The input to a path-planning algorithm is a geometric map of the environment (and possibly the material composition of the environment in more advanced path planners), where positive and negative obstacles have been identified, and the starting and goal points are given. The output of the path-planning algorithm is a set of waypoints that specify the trajectory the vehicle must follow. For a completely known environment, path planning need only be performed once before the motion begins. However, the environment is often only partially known, and as obstacles are discovered they must be entered into the map and the path must be replanned.

State of the Art

Academic. Research in planning robotic motion dates back to the late 1960s, when computer-controlled robots were first being developed. Most advances were made in the 1980s and 1990s as computers became readily available and inexpensive. An excellent reference that summarizes many of these algorithms is Latombe (1991).

The notion of configuration space is commonly used to represent obstacles in the robot's environment (Lozano-Perez, 1983). Configuration space for a mobile robotic vehicle is typically a two- or three-dimensional space representing the x and y position and possibly the orientation in the x–y plane. The vehicle and obstacles are simplified in shape to polygons, and the vehicle itself is further simplified to a point by "growing" the obstacles by the vehicle silhouette.

As described in Latombe (1991), there are three computational approaches to path (or motion) planning. The first is a roadmap approach where regions of free space are linked together by a network of one-dimensional curves. Once a roadmap is constructed, path planning is reduced to searching for roads that connect the initial starting point to an ending point.

The second approach involves decomposing the free space into nonoverlapping regions called cells. This cell decomposition may be exact or approximated by a prespecified geometric shape, typically a square. A connectivity graph represents the adjacency relation among the cells. The graph is searched to determine a sequence of cells that link the starting position to the goal position.

The third approach is a potential field method where the direction and speed of motion are specified by the gradient of a potential field function that is minimized when the vehicle reaches the goal point. Obstacles are avoided by adding in repulsive terms that move the vehicle away from the obstacles. The potential fields algorithm is computationally efficient and easy to implement in real time. As obstacles are discovered, repulsive vectors are easily added. One disadvantage of a potential field algorithm is that it is possible for the local minimum to occur and for the resultant gradient vector to be zero before the vehicle reaches the goal. Another disadvantage is that potential fields may also cause instability (i.e., oscillatory motion) at higher speeds (Koren and Borenstein, 1991). In these cases the first two approaches must be used to drive the vehicle away from the local minimum.

From the systems perspective, path-planning tools depend to a great extent on the quality of the perception and map-building tools. Recent perception advances in the area of simultaneous localization and mapping (SLAM) (Thrun et al., 1998; Choset and Nagatani, 2001; and Dissanayake et al., 2001) will greatly benefit path planning. This technology is potentially very useful for the Army's UGV program, as it allows a vehicle to estimate where it is located as well as build a map of the environment. DARPA has funded much of the work found in Thrun et al. (1998) through its Tactical Mobile Robotics program.

Commercial. Many of the commercial robot simulation packages such as Simstation and RoboCad contain general six-degrees-of-freedom path planners for industrial robot manipulators. They are used to plan the free space motion of the manipulator in a cluttered factory environment. Some of the commercial mobile robotic vehicles by companies, such as the i-Robot Corporation, contain a simple two-dimensional path planner that is specific for their vehicles. There are no general purpose path planners for UGVs on the commercial market.

Current Army Capabilities. The Demo III experimental unmanned vehicle (XUV) contains an advanced and sophisticated path planner that combines world modeling, optimization, and computational searching algorithms. Planning is performed at several levels based on the time horizon (e.g., 500-ms, 5-s, 1-min, and 10-min plans) and the spatial resolution (0.4-meter, 4-meter, and 30-meter grid spacings). Path segments are weighted based on path length, offset from reference path, detected obstacles, terrain slope, and so on. The segments are stored in a graph and Dijkstra's algorithm is used to search the graph for the optimal solution. In addition to shortest-path-length plans the planner can also compute road-only paths, and tree line tracking, low-visibility paths. Updating the vehicle path is currently performed at 4 Hz using a 300-MHz Motorola G3 processor.

The success of path planning generally depends on reliable sensor measurements. There are path planners that take uncertainty in sensor measurements into account, but even these planners perform poorly if the sensor measurements are outside assumed statistical limits (Thrun et al., 1998; Choset and Nagatani, 2001; Dissanayake et al., 2001). Although specific details were not given at Demo III, the National Institute of Standards and Technology (NIST) developers of the path planner appear to use a recursive estimator such as a Kalman or Information Filter to improve estimates of the vehicle's current location and its surrounding obstacles. The new LADAR system on the Demo III XUV vehicle is certainly a big improvement over previous systems using stereo-video alone. The LADAR system is able to detect obstacles regardless of their contrast in the environment. The inertial navigation unit in the Demo III XUV greatly aids in localization of the vehicle in the environment.

Technology Readiness

Path planning for an individual UGV is relatively mature, but mission planning and multiple UGV and UAV planning are relatively immature. Path-planning technology is highly dependent upon both Perception and Navigation technologies for success. As demonstrated in Demo III, the state of the art in path planning for an individual UGV, such as the Donkey and Wingman examples, is estimated at TRL 5, because of limited testing in relevant environments.

The technology readiness level of multiple UGV and UAV path planning is currently TRL 3 for multiple UGVs and TRL 1 for multiple UGVs and UAVs. Under DARPA's Tactical Mobile Robotics program, path planning for multiple UGVs was demonstrated in Feddema et al. (1999) for six "bread-box-sized" robotic vehicles; in the DARPA PerceptOR program path planning for a combined UAV and UGV has just recently been simulated.

When planning the path of multiple unmanned systems, such as would be the case for the Hunter-Killer example, the communications bandwidth between vehicles is a very important factor. The less communications bandwidth there is, the less coordination between vehicles. Available communications bandwidth also depends on the mission, with more covert missions having less bandwidth. Trade studies need to be performed to determine how much bandwidth is available for each mission. Once this is determined it is possible to develop the appropriate path planners for multiple coordinated vehicles. The budget requirements necessary to bring path planning for multiple UGVs and UAVs up to a TRL 6 is substantially more than for the mission planning aids, and the time horizon could be 10 to 15 years away.

Salient Uncertainties

Algorithm development for multiple vehicle path planners is a relatively low-risk but time-consuming effort. Path planning is a software technology that will most likely be upgraded as perception sensors and mobility platforms are upgraded. As with any software in critical systems, the software must go through a stringent, structured design review, and all branches of the code must be thoroughly tested and validated before being installed. Most planning efforts for robots are in the area of path planning. The major gaps with which the Army should be concerned are in the mission-planning area as discussed below.

Recommended Areas of Research

The following are recommended areas of research and development:

1. The trade-off in computational space and time complexity for multiresolution map generation needs to be further evaluated.

2. Planning for sensor acquisition based on mapping and localization requirements should be evaluated.
3. A hierarchy of path-planning algorithms should be employed. The fastest, most efficient algorithms should be used whenever possible. If these algorithms fail, more sophisticated algorithms should be used. This should provide for graceful degradation of performance (e.g., loss of speed).
4. Most UGV path planning to date has used only geometric and kinematic reasoning. An important next step is to include models of vehicle dynamics, terrain compliance, and dynamics of vehicle and terrain interaction into future planners. Dynamic programming techniques have been successfully applied to trajectory planning of rockets (Dohrmann et al., 1996), and they may also be successfully applied here.
5. The weights used to determine the optimal path are heuristically defined by the developer. Much experimentation is necessary to determine the appropriate weights for all variations in terrain and weather.
6. Simultaneous planning for multiple UGVs and UAVs is still in its infancy. Although not demonstrated for the committee, the Demo III software was capable of controlling up to four UGVs simultaneously. The planning for these vehicles was performed independently with the operator specifying phase transition points to coordinate the vehicles at key locations. In the future a more advanced planner will be needed to control the positions of tens of UGVs and UAVs without having to plan each path individually. This capability may not be needed for initial deployment of UGVs, but it will certainly be needed to meet FCS objectives.

Mission Planning

This section defines the scope of autonomous mission-planning technology. It describes the state of the art and estimates the levels of technology readiness.

Definition

From a military perspective, autonomous mission planning goes well beyond path planning. It is the ability of the autonomous UGV to determine its best course of military action, considering synergistically the mission being supported by the UGV; enemy situation and capabilities; terrain, features, obstacles, and weather conditions; the UGV's own and friendly force situation and vulnerabilities; noncombatant information; time available; knowledge of military operations and procedures; and unique needs of the integrated mission package. All but the last few items directly support the development of a warfighter's situational awareness.

The military knowledge base includes tactics, techniques, and procedures as defined in tactical fighting documents and standard operating procedures; and information from unit operations orders, including friendly force structure, detailed mission execution instructions, control graphics, enemy information, logistics (e.g., when and where to refuel), priority for supporting fires, and special instructions. The special instructions can include rules of engagement, communications protocols, and information needed about the enemy.

With enhanced military SA, understanding of its mission, and knowledge of military operations, the UGV can execute its specific mission tasks. To support this execution the UGV will identify individual and unit maneuver needs, covered and concealed routes and battle positions, tactically significant observation and firing positions for both itself and the enemy, and what and when to report and engage.

Assuming the appropriate information is available, the critical technologies needed to provide the above capabilities are software; highly efficient processing capability; and rapid-access, high-capacity, low-power storage devices for real-time cognitive processes.

State of the Art

Mission-planning capabilities for robots are very immature. Most research in the perception and planning areas is focused on path planning, with little to no efforts in mission planning. Little work is being done to develop the cognitive algorithms and knowledge bases needed to support autonomous mission planning (Meyrowitz and Schultz, 2002; U.S. Army, 2001). Most mission-planning technology efforts are in the area of developing mission-planning aids for humans using command and control systems. The only autonomous mission planning appears to be in modeling and simulation technology development efforts in support of developing more realistic decision-making capabilities within simulations (Toth, 2002).

Much work has begun in the area of cognitive modeling. Cognitive models based on neural networks, Bayesian networks, case reasoning, and others are being considered for supporting the decision-making capabilities of synthetic entities. These modeling and simulation efforts could be leveraged by robotic development programs. The Army and Navy have recently begun to exploit this opportunity (U.S. Army, 2002; Toth, 2002).

Needs of Example UGV Systems

The mission-planning needs of the four notional applications vary significantly. The basic Searcher and Donkey will not require mission-planning capabilities. The human lead will plan most of the mission for the Wingman; however, the Wingman will need some mission-planning capability to react to changes in mission while on the move. The Hunter-Killer team will require very complex autonomous mission-planning capabilities. More advanced versions of the first three applications would require mission-planning capabilities of varying levels of complexity.

Execution of the missions will vary. The teleoperator will execute most of the Searcher's mission. The Donkey mostly will follow its leader's execution of the mission. The Wingman and Hunter-Killer team execute most of their missions on their own. Various aspects of the example missions will be executed by specific software developed for tactical behaviors and cooperative behaviors as discussed in the section on Behaviors and Skills.

Technology Readiness Estimate

While the mission planning in Demo III was very good (possibly TRL 5), it was very limited in the number and scope of RSTA mission functions attempted. Much more complete mission- and path-planning algorithms will be required for such missions as counter-sniper, indirect fire, physical security, logistics delivery, explosive ordnance disposal, and military operations in urban terrain (MOUT). Mission-planning aids for human command and control of each of these missions could be brought up to TRL 6 in a few years with moderate funding.

Autonomous mission-planning technologies that would be needed for Wingman and Hunter-Killer systems are at TRL 3. TRL 6 may not be achieved for another 10 years.

Salient Uncertainties

The largest capability gap is in mission planning that is specific to the Army doctrine. This is a chicken-and-egg problem in that the use of robot vehicles is not defined in the Army doctrine, and therefore the doctrine will also have to be developed. Computer scientists and soldiers need to work together to understand the capabilities of UGVs and the needs of the soldier. The soldiers will write the doctrine, while the computer scientists will develop the mission-planning tools.

Automated mission planning is being addressed in the modeling and simulation community, but capabilities are still very immature. Advances are being made in software; highly efficient processing capability; rapid-access, high-capacity, low-power storage devices for real-time cognitive processes; local or distributed high-fidelity knowledge bases; high bandwidth, mobile communications networks; perception technologies; mobility systems for complex terrain, and natural language and gesture recognition technologies that understand military language and gestures. However, these advances are not being integrated into a mission planning technology for robot systems.

Feasibility and Risks

Near-term success in autonomous mission planning does not seem feasible. Modeling and simulation has not had the success in modeling decision making within simulations that was anticipated. Within DOD there has been a change in thrust in human behavior representation efforts. This will greatly impact the development of mission-planning capabilities of synthetic entities. The reason for this change is that past efforts seem to have reached a plateau, with the belief that adding another million lines of code to these past efforts would not get human behavior representations any closer to the simulation needs of the warfighter (Numrich, 2002; Toth, 2002). It is not known how difficult it will be to transfer successful modeling and simulation efforts to a UGV program.

Areas of R&D

The Demo III system demonstrated several preliminary mission-planning capabilities for the scout (RSTA) mission. These will need to be enhanced and other missions added as the operational requirements for FCS demand.

The UGV community should work closely with the modeling and simulation community to leverage algorithms and benefit from lessons learned. The Army should focus on developing cognitive models and knowledge bases for support of mission-planning capabilities.

Salient Uncertainties in Planning

Autonomous path and mission-planning technologies will require highly skilled personnel to develop the knowledge base and for configuration management and maintenance of mission-planning subsystems.

The foregoing assessment of path- and mission-planning technologies provide the basis for answering Task Statement Question 4.a as it applies to the Planning component of "intelligent" perception and control. See Box 4-3.

BEHAVIORS AND SKILLS

Behaviors and skills software combines the inputs from perception, navigation, and planning and translates them into motor commands. Specific effector behaviors depend upon the particular mission equipment installed on the UGV and will vary with the operational requirements.

A behavior is coupling of sensing and acting into a prototype, observable pattern of action. It can be innate, learned, or strictly a stimulus response (e.g., ducking when something is thrown at you). A skill is a collection of behaviors needed to follow a plan or accomplish a complex task (e.g., riding a bicycle). This section is concerned with behaviors and skills that enable tactical responses, including survival and goal-oriented behaviors, as well as behaviors specific to combat tactics.

Tactical behaviors, especially those associated with military skills, are essential to enable the UGV to perform in a battlefield environment. Another area, cooperative behaviors, will enable UGVs to accomplish missions involving

> **BOX 4-3**
> **Task Statement Question 4.a**
> **Planning Component of "Intelligent" Perception and Control**
>
> **Question:** What are the salient uncertainties in the "intelligent" perception and control components of the UGV technology program, and are the uncertainties technical, schedule related, or bound by resource limitations as a result of the technical nature of the task to the extent it is possible to enunciate them?
>
> **Answer:** Algorithm developments for mission planning and for multiple UGV and UAV path planning are relatively immature. When planning the path of multiple UGVs and UAVs, the communications bandwidth between vehicles is a very important factor. Trade studies need to be performed to determine how much bandwidth is available and how the requirements will vary for specific missions. Once this is determined, it is possible to develop the appropriate path planners for multiple coordinated vehicles.
>
> The budget requirements necessary to bring path planning and automated mission planners for multiple UGVs and UAVs up to a TRL 6 is substantially more than for single mission planners, and the time horizon could be 10 to 15 years away.
>
> These uncertainties are bound by resource limitations and result from the technical nature of the task.

other unmanned systems, including UAVs. Both of these areas are discussed in the context of the four example systems postulated in Chapter 2.

Tactical Behaviors

This section defines the scope of tactical behavior technology. It describes the state of the art, estimates technology readiness, and discusses capability gaps.

Definition

The term "tactical" has double meanings for military robots. For commercial robots tactical behaviors include software to perform and control mission functions, such as manipulating tools or sensors. For military UGVs the definitions must be expanded to include behaviors based on military protocols used in tactical combat.

Tactical behaviors may need to be modified by the mission or by the influence of cultural, political, or economic concerns. They affect how UGVs will maneuver, engage, communicate, take evasive actions, and learn. Most of these behaviors are defined by skills and tasks identified in fighting manuals, Army Readiness Training Evaluation Program (ARTEP) tasks, standard operating procedures (SOPs), and so on. Several of these skills and tasks are discussed below.

Tactical maneuver includes movements both as an individual UGV and as part of a unit formation. With access to detailed terrain and feature information and friendly or enemy situations, individual movement behaviors include using folds in terrain and thick vegetation for cover and concealment, gaining and maintaining contact with an enemy entity without being detected, and occupying static positions that provide optimal line of sight for communications or for engaging the enemy. UGV tasks as a member or members of a maneuvering unit include moving in an appropriate position, at a particular distance, and with a specific mission package orientation with respect to other entities in the formation. Both the Wingman and Hunter-Killer examples depend upon such military tactical maneuver skills.

For UGVs equipped with or access to weapons or linked to weapons effects, tactical behaviors must include targeting, engaging, and assessing damage. Engagements can be accomplished with onboard weapons, perhaps including nonlethal weapons, or with munitions on other ground, air, or sea weapons, or soldier platforms. Battle damage assessments determine whether the target is still a threat to the UGV or other friendly forces. Targeting includes finding, identifying, and handing off targets. An engagement skill required by the Wingman UGV, for example, would be to draw lethal fire away from accompanying manned platforms.

The UGV must know when and how to communicate. It must know when to report crossing control measures, enemy contact, and chemical alerts. It must also know how to request orders and supporting fires. For the Wingman and Hunter-Killer, communicating includes cooperating or collaborating on teams of UGVs, or UGVs and humans, as well as interfacing with personnel outside its organization. It must be able to coordinate or facilitate operations with other friendly forces by radio and through natural language and gestures.

Physical security is another consideration. Tactical evasive actions primarily relate to actions that support self-preservation from enemy personnel, enemy weapons, and natural dangers. Survivability techniques include changing battle positions, dashing from one point to another, hiding, firing, and calling for additional support. The usefulness of a UGV to the enemy can be significantly reduced by the ability of the friendly force to command the UGV even while the system is under enemy control. If captured, a UGV might not hesitate to call in artillery on itself. To ensure that the UGV does not become a Trojan horse the enemy would need to render it practically useless. If cornered and faced with imminent destruction, it is possible that an appropriate tactical behavior for a robot might be to self-destruct.

The UGV must also know how to protect itself from natural dangers like precipitation, heat, water obstacles, and steep drops (negative obstacles). It has to know when it can navigate over, through, and around and estimate compliancy of an obstacle. Finally, it must know what actions to take when it is low on ammunition, fuel, oil, energy, or platform states (upside down, track thrown) warrant assistance.

The UGV must also know how to react to noncombatants and potential militant forces. It must abide by rules of engagement. It should be able to communicate with or at least make announcements to civilians in the local language.

A critical tactical behavior capability is for the UGV to know how its actions impact the plan or the commander's intent; when to call for assistance; when to override or update behavior; and when is the appropriate time to engage or disengage its behaviors. Finally, the UGV must be able to learn by adjusting its knowledge base of tactical behaviors as it experiences repeatable enemy actions or other learning events.

The military knowledge base needed to support these behaviors is similar to those needed for mission planning, including tactics, techniques and procedures as defined in tactical fighting documents and SOPs; and information from unit operations orders, including friendly force structure, detailed mission execution instructions, control graphics, enemy information, logistics (e.g., when and where to refuel), priority for supporting fires, and special instructions. The special instructions can include rules of engagement, communications protocols, and lists of needed information about the enemy.

The mobility platform and mission equipment are critical supporting technologies for the performance of tactical behaviors. Other technologies for supporting tactical behaviors include stealth technologies and human–robot interface technologies, especially natural language and gesture recognition and transmission of English and foreign languages. This means that the development of tactical behaviors must be coordinated and synchronized with development of these supporting technologies.

State of the Art

Tactical behavior capabilities for robots are very immature. There appears to be little being done in the area of tactical behaviors. Little work is being done to develop the cognitive algorithms and knowledge bases needed to support autonomous tactical behavior (U.S. Army, 2001; Meyrowitz and Schultz, 2002). Most UGV efforts in developing tactical behaviors appear to be in self-preservation actions. The larger set of tactical behaviors is being addressed in modeling and simulation technology development efforts in support of developing more realistic behaviors of entities within simulations (Toth, 2002). Much work has begun in the area of cognitive modeling. Cognitive models based on neural networks, Bayesian networks, case-based reasoning, and others are being considered for supporting the decision-making capabilities of synthetic entities. These modeling and simulation efforts could be leveraged by robotic development programs. The Army and Navy have recently begun to exploit this opportunity (U.S. Army, 2002; Toth, 2002). Army UGV tactical behavior programs should also leverage work on past mobile minefield efforts and DARPA's Tactical Mobile Robotics program.

Technology Readiness

The need for tactical behavior software varies significantly for the applications of the four example systems. The basic Searcher has no requirement, but Donkey will need to move in a way that minimizes the chance for detection by the enemy. Wingman and Hunter-Killer will require active behaviors as nonlethal self-protection, avoiding enemy observation, and fleeing and hiding. The human lead will control or execute most of the tactical behaviors of the Searcher and Donkey. The Wingman and Hunter-Killer team will both require the essential military tactical behavior capabilities described in this section, with those needed by the Hunter-Killer being much more complex. Advanced versions of all applications would require tactical behaviors of increased complexity.

Simple evasive actions will reach TRL 6 in the near term. Self-preservation actions, like avoiding enemy detection, should reach TRL 6 in 3 to 5 years. Complex tactical behaviors are at TRL 3, and TRL 6 may not be achieved for another 10 years.

Salient Uncertainties

Most tactical behaviors development efforts are focused on self-preservation. As stated earlier, tactical behaviors are being addressed in the modeling and simulation community, but capabilities are still very immature. Advances are being made in software; highly efficient processing capability; and rapid-access, high-capacity, and low-power storage devices for real-time cognitive processes. However, these advances are not being integrated into tactical behavior technologies for UGV systems.

Near-term success in autonomous, complex tactical behaviors does not seem feasible. The feasibility and risks described in the mission-planning section hold true for tactical behavior technology. Additionally, mobility for UGVs that requires near-human capabilities to negotiate complex terrain is at very high risk. Stealth technologies are also at high risk.

The foregoing provides the basis for the answer to Task Statement Question 4.b as it pertains to tactical behaviors. See Box 4-4.

Recommended Areas of R&D

As with mission-planning technology, the committee recommends that the UGV community work closely with the modeling and simulation community to leverage algorithms and benefit from lessons learned. UGV efforts should focus on developing the cognitive models and knowledge bases for supporting tactical behaviors as they are defined.

> **BOX 4-4**
> **Task Statement Question 4.b**
> **Tactical Behaviors**
>
> **Question:** What are the salient uncertainties for the other main technology components of the UGV technology program (e.g., adaptive tactical behaviors, human–system interfaces, mobility, communications)?
> **Answer:** Near-term success in developing autonomous, complex tactical behaviors does not seem feasible. The feasibility and risks described for mission planning also hold true for tactical behavior technology. Additionally, the aspects of A-to-B mobility that require near-human capabilities to negotiate complex terrain are at very high risk. Stealth technologies for UGVs are also at high risk.

Cooperative Robot Behaviors

This section defines the scope of the cooperative behaviors technology area. It describes the state of the art, estimates technology readiness, and identifies the impact on Army operations or logistics.

Definition of Cooperative Behaviors

In the field of psychology the word "behavior" is defined as "the aggregate of observable responses of an organism to internal and external stimuli." In robotics, behavior is often used to describe the observable response of a single robot vehicle to internal and external stimuli. When multiple vehicles are involved, the terminology "cooperative behavior" is often used to describe the response of the group of vehicles to internal and external stimuli.

State of the Art

Academic. In recent years there has been considerable interest in the control of multiple cooperative robotic vehicles, the vision being that multiple robotic vehicles can perform tasks faster and more efficiently than a single vehicle. This is best illustrated in a search-and-rescue mission when multiple robotic vehicles would spread out and search for a missing aircraft. During the search the vehicles share information about their current location and the areas that they have already visited. If one vehicle's sensor detects a strong signal indicting the presence of the missing aircraft, it may tell the other vehicles to concentrate their efforts in a particular area.

Other types of cooperative tasks range from moving large objects (Kosuge et al., 1998) to troop hunting behaviors (Yamaguchi and Burdick, 1998). Conceptually, large groups of mobile vehicles outfitted with sensors should be able to automatically perform military tasks like formation-following, localization of chemical sources, de-mining, target assignments, autonomous driving, perimeter control, surveillance, and search-and-rescue missions (Noreils, 1992; Hougen et al., 2000; Brumitt and Hebert, 1998; Kaga et al., 2000). Simulation and experiments have shown that by sharing concurrent sensory information, the group can better estimate the shape of a chemical plume and therefore localize its source (Hurtado et al., 1998). Similarly, for a search-and-rescue operation a moving target is more easily found using an organized team (Jennings et al., 1997; Goldsmith et al., 1998).

In the field of distributed mobile robot systems much research has been performed, and summaries are given in Cao et al. (1995) and Parker (2000). The strategies of cooperation encompass theories from such diverse disciplines as artificial intelligence, game theory and economics, theoretical biology, distributed computing and control, animal etiology, and artificial life.

Much of the early work focused on animal-like cooperative behavior. Arkin (1992) studied an approach to "cooperation without communication" for multiple mobile robots that are to forage and retrieve objects in a hostile environment. This behavioral approach was extended in Balch and Arkin (1998) to perform formation control of multiple robot teams. Motor schemas such as avoid static obstacle, avoid robot, move to goal, and maintain formation were combined by an arbiter to maintain the formation while driving the vehicles to their destination. Each motor schema contained parameters such as an attractive or repulsive gain value, a sphere of influence, and a minimum range that were selected by the designer. "When inter-robot communication is required, the robots transmit their current position in world coordinates with updates as rapidly as required for the given formation speed and environmental conditions" (Balch and Arkin, 1998).

Kube and Zhang (1994) also considered decentralized robots performing tasks "without explicit communication." Much of their study examined comparisons of behaviors of social insects, such as ants and bees. They considered a box-pushing task and utilized a subsumption approach (Brooks and Flynn, 1989; Brooks, 1986), as well as ALN (adaptive logic networks). Similar studies using analogs to animal behavior can be found in Fukuda et al. (1999). Noreils (1993) dealt with robots that were not necessarily homogeneous. His architecture consisted of three levels: functional level, control level, and planner level. The planner level was the high-level decision maker. Most of these works do not include a formal development of the system controls from a stability point of view. Many of the schemes, such as the subsumption approach, rely on stable controls at a lower level while providing coordination at a higher level.

More recently researchers have begun to take a system controls perspective and analyze the stability of multiple vehicles when driving in formations. Chen and Luh (1994) examined decentralized control laws that drove a set of holonomic mobile robots into a circular formation. A conservative stability requirement for the sample period is given in terms of the damping ratio and the undamped natural frequency of the system. Similarly, Yamaguchi studied line formations (Yamaguchi and Arai, 1994) and general formations (Yamaguchi and Burdick, 1998) of nonholonomic vehicles, as did Yoshida et al. (1994). Decentralized control laws using a potential field approach to guide vehicles away from obstacles can be found in Molnar and Starke (2000); and Schneider et al. (2000). In these studies, only continuous time analyses have been performed, assuming that the relative position between vehicles and obstacles can be measured at all times.

Another way of analyzing stability is to investigate the convergence of a distributed algorithm. Beni and Liang (1996) prove the convergence of a linear swarm of asynchronous distributed autonomous agents into a synchronously achievable configuration. The linear swarm is modeled as a set of linear equations that are solved iteratively. Their formulation is best applied to resource allocation problems that can be described by linear equations. Liu et al. (2001) provide conditions for convergence of an asynchronous swarm in which swarm cohesiveness is the stability property under study. Their paper assumes position information is passed between nearest neighbors only and proximity sensors prevent collisions.

Also of importance is the recent research combining graph theory with decentralized controls. Most cooperative mobile robot vehicles have wireless communications, and simulations have shown that a wireless network of mobile robots can be modeled as an undirected graph (Winfield, 2000). These same graphs can be used to control a formation. Desai et al. (1998, 2001) used directed graph theory to control a team of robots navigating terrain with obstacles while maintaining a desired formation and changing formations when needed. When changing formations, the transition matrix between the current adjacency matrix and all possible control graphs are evaluated. In the next section the reader will notice that graph theory is also used in this paper to evaluate the controllability and observability of the system.

Other methods for controlling a group of vehicles range from distributed autonomy (Fukuda et al., 1998) to intelligent squad control and general purpose cooperative mission planning (Brumitt and Stentz, 1998). In addition, satisfaction propagation is proposed in Simonin et al. (2000) to contribute to adaptive cooperation of mobile distributed vehicles. The decentralized localization problem is examined by Roumeliotis and Bekey (2000) and Bozorg et al. (1998) through the use of distributed Kalman filters. Uchibe et al. (1998) use canonical variate analysis (CVA) for this same problem.

Feddema and Schoenwald (2001) discussed models of cooperation and how they relate to the input and output reachability and structural observability and controllability of the entire system. Whereas decentralized control research in the past has concentrated on using decentralized controllers to partition complex physically interconnected systems, this work uses decentralized methods to connect otherwise independent nontouching robotic vehicles so that they behave in a stable, coordinated fashion. These methods allow the system designer to determine the required sampling periods for communication and control and the theoretical limits on the interaction gains between each vehicle. Both continuous time and discrete time examples are given with stability regions defined for up to 10,000 vehicles. The results of this stability analysis have been applied to several missions: formation control, robotic perimeter surveillance, facility reconnaissance, and a self-healing minefield. Figures 4-4, 4-5, and 4-6 show the types of user interfaces used to control a formation, guard a perimeter, and surround a facility (Feddema et al., 1999).

Automated Highway Systems. The University of California Partners for Advanced Transit and Highways (PATH) program has been developing concepts and technologies for cooperative control of automated highway vehicles since 1990. Cooperation between vehicle and roadway systems and between individual vehicles has been emphasized in this research in order to enable higher performance of the vehicle control systems and to reduce their vulnerability to sensor imperfections. Protocols have been designed for vehicle–vehicle cooperative maneuvering (Hsu et al., 1991), and experimental implementations on passenger cars have demonstrated the improvements that can be achieved in vehicle-following accuracy and ride quality when vehicles share their state information over a wireless communications link rather than relying only on autonomous sensing (Rajamani and Shladover, 2001).

Military. The Demo II and III projects demonstrated a simple follow-the-leader cooperative behavior where the lead vehicle records GPS waypoints as it moves to a goal, and then transmits the GPS waypoints to the following vehicle, which then traverses the same path. The follower vehicle uses its own local perception sensor data to keep the vehicle on the road while following the GPS waypoints. The follower ATD STO project is extending this capability to meet more difficult requirements in terms of separation distance, delayed travel time, speed, and difficulty in terrain. The follower vehicle is to follow the path of the lead vehicle up to 200 kilometers and 24 hours later. Maximum speeds are to increase to 65 km/h on a primary road and 30 km/h over rough terrain.

Ideally the follower vehicle should be able to tolerate GPS drop-outs or jamming. The proposed schemes for navigation in the case of GPS drop-outs include using hand- or vehicle-emplaced transponder beacons or stored images at waypoints to guide the vehicle. The transponder beacon ap-

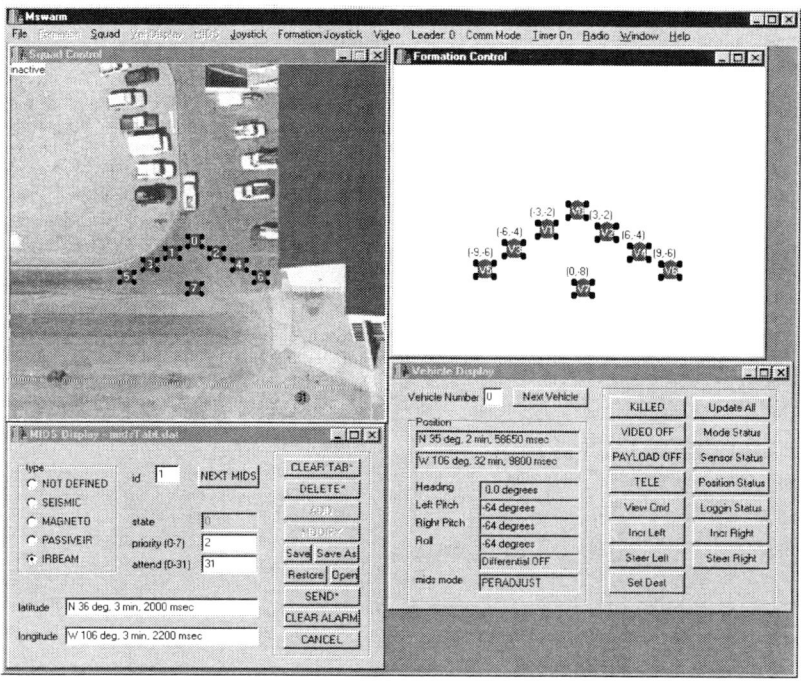

FIGURE 4-4 User interface for controlling a formation of robot vehicles. On the left the current vehicle locations are displayed on an aerial photograph. On the right the user may drag and drop vehicle icons to arrange in any desired formation. SOURCE: Feddema et al. (2002). © 2002 IEEE.

pears to be a technically feasible approach assuming that the enemy does not find and remove the beacons. The stored-images approach is attractive because the enemy does not know what the "bread crumb" image looks like and thus it is more difficult to foil. Unfortunately the approach may be technically impossible given the current limitation of machine vision. The perspective view of a camera changes considerably with orientation. A rock or tree when viewed from one angle looks completely different from a rock or tree when viewed from another angle. The camera images will not be sufficient in an environment lacking distinctive features, such as the desert, where all plants look the same, or the forest, where all trees look the same.

The success of the follower will depend on several factors including: the presence of GPS (or beacons) and obstacles, the level of object recognition, and the weather conditions. Little is known about how such factors might affect the success of a follower mission. A chart similar to Figure 4-7 would help the Army to understand how such conditions might affect the leader-follower concept.

Technology Readiness

The technology readiness level of basic leader-follower cooperative behavior, such as might be exhibited by the Donkey and Wingman examples, is already TRL 6, but cooperative robot behavior, such as needed by the Hunter-Killer is still in a state of infancy. This is currently an area of much research and is no more than TRL 2 or 3. Although some simple cooperative control strategies have been demonstrated at universities and at the national laboratories, a basic understanding of how to design cooperative behaviors

FIGURE 4-5 User interface for perimeter surveillance. The perimeter is marked in blue and miniature intrusion detection sensors are marked by circular numbered icons. An alarm is identified when the icon turns red. The vehicles closest to the intrusion attend to the alarm, while the others adjust their position around the perimeter to prepare for other possible alarms. SOURCE: Feddema et al. (2002). © 2002 IEEE.

FIGURE 4-6 User interface for a facility reconnaissance mission. The initial positions of the vehicles were at the lower left corner of the screen. The vehicles first follow their assigned paths (drawn in black). Once they reach the end of their paths, the vehicles use a potential field path planner to avoid obstacles (drawn in red) and navigate towards goal attractors (drawn in green). To avoid collision between the vehicles and to uniformly cover the goal attractors, repulsive forces push the vehicles away from each other. The path plan is first previewed by the operator, after which the goal and obstacle polygons are downloaded to the vehicle and the same potential field path planner drives the vehicles toward the goal polygons while avoiding obstacle polygons, unplanned obstacles, and neighboring vehicles. The final path is not necessarily the same as the previewed paths since the potential field path planner will avoid sensed obstacles that were not in the original map and the real position of neighboring vehicles may be different at the time of execution. SOURCE: Feddema et al. (2002). © 2002 IEEE.

is still not understood. The budget requirements necessary to bring cooperative behaviors for multiple UGVs and UAVs up to a TRL 6 could be several million dollars, and the time horizon could be 10 to 15 years away.

Salient Uncertainties

There are many possible Army missions for which cooperative behavior will be important, including:

- Perimeter surveillance
- Facility reconnaissance
- Plume localization
- Distributed communication relays
- Distributed target acquisition
- Explosive ordnance detection and
- Building a camera collage.

In each of these missions the relative position between adjacent vehicles is a primary control variable. For example, when guarding a perimeter or surrounding an enemy facility, it is desirable for robotic vehicles to be spread evenly around the perimeter. Unfortunately, current research on cooperative robotic vehicle systems assumes that the relative position between vehicles can be measured either with GPS or with acoustic or visual sensors. For military applications GPS may not always be available and acoustic and visual sensors are not covert and are limited to line-of-sight situations. A new means of measuring the relative position between vehicles is needed. For instance, a radio frequency-ranging system such as that proposed by Time Domain, Inc., may be one way of solving this problem. Without a robust means of determining the relative position of another vehicle over a significant range, cooperative robotics may not be possible for military missions.

For cooperation to occur either the robot vehicle must sense the state of another robot vehicle or the state must be communicated by another means, such as RF radios. In most cases considerable perception capabilities are required to perceive the state of another vehicle. It is often much simpler to communicate the state of the vehicle with RF radios. In these cases the feasibility of performing a cooperative task depends on the communication range and bandwidth of the radios onboard the vehicles.

Cooperative behavior is a software technology area that will most likely be upgraded as communication and perception sensors are upgraded. As with any software in critical systems the software must go through a stringent structured design review and all branches of the code must be thoroughly tested and validated before being installed. As the number of vehicles involved in the cooperative system increases there will be a possible combinatorial explosion of cases to test, since each vehicle could be executing a different branch. Simulation may be the only possible way to test all cases efficiently.

The foregoing provides the basis for the answer to Task Statement Question 4.b as it pertains to cooperative robot behaviors. See Box 4-5.

Areas of Research and Development

Research is needed in the following areas:

- Simulation tools for testing cooperative behaviors should be developed. Similar to the robosoccer simu-

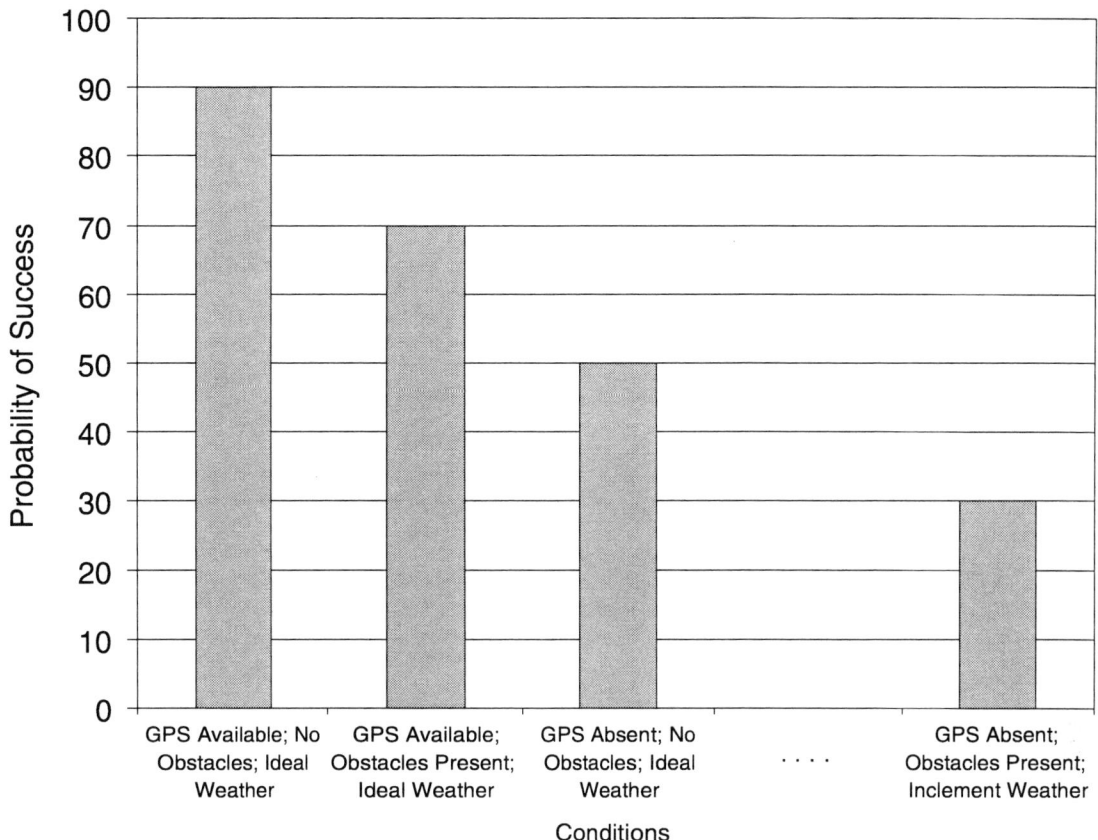

FIGURE 4-7 Probability of success depends on the presence of GPS and obstacles, the level of object recognition, and the weather conditions.

lator, these simulations can be used to compare and contrast competing cooperative behaviors. Two opposing teams executing competing cooperative behaviors can battle each other in cyberspace.
- A playbook of cooperative behaviors needs to be tested and evaluated for their usefulness on real hardware. This playbook might include perimeter surveillance, communications relay, search and rescue, building infiltration, and de-mining.
- Many of the cooperative tasks will require high-bandwidth communications. A trade study is needed to evaluate the trade-off between communication bandwidth and responsiveness of the system to perform the task. Will the bandwidth requirements of secure communications substantially limit which cooperative behaviors are feasible?
- Detailed mathematical modeling of cooperative behavior is needed to fully understand how to design a local individual behavior that, combined with others executing the same local behavior, results in a useful global behavior. Currently these design tools do not exist.

LEARNING/ADAPTATION

This section outlines the state of the art in machine learning, including adaptive control. The committee considered "machine learning" to be synonymous with what is commonly called "soft computing." Several briefings used the term "learning" in the sense of "perception" or "sensing," both of which are covered in the opening section of this chapter. The Learning/Adaptation technology area includes neural networks, fuzzy logic, genetic algorithms, and adaptive controls. The first three are typically associated with artificial intelligence while the fourth is associated with control theory.

State of the Art

The primary characteristic of soft computing and other algorithmic approaches is that they are based on heuristics instead of provably correct mathematical properties. Thus, while results cannot be proven, the approaches have two primary advantages. First, they are not model-based and, second, they yield "good" solutions that would take a prohibi-

> **BOX 4-5**
> **Task Statement Question 4.b**
> **Cooperative Robot Behaviors**
>
> **Question:** What are the salient uncertainties for the other main technology components of the UGV technology program (e.g., adaptive tactical behaviors, human–system interfaces, mobility, communications)?
>
> **Answer:** For cooperation to occur either the robot vehicle must sense the state of another robot vehicle or the state must be communicated by another means, such as RF radios. In most cases considerable perception capabilities are required to perceive the state of another vehicle. It is often much simpler to communicate the state of the vehicle with RF radios. In these cases the feasibility of performing a cooperative task depends on the communication range and bandwidth of the radios onboard the vehicles.
>
> Cooperative behavior is a software component that will mostly likely be upgraded as communication and perception sensors are upgraded. As with any software in critical systems the software must go through a stringent structured design review and all branches of the code must be thoroughly tested and validated before being installed. As the number of vehicles involved in the cooperative system increases there will be a possible combinatorial explosion of cases to test, since each vehicle could be executing a different branch. Simulation may be the only possible way to test all cases efficiently.
>
> Cooperative behavior is still in a state of infancy. Although some simple cooperative control strategies have been demonstrated at universities and at the national laboratories, the design of cooperative behaviors is still not understood. The budget requirements necessary bring cooperative behaviors for multiple UGVs and UAVs up to a TRL 6 could be several million dollars, and the time horizon could be 10 to 15 years away.
>
> These uncertainties are bound by resource limitations and result from the technical nature of the task.

tively long time using rigorous mathematical approaches (see Gad-el-Hak, 2001).

Neural Networks

Neural networks are collections of simple, interconnected, parallel-processing units roughly equivalent to the structure and operation of biological brains. Individual processing units or neurons are interconnected with synapses. If the accumulation of inputs in a particular neuron exceeds a particular threshold, the neuron fires by sending an electrical signal along its axon (output connection), which is connected to other neurons. These neurons in turn will fire if the accumulation of their input signals exceeds a certain threshold. This process, which generally occurs through several layers of cooperative behavior, leaves the system in a state that can be related to the input that created it. Learning, or the generation of a particular final state for a given input state or range of input states, occurs by appropriate adjustments to the synaptic connections between neurons (Schalkoff, 1997; Haykin, 1999; Chen, 1996).

The neural network technology application most relevant to autonomous UGVs would enable a vehicle to drive along a road. Another relevant application is image-segment identification needed for navigation or target recognition. On-road driving has been successfully implemented by academic researchers at Carnegie Mellon University (Baluja, 1996; Jochem et al., 1995a,b; Hancock and Thorpe, 1995; Pomerleau, 1992). Successful road-following was accomplished by training an artificial neural network using reduced resolution vision inputs and steering outputs to follow a road. To our knowledge artificial neural networks have not been used successfully for off-road navigation, probably because the highly unstructured nature of the off-road environment would make it very difficult to train the network to handle all possible likely scenarios.

Fuzzy Control

Fuzzy control is a design technique that is based upon mathematics concepts from fuzzy logic, which is an extension of classical logic, which in turn is based upon an extension of classical set theory. In classical set theory an element is either a member of a set or not. In fuzzy sets an element can have a fractional "degree of membership" (Zadeh, 1965; 1968a,b; 1971). The main advantages of fuzzy logic in controls applications are that (1) it provides a nonmodel-based means to synthesize controllers (i.e., the equations of motion do not need to be derived) and (2) it provides a structure for translating human knowledge or intuition about a complex system to a computer controller. Possible UGV applications include an alternative mechanism by which to effectively control a vehicle in situations where traditional control methodologies fail or are impossible to implement due to unknown modeling aspects of a complex system (Dubois et al., 1997; Tunstel et al., 2001; Kim and Yuh, 2001; Kadmiry et al., 2001; Wang and Lee, 2001; Howard et al., 2001).

Genetic Algorithms

Genetic algorithms represent an optimization technique based upon concepts from biological evolution. They work well when a global optimization cost function is discontinuous and for finding "good" solutions where more mathematically rigorous algorithms will fail to find a solution in a reasonable amount of time. Here candidate solutions are generated randomly, with variable values represented as genes in a chromosome (a string of ones and zeros). Successive generations are obtained by "mating" pairs of members where the "parents" are selected with a bias toward those with better values of the merit function. The combination ensures that the "offspring" inherit information from both parents. Overviews include those by Goldberg (1989), Michalewicz (1992), Mitchell (1997), and Man et al. (1999). Regarding autonomous vehicles, one possible use would be to find an optimal (preplanned or re-preplanned) path between two points. Given that there may be an infinite number of paths with random obstacles the genetic algorithm approach is one method of finding the "best" path.

Adaptive Control

Adaptive control is a form of machine learning in that sensory information is used to modify either an internally stored system model or parameters in a controller. The class of systems to which an adaptive controller can be applied is limited relative to systems that can be controlled by soft computing. Overviews include those by Kaufman et al. (1994), Landau et al. (1998), Steinvorth (1991), and Astrom and Wittenmark (1989). While adaptive control is mature within control theory, the vast majority of results are limited to linear systems. Therefore, the more difficult aspects of autonomous navigation, which may be highly nonlinear, cannot be addressed by adaptive control. On the other hand, it may be entirely appropriate for lower-level, interior control loops within the operating mechanism itself.

Learning Applications

Learning may be applied to any of the software technologies of the UGV; however, its use has been limited in part by lack of computational resources. In perception, for example, learning is used extensively in feature classifiers (i.e., to classify regions as road or non-road or for terrain classification) (see Appendix C). Classifiers typically are based on neural-network or other statistical techniques. An issue is the extent to which models appropriate to off-road scenes could be developed. More generally the issue is how to provide performance evaluation functions so the system can self-assess its performance.

Technology Readiness

Machine learning is in a state of infancy, TRL 1 or 2 at best. Adaptive-learning algorithms, such as might be used by the Donkey example, are much more mature, TRL 3 or 4. While the latter may suffice for many Army requirements, the degree of machine learning that will be required for UGVs such as Wingman and Hunter-Killer to perform military missions and tasks is presently unknown.

Salient Uncertainties

It is widely accepted that learning is an essential element of unmanned systems; yet learning is not necessary and in many cases may be undesirable for military applications. Emergent, unpredictable behaviors are not desirable in soldiers, so why should they be desirable in unmanned systems? Would anyone board an airplane whose pilot was just learning the flight controls? Would you trust a robot with a weapon that is learning to use it? Even worse, would you trust a group of robots with weapons that they learned from each other how to handle?

There are at least three problems associated with learning: (1) Many of the current learning algorithms come up with "black-box" solutions that cannot be analyzed. Genetic programs now generate code that is incomprehensible. How can the Army verify and validate the algorithm that a learning system has generated? Presently there is no way to know if the system has learned a harmful side effect that will show up at the wrong time. (2) Current learning algorithms also require considerable training (millions of trials) and typically fail during the training run. Would such training and failure be acceptable for military applications? (3) Learning is not nearly as advanced, as some people believe. This technology is still in its infancy, and while it should be pursued at the academic level, it will take much longer than 20 years to reach the level where it should be incorporated into a military system with a weapon.

Critical elements missing from the Army's approach are recognition of and a measure of the complexity of the environment in which UGVs will have to operate and a comparison with the level of complexity with which high-level control algorithms (i.e., current decision-making algorithms) can effectively handle. Methods from soft computing have been demonstrated in laboratory environments (i.e., very highly structured environments), but rarely if ever have they displayed even a fraction of the degree of robustness necessary to handle the complex environments envisioned by the Army for UGVs. It is still an open question whether these techniques ultimately will provide the solution to allow a UGV to operate effectively in a highly unstructured and uncertain environment or whether more standard but "brute force," (i.e., computationally intensive) approaches will provide the solution.

Uncertainty exists concerning the degree to which methods from machine learning will ultimately provide solutions to complex real-world problems. While all the technology areas described have great potential and seem to display continued evolution, a true "learning machine" that could display sufficiently adaptive behavior (to include, for example, adaptive reasoning and reasoning under uncertainty to deal with UGV combat environments) is far from reality. Resource limitations are not relevant given the intensity and amount of attention given to machine-learning paradigms, particularly in academia.

The foregoing provides the basis for the answer to Task Statement Question 4.a as it pertains to learning/adaptation. See Box 4-6.

Areas of Research

The most promising areas of machine-learning algorithms is in perception and signal processing. Neural networks are already used in optical character recognition and handwriting recognition. The analogous application for UGVs is to infer high-level information from sensor data. Genetic algorithms are well established as optimization techniques.

> **BOX 4-6**
> **Task Statement Question 4.a**
> **Learning/Adaptation Component of "Intelligent" Perception and Control**
>
> **Question:** What are the salient uncertainties in the "intelligent" perception and control components of the UGV technology program, and are the uncertainties technical, schedule related, or bound by resource limitations as a result of the technical nature of the task, to the extent it is possible to enunciate them?
>
> **Answer:** Significant uncertainty exists concerning whether methods from machine learning will be essential to the successful development of UGVs. While the technology areas described have potential to provide solutions, a true "learning machine" that could display sufficiently adaptive behavior to deal with the complexities of the UGV combat environment is far from reality. Methods from soft computing have been demonstrated in highly structured laboratory environments, but rarely if ever have they displayed even a fraction of the robustness necessary to handle the complex environments envisioned by the Army for UGVs. This uncertainty extends to include the break point between adaptive control solutions and artificial intelligence solutions for each of the "intelligent" components of the autonomous system.
>
> Missing elements include a recognition of and a measure of the complexity of the environment in which UGVs will have to operate to compare with the level of complexity that high-level control algorithms (i.e., current decision-making algorithms) can handle effectively.
>
> Resource limitations are not relevant given the intensity and amount of attention given to machine-learning paradigms, particularly in academia.

Identifying components of the overall control and decision-making strategy of a UGV that require optimization warrants near-term attention. Using methods from soft computing and adaptive control for higher-level decision making should be incrementally pursued.

Given current shortcomings, all four of the technology areas merit far-term investigation. In particular, each should be applied to progressively more complex problems in an effort to determine (1) algorithmic modifications and/or evolution necessary to handle increasingly complicated and uncertain problems and (2) how the amount of computing power necessary to effectively implement such algorithms in real-world situations scales with the complexity of the problem.

For FCS the Army should focus on use of learning technologies to resolve A-to-B mobility issues and on adaptive learning algorithms to develop tactical behaviors.

SUMMARY OF TECHNOLOGY READINESS

Table 4-9 summarizes the technology readiness level assessments made in each of the preceding sections vis-à-vis the four example UGV systems defined in Chapter 2. The table shows the time frame that the committee believes is appropriate for achieving TRL 6.

Capability Gaps

Each of the chapter sections identified salient uncertainties and technology and capability gaps that must be filled by the Army to support development of the four example systems. These are summarized in Table 4-10. For each gap listed, the committee estimated a degree of difficulty/risk (indicated by shading) according to the following criteria:

TABLE 4-9 Estimates for When TRL 6 Will Be Reached for Autonomous Behavior Technology Areas

Technology Areas	Searcher	Donkey	Wingman	Hunter-Killer
Perception				
For A-to-B mobility	Near-term	Mid-term	Mid-term	Far-term
For situation awareness	Mid-term	Mid-term	Mid-term	Mid-term
Navigation	Near-term	Near-term	Mid-term	Mid-term
Planning				
For path	Near-term	Near-term	Mid-term	Mid-term
For mission	Near-term	Near-term	Mid-term	Far-term
Behaviors and skills				
Tactical behaviors	Near-term	Mid-term	Far-term	Far-term
Cooperative behaviors	Near-term	Near-term	Near-term	Far-term
Learning/adaptation	Near-term	Mid-term	Mid-term	Far-term

Near-term
Mid-term (2006–2015)
Far-term (2016–2025)

TABLE 4-10 Capability Gaps in Autonomous Behavior Technologies

Degree of Difficulty/Risk

Low
Medium
High

Technology Areas	Capability Gaps			
	Searcher	Donkey	Wingman	Hunter-Killer
Perception				
A-to-B mobility on-road		Algorithms and processing fast enough to support 40 km/h (road-following, avoidance of moving and static obstacles).	Algorithms and processing fast enough to support 100 km/h (road-following, avoidance of moving and static obstacles). Sensors with long range.	Algorithms and processing fast enough to support 120 km/h (road following, avoidance of moving and static obstacles). Sensors with long range.
A-to-B mobility off-road	Algorithms for real-time two-dimensional mapping and localization. Miniature hardened range sensors. All-weather sensors.	Detect and avoid static obstacles (positive and negative) at 40 km/h day or night. Classify terrain (traversable at speed, in low visibility). Classify vegetation as "push through" or not, detect water, mud, and slopes. Algorithms for GPS mapping and corrections.	Sensors and strategies for fine positioning in bushes. Detect and avoid obstacles at 100 km/h. Classify terrain and adapt speed, control regime. Continually assess terrain for potential cover and concealment. Multiple sensor fusion.	Algorithms for multiple sensor and data fusion. Detect and avoid static obstacles at 120 km/h. Classify terrain and adapt speed, control regime. Continually assess terrain for cover and concealment.
Situation awareness		Algorithms for detecting humans (even lying down, versus other obstacles). Sensors and algorithms for detecting threats.	Track manned "leader" vehicle. Select suitable OP (provides LOS cover and concealment). Detect, track, and avoid other vehicles or people. Distinguish friendly and enemy combat vehicles. Detect unanticipated movement or activities. Acoustic, tactile sensors for recognition.	Algorithms and sensors to recognize movement and identify source. Select suitable OP (provides LOS cover and concealment). Detect, track, and avoid other vehicles or people. Distinguish friendly and enemy combat vehicles. Detect unanticipated movement or activities. Detect potential human attackers in close proximity. Sensors while concealed (indirect vision).

continues

TABLE 4-10 Continued

Technology Areas	Capability Gaps			
	Searcher	Donkey	Wingman	Hunter-Killer
				Localization to coordinate multirobots.
				Identify noncombatants.
Navigation		Relative navigation utilizing communications and GPS.	Integration of GPS, digitized maps, and local sensors.	Error detection and correction.
Planning Path		Use DTED maps; 1-km replanning for obstacle avoidance.	Plan relative to leader; reason about overlapping views.	Tactical formation planning.
		Electronic "breadcrumbs."	Plan to rejoin or avoid team; use features other than terrain.	Adjust route based on external sensor inputs.
		Decision template for alternative routing.	Reasoning algorithms to identify and use concealment.	Plan to optimize observation points, target kill arrays, and communication links.
				Multiobject and pursuit-evasion path planning for multiple UGVs.
Mission			Mimic leader actions.	Plan for complex missions including combat survival.
			Independent actions.	Plan for team and marsupial operations.
				Independent actions.
Behaviors and skills Tactical skills	Basic nonlethal self-protection if touched or compromised.	Avoid enemy observation.	Hooks for specialized mission functions (e.g., RSTA, indirect fire).	Independent operations; fail-safe controls for lethal missions.
		"Flee and hide."	Self-protection.	Self-preservation and defensive maneuvers.
			Complex military operational behaviors.	Complex military operational behaviors.
Cooperative robots			Formation controls of multiple UGVs.	Formation controls of multiple UGVs and UAVs.
			Cooperation for such tasks as hiding in bushes.	
Learning/adaptation		Basic learning for survivability.	Advanced terrain classification.	Advanced fusion of multiple sensor and data inputs.
			Basic machine learning augmentation of behaviors.	Advanced machine learning augmentation of behaviors.

- Low Difficulty/Low Risk—Single short-duration technological approach needed to be assured of a high probability of success
- Medium Difficulty/Medium Risk—Optimum technical approach not clearly defined; one or more technical approaches possible that must be explored to be assured of a high probability of success
- High Difficulty/High Risk—Multiple approaches possible with difficult engineering challenges; some basic research may be necessary to define an approach that will lead to a high probability of success

Tables 4-9 and 4-10 provide the basis for answers to Task Statement Questions 3.d and 4.c. See Boxes 4-7 and 4-8.

BOX 4-7
Task Statement Question 3.d
Autonomous Behavior Technologies

Question: What technology areas merit further investigation by the Army in the application of UGV technologies in 2015 or beyond?

Answer: The committee postulated operational requirements for four example UGV systems and determined critical capability gaps in multiple UGV technology areas merit further investigation by the Army. The technology areas and respective capability gaps are listed in Table 4-10.

BOX 4-8
Task Statement Question 4.c
Autonomous Behavior Technologies

Question: Do the present efforts provide a sound technical foundation for a UGV program that could meet Army operational requirements as presently defined?

Answer: Operational requirements are not clearly defined, and the technological base has consisted of diffuse developments across multiple potential missions. While relevant technologies will be enabled in the present program, the lack of user pull is a major detriment to achieving timely integration of a UGV system into the FCS. Furthermore, unless funding of the UGV technology base is significantly enhanced, the simplest of semiautonomous battlefield systems is not likely to be achieved before the 2010 time frame.

5

Supporting Technologies

An unmanned ground vehicle (UGV) system encompasses the broad technology areas shown in Figure 5-1. Enabling technologies for core autonomous behavior were reviewed in Chapter 4. This chapter assesses the state of the art in UGV supporting technologies for human–robot interaction, mobility, communications, power/energy, and health maintenance. Each section describes the scope of the technology areas, estimates technology readiness levels, and identifies technology gaps.

HUMAN–ROBOT INTERACTION

This section defines the scope of the human–machine interaction technology area. It describes the state of the art, estimates technology readiness, describes capability gaps, and identifies salient uncertainties.

Definitions

Human–robot interaction (HRI) covers the macrocosm of how intelligent agents work together in a system. It encompasses human–robot interfaces, which are specialized human–computer interfaces for the particular needs of HRI in a defined system but is much broader. Human–robot interaction is not synonymous with human-centered computing, whereby computers augment human ability, but it is assumed that the principles of human-centered computing or design will be applied to HRI systems when appropriate.

HRI has recently emerged as a topic of research, in part due to the shift in the AI community from a goal of fully autonomous robots operating in isolation (e.g., a planetary rover) to service robots operating side by side with humans under some form of semiautonomy. HRI appears on the Department of Energy (DOE) Robotics and Intelligent Machines Roadmap as one of the four basic areas (DOE, 1998a).

HRI is particularly relevant to the Future Combat Systems (FCS) concept because it addresses how humans will interact with multiple robots (particularly in times of stress and cognitive fatigue), how responsibilities will be dynamically allocated between humans and robots based on the context, and how the impact of uncertainty and information overload can be mitigated. HRI is instrumental in reducing training times and providing a common interaction mode if users are expected to control UGVs and unmanned aerial vehicles (UAVs).

State of the Art

Human–robot interaction has been gaining momentum as a separate field of study either directly or as a conclusion since 1996. HRI is a cross-disciplinary area, populated by the robotics, artificial intelligence, cognitive science, and human–computer interface (HCI) communities. The motivation for HRI has stemmed from a variety of sources.

HRI studies and technology are clearly needed to minimize the hidden costs of robot systems before they can be fielded. The state of the practice for unmanned systems to date is multiple operators per system. This is true for all fielded UAVs (such as Pioneer and Predator) and prototypical UGVs (including XUV and TMR robots). The TMR (tactical mobile robots) program thus far, for example, has demonstrated several prototype component systems, but none has a 1:1 human-to-robot operational ratio. Although designed to be controlled by a single person, the Urbans, Solem, and Packbot robots require two people for carrying to the field and others to perform planning and maintenance chores. Additional operators are needed to trade off controller duties for extended-duration missions.

Health care robots and toys have brought robots to a different class of end user who does not have and does not want to obtain specialized robotic skills, necessitating new modes of interaction. Work in HCI has shown that people often work better with more naturalistic or social interfaces; this assessment is expected to transfer to robots, with people

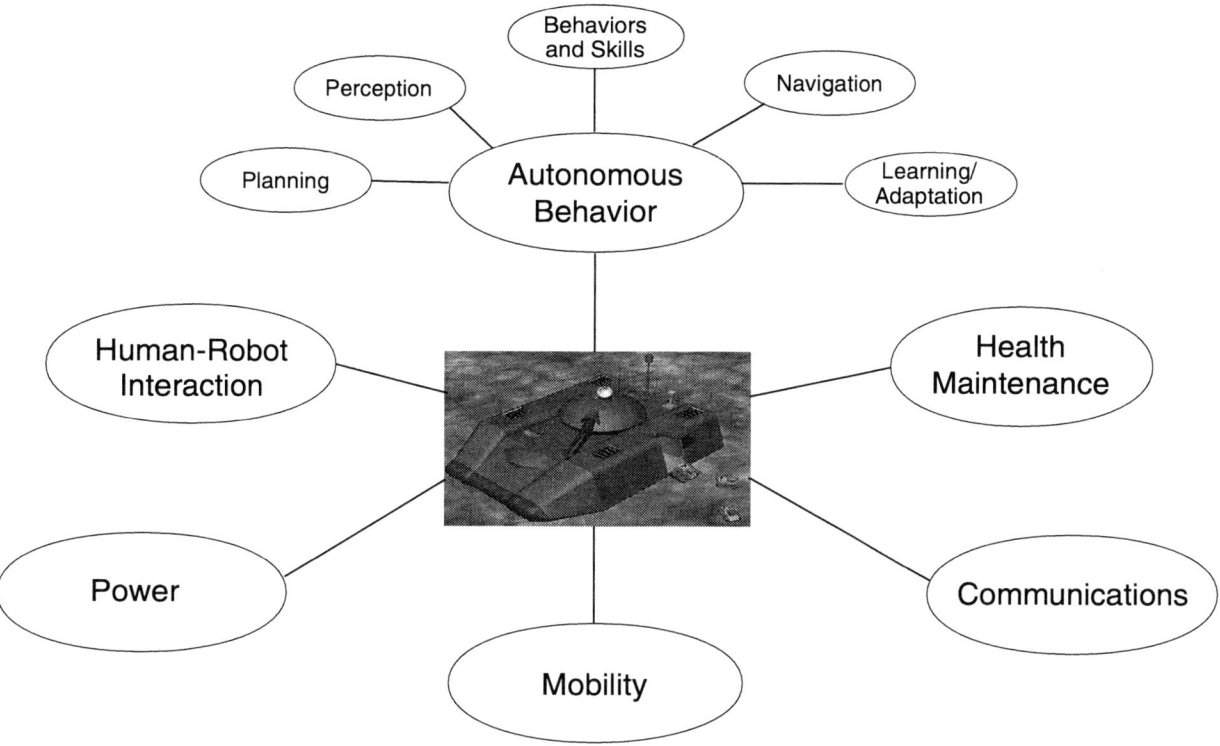

FIGURE 5-1 Areas of technology needed for UGVs.

expecting to deal with robots as creatures with their own personality and motivations.

The HRI literature is limited and mostly associated with a series of workshops sponsored by DOE, the National Science Foundation, and Defense Advanced Research Project Agency (DARPA) (DOE, 1998a; Murphy and Rogers, 2001). One outcome of the literature is a preliminary taxonomy of the issues in HRI proposed in the 2001 DARPA/National Science Foundation Study on Human–Robot Interactions (Murphy and Rogers, 2001). The taxonomy consists of six issues in HRI: communications, modeling, teamwork, usability and reliability, application domains, and representative end users. Of these, communications, modeling, teamwork, and usability and reliability are relevant for FCS vehicles. Application domains and representative end users are not of interest in this study because the FCS program specifies the application and user.

Murphy and Rogers (2001) describe the many facets of communications in HRI:

> *Direct human–robot communication* is possibly the most obvious issue. Modalities include speech, vision, gestures, and teleoperation, though there may be other forms. *Mediated human–robot communication* is another topic. This arises from virtual environment, graphical user interfaces, and can be enacted by collaborative software agents. The physical *interaction and interfaces* impact communication. These methods include physical interaction between robots and human, mixed-initiative interactions between humans and robots, and dialog-based interaction.

Murphy and Rogers (2001) also describe the scope of modeling.

> Modeling issues spanned traditional concerns (cognitive, task and environment modeling) to more HRI specific concerns. Cognitive modeling of human reasoning, behavior, intention and action is needed for imitation (i.e., robot learns how to behave from the human) and for collaboration (i.e., robot understands what the human is doing within the context of the task) [as well as the human understanding of robotic behavior to ensure that the behavior fits the human's mental model and avoid what might be characterized as "clumsy automation"]. Task and environment modeling is needed as a basis for performance. Other modeling issues are social relations, learning, and methods.

Teamwork is particularly relevant for control of unmanned platforms for the military. Teamwork issues can be divided into architectures and task allocation. Architectures focus on the optimal organization of teams (e.g., multiple robots and a single human, multiple humans and a single robot, multiple robots–multiple humans). Research into architectures is expected to determine the situations that require authoritarian, hierarchical relationships and those that require more democratic structure. While architectures are being investigated by the multiagent community, they often neglect questions of how a single robot can work with more than one human, balancing multiple demands, and how tasks can be traded between humans and robots as needed to prevent human mental or sensory overload.

Task allocation in human–robot teams is important because each partner in the team has skills that the other lacks, including intelligence skills. One example is the correct partitioning of skills. In surgery robotics it may be easy to determine a human's handshake and the limits of visual acuity, but it is much more difficult to detect deficiencies in spatial reasoning. It is not clear what we need to know about environments, tasks, humans, and robots to be able to optimize mission performance even if we knew the capabilities of the human and robot (Murphy and Rogers, 2001).

A major issue is the overall utility of HRI systems. What are the appropriate metrics for evaluating the success, effectiveness, and quality of human–robot teams and can such metrics be task independent? The need for metrics also emphasizes the need for benchmarks to directly compare work in the HRI arena with other aspects of performance and to determine where the effectiveness of different human–robot interfaces can be measured. Usability studies involving the task analysis of users and measures of utility for the different types of human–robot relationships are also warranted (Murphy and Rogers, 2001). This taxonomy of communications, modeling, teamwork, and usability and reliability emphasizes HRI as a systems-level design set of issues, not a specific enabling technology. Advances are needed in all elements of the taxonomy, particularly modeling. Of these elements, social informatics (the number and organization of the agents, the roles they play, how they transition between roles, how they collaborate and how they interact safely) is probably the area must familiar to the public, since it includes robots that express emotions and can interpret and respond to the emotions of end users. This type of social informatics is of interest to the entertainment industry and has captured media attention, but it should not be considered the ultimate definition or goal of HRI.

Foreign Technology Development

European and Japanese interest in HRI appears to be driven by the service robot sector, with no direct relevance to military operations. In Europe the Royal Institute of Technology in Stockholm is the leader in HRI, whereas in Japan work is distributed. The work out of Sweden (Kerstin Severinson-Eklundh) is possibly the most advanced in general human–robot interaction, but the field is still immature and no one has a lead.

Current Army Capabilities

The only aspect of HRI addressed by current Army UGV programs is communications to provide the needed operator interfaces. While these programs have incorporated HCI design principles, it is clear that the larger scope of HRI has not been covered. HCI is one subset of HRI, and HRI is often confused with HCI. HRI is concerned about the entire information process and groups of people. HRI focuses on such issues as visualization of information (e.g., whether a polar plot or a three-dimensional surface plot is a better representation for the user's needs) rather than the HCI aspect of whether the display controls, for example, are incorporated with radio buttons. To illustrate, an encouraging attempt was made to resolve an HRI issue in the DARPA TMR program when a project was undertaken to assess which operator skills were necessary to control a robot effectively for significant periods of time under heavy cognitive loading using existing interfaces. Unlike the HCI, which can be added later (although at some cost penalty), a good HRI design must be integral to the entire specification, design, implementation, and evaluation process for the UGV system. A significant investment in testing will be needed just to ascertain the realistic possibilities for reducing human attention and errors during operations of FCS UGVs.

HRI is nascent and has not been systematically explored or applied in military UGV applications. This is likely due to the Army's emphasis on demonstrations rather than fieldable UGV systems. Although not included in past UGV developments, research in HRI is planned in the emerging Army Collaborative Technology Alliance (CTA) for robotics. This is important because interfaces designed independently of the other elements of HRI are not as likely to be effective. It is not surprising that the other elements of HRI (modeling, teamwork, usability and reliability) have also been overlooked.

The Donkey, Wingman, and Hunter-Killer example systems all require levels of interaction that have not been previously investigated. Donkey must be capable of asking for guidance and Wingman must be capable of close (but minimal) interface with soldiers. Wingman must also be capable of multimodal interfaces (optional teleoperation).

Natural user interfaces are assumed essential for battlefield scenarios and are the hardest gap to fill. Aside from degree and mode of interface, interaction requirements for Wingman and Hunter-Killer will not be completely determined without some experimentation. It is believed that Hunter-Killer will require methods for interacting and intervention by soldiers under stress, and it will probably need near-autonomous control algorithms to support multiple operators (and/or robots) on the battlefield. Training for the soldier must also be comprehensive enough to include a complete understanding of the robots's repertoire of behaviors and when they will be executed especially in the more autonomous modes.

Technology Readiness

Except for the Searcher, which has no interaction requirements beyond interface controls, the technical readiness level of HRI technologies vis-à-vis the example UGV systems is at TRL 1 (technology readiness level) or less. Telesystem control algorithms to support one operator per robot (as needed by the Searcher) are at TRL 5. Semiautonomous control software to support multiple robots (as needed for Donkey systems) has been developed and is at TRL 4 or 5.

Salient Uncertainties

There are two types of gaps between the current state-of-the-art HRI capabilities and the projected needs: interfaces (or the mechanisms by which the human communicates with the robot and interprets the output) and control scheme (or the partitioning of roles and responsibility among the human and robotic agents).

The interface needs are largely high risk, because it is assumed they will require more naturalistic interfaces. One example of a naturalistic interface is speech, also known as natural language processing (NLP), or the use of gestures when speech is not permitted. These interface needs are complicated by the remote operation of a robot, which will normally mean that an operator is not within the physical line of sight, unlike remote control of a television, which can be performed only when the remote control shares a physical line of sight with the television.

Speech and gesture interfaces are currently being explored and appear promising, though much work needs to be done for application to noisy environments. NLP and hand gestures represent only a fraction of the ways information can be communicated, and it is expected that perceptual user interfaces that cover multiple modalities will become of increasing importance.

The control scheme is related in that a poorly designed interface can make a system hard to control. The control scheme refers to the underlying organization of responsibilities. In a teleoperated system the human plays an active role, usually consuming 100 percent of the available time. In practice, teleoperated robots often require multiple operators, making the 1:1 ratio ambitious. In the Donkey scenario the goal is one operator per herd. This requires that the Donkeys be given some autonomy, similar to a horse, in order to make the low human-to-robot ratio feasible. In the Hunter-Killer example the control is almost fully autonomous for many reasons. While the 1:6 ratio seems only an incremental advance over the Donkey 1:5 ratio, the Hunter-Killer requires heterogeneous robots: robots that have inherently different physical or software capabilities. As a result it becomes increasingly harder for an operator to keep up with the differences and the ramifications of those differences.

Perhaps the most glaring uncertainty is what will be discovered in field experiments. HRI is resource intensive and requires that many people be tested in realistic conditions over a significant period of time. It will be hard to fill all of the gaps in technology; therefore the Army should begin with investigations that are systemically planned to meet the operational requirements of the UGV systems. These include:

**BOX 5-1
Task Statement Question 4.b
Human–Robot Interaction**

Question: What are the salient uncertainties for the other main technology components of the UGV technology program (e.g., adaptive tactical behaviors, human–system interfaces, mobility, communications)?

Answer: HRI requirements for Army UGVs will be much more demanding than those for commercial developments. Advances in HRI technology will enable reductions in manpower needed to control UGVs of all classes, so satisfaction of these requirements is key to the acceptability of robots on the battlefield. Past research in implementing AI-based tools in operational settings has demonstrated that understanding how the human will use and respond to, as well as what is expected from, the tools is paramount for success. The most glaring uncertainty is which HRI technologies will be needed to meet the Army's requirements once it settles on operational requirements and begins to perform field experiments. HRI research is resource intensive and will require many tests under realistic conditions over a significant period of time.

---

basic interfaces to minimize soldier-operator fatigue; NLP, natural-user, and multimodal interfaces; algorithms for multiple operators; and HRI requirements for employment of heterogeneous robots.

The foregoing provides the basis for the answer to Task Statement Question 4.b as it pertains to human–robot interaction. See Box 5-1.

Recommended Research and Development

The HRI areas in which the Army is most vulnerable are modeling and usability and reliability. The robotics community is addressing teamwork and that work should be transferable; however, modeling of the task and the capability of the agents are highly domain specific, as is means of ensuring usability and reliability. Therefore, the Army cannot rely on academia to provide this important component. The failure to have a good understanding of the HRI requirement will likely result in a mismatch of human–robot abilities. As a consequence, for example, there may be too few humans allocated per group of robots to handle unexpected problems in real time, jeopardizing the mission. HRI should provide a realistic assessment of the proper human to robot ratio and ways to streamline the interactions. The Army should do the following:

- Integrate HRI into FCS design concept, since it is pervasive.
- Initiate an applied research program to study and model roles, cognitive abilities, team organization,

responsibilities of team members, and robotic behaviors that are consistent with soldier expectations and mission success. In particular, model the transitions in responsibility and control (e.g., rather than wait, robot will assume initiative in Case X).
- Initiate an applied research program to model dynamic allocations of responsibility, particularly interruptions while the human is doing one task and must suddenly respond to a crisis detected by the robot. The modeling should focus on how to provide relevant, timely information to the human to mitigate the notoriously poor quality of decisions made by humans in interruption studies.

Useful HRI studies should be possible within the next three years. These do not require the development of new hardware but rather require access to the FCS concept and intended users to determine technical requirements for HRI architectures and task allocation.

MOBILITY

This section defines the scope of the mobility technology area. It describes the state of the art, estimates technology readiness, and identifies salient uncertainties.

Definition of Mobility

"Mobility" is the term used to describe the ability of the robotic vehicle to traverse a rough terrain without any perception. The mobility of a UGV is often expressed in terms of the size of an obstacle (both negative and positive) it can negotiate and still continue along a specified path. As pointed out in U.S. Army (1998), for several reasons a UGV must have a high degree of mobility:

- A high degree of mobility minimizes the perception burden.
- Timely mission accomplishment cannot be achieved if the platform has to spend its time searching for an easy path through difficult terrain.
- The best route for covert missions will mostly likely not coincide with the easiest mobility route.
- A high degree of mobility will keep the vehicle from becoming stuck, thus requiring human assistance.

State of the Art

Most UGVs can be categorized in one of three forms: wheeled, tracked, or hybrid (combination of wheeled and tracked). Wheeled vehicles are the simplest, quietest, and most reliable. Tracked vehicles are known to have better traction and flotation than wheeled vehicles on such slippery surfaces as mud fields, rice paddies, and snow; unfortunately the tracks are more susceptible to breakage and are noisier in general.

As explained in U.S. Army (1998), there are several criteria used to evaluate the mobility of a UGV. For discrete obstacle negotiation, the criteria include tree and stump knock-over, gap crossing, fording water, vertical step crossing, and tree and stump avoidance. For all-terrain mobility the criteria include horsepower per ton, axial twist, ground pressure, vehicle cone index (VCI), forward/reverse slope, side slope operation, side slope stability margin, width for rollover resistance, side step clearance height, high-low speed range, and ground clearance. Table 5-1 shows the desired criteria for a high-mobility UGV weighs less than 2,000 lb. and can be transported in an high-mobility multi-purpose wheeled vehicle (HMMWV) (U.S. Army, 1998).

Table 5-2 shows the specifications of the Demo III XUV (experimental unmanned vehicle) and several U.S.-produced teleoperated UGVs that have been developed for explosive handling, SWAT (special weapons and tactics), HAZMAT (hazardous materials) response, nuclear power plant surveillance and maintenance, and airport security (Shephard's, 2001). The Gecko and Mini-Flail were designed and built for the Department of Defense UGV Joint Program Office. There are also foreign platform options from Canada, England, and Ireland (Shephard's, 2001). As shown in the table, with the exception of the Demo III XUV, these UGV platforms are skid-driven systems that have been designed for much slower speeds than are needed for the Wingman

TABLE 5-1 Desired Criteria for a High-Mobility UGV Weighing Less Than 2,000 Pounds

Discrete Obstacle Negotiation	Desired Criteria
Tree and stump knockover	2–3 ft stump
Gap crossing	1–2 m
Vertical step crossing	18–24 in.
Fording water	4–5 ft deep
Tree and stump avoidance	9–12 in. random spacing
Berm climbing	39 in. desired
Mobility	
Hp/ton	20–40
Axial twist	±10 degrees
Ground pressure	1–2.5 psi
Vehicle cone index	Less than 12 with 6 being desirable
Forward/reverse slope	60%
Side slope operation	60%
Side slope roll stability margin	0.3
Width for rollover resistance	T/2H >= 1.1 required with T/2H = 1.3 desired
Side step clearance height	20 in. goal with 24 in. desired
High-low speed range	1–30 mph
Ground clearance	10–12 in. nominal with variable 4–24 in. desired

SOURCE: Data from U.S. Army (1998).

TABLE 5-2 Current Options for Army UGV Mobility Platforms

Model	Demo III XUV	Mini Andros II	Andros Wolverine	Gecko	Mini-Flail	MPR-150	RATS
Manufacturer	General Dynamics	REMOTEC	REMOTEC	AmDyne Corporation	Marion Metal Works	OAO Robotics	OAO Robotics
Power plant	Diesel engine	2 12-volt DC batteries	Four 12-volt DC batteries	Gasoline ICE (diesel option)	Diesel engine	24-volt DC batteries	Diesel engine
Dimensions (l × w × h in inches)	118 × 66 × 48	42 × 24 × 37	58 × 28 × 40	119 × 57.5 × 120	120 × 50 × 43	38 × 23.5 × 31	100 × 52 × 110
Number of wheels/tracks	4 wheels	Hybrid: 4 wheels plus 2 articulated tracks	Hybrid: 6 wheels covered by tracks	8 wheels	4 wheels	Tracks	4 wheels
Steering	4-wheel Ackerman	Skid	Skid	Skid	Skid	Skid	Skid
Turning radius	128 in.	0 in.	0 in.	0 in.	0 in.	0 in.	0 in.
Weight	3,400 lb	190 lb	600 lb	1,000 lb with max payload	2,450 lb	217 lb	4,000 lb
Speed	0–40 mph	0–1.1 mph	0–2 mph	0–20 mph	3–5 mph	0–1.5 mph	0–5 mph
Performance	Traverse moderate terrain at 2/3 speed of HMMWV	Climb 45-degree slope, 16-in ledge, 21-in-wide ditch	Climb 38-degree slope and 24-in-wide ditch	Climb 45-degree slope, amphibious, low pressure footprint	Flail head used to destroy ordnance on relatively flat terrain	Climb 45-degree slope	Climb 30-degree slope
Manipulator	No	15-lb payload at full reach	3 DOF 60-lb payload at full reach, 100 lb payload at 18 in.	No	No	5 DOF	6 DOF 200-lb lift
Intended use	Scout Mission	Explosive handing, SWAT operations, HAZMAT response, nuclear surveillance/maintenance, airport security	Explosive handing, SWAT operations, HAZMAT response, nuclear surveillance/maintenance	Remote "truck" for hazardous situations	Destroy and neutralize explosive ordnance	Law enforcement, explosive ordnance disposal, nuclear, emergency response, firefighting	Law enforcement, explosive ordnance disposal, nuclear, emergency response, firefighting

DOF = Degrees of freedom. SOURCE: U.S. Army (2001); Shephard's (2001).

and Hunter-Killer example applications. Speed considerations have not been emphasized in the mobility designs for existing systems, because existing interface technologies cannot support teleoperation of UGVs at high speeds.

The Army is pursuing platforms to support UGV requirements of the science and technology objective (STO) programs and demonstrations and the Joint Robotics program. The Army Tank-Automotive and Armaments Command (TACOM) is currently contracting with General Dynamics to have a new 6 × 6 vehicle built as part of the Joint Robotics Program/Mobility Enhancement program (JRP/MEP).

The study provided to the committee by TACOM (U.S. Army, 1998) suggests that a 6 × 6 vehicle would have improved mobility over the Demo III XUV. The study states that "the best 4 × 4s do not do well in mobility and immobilization resistance. The proposed 6 × 6 vehicle will have "better mobility than the 4 × 4, slightly better immobilization resistance, but much poorer stealth compatibility and utility." The gap crossing of the 4 × 4 vehicle is approximately two-thirds the diameter of a tire (21 inch for a 32 inch diameter wheel), while the gap crossing of the 6 × 6 vehicle will be approximately the length between two of its wheels (approximately 34 inches). The 4 × 4 vehicle has a VCI of 12 while the 6 × 6 vehicle will have a VCI of 6. The U.S. Army report (1998) also suggests that an even better technical approach is "an 8 × 8 configuration with two swiveling halves and 'mesh' track overlays, achieving the mobility and immobilization resistance of tracks, the signature mitigation of wheels, and the best maneuverability and obstacle avoidance features of both."

Technology Readiness

Commercial and academic prototypes of UGV platforms, some built as part of the DARPA TMR program, are adaptable to mobility requirements of the Searcher example. Several of these were successfully demonstrated during recovery operations at the World Trade Center site and in the caves of Afghanistan. These platforms are at TRL 6.

Two prototype vehicles are being developed under the DARPA UGCV program discussed in Chapter 3. The Carnegie Mellon Spinner prototype vehicle will be a 6-ton vehicle with a mobility subsystem incorporating a 6-wheel drive designed to be capable of off-road operation; the Lockheed Martin Retauris prototype vehicle will be a 1,300-pound hybrid-electric vehicle with a mobility drive system consisting of six suspension arms capable of 360-degree rotation with electric-motor wheel drives. A 12-month testing phase is intended to help identify mobility technologies for potential application to the Army's FCS program and to evaluate characteristics such as air deployability and resilience to terrain-induced damage. Both prototypes have possible applicability to the Hunter-Killer example system and are estimated to be at TRL 1.

The technology readiness level of a 4 × 4 mobility platform such as the Demo III XUV vehicle is TRL 7. This mid-sized platform has been demonstrated in field environments and might be adapted to the Donkey example applications. A new 6 × 6 mobility platform, possibly adaptable to a prospective Wingman or Hunter-Killer platform, is currently at TRL 2. Building on past experience with designing and building the XUV, the 6 × 6 platform could be at TRL 6 in 2 years; however, it is doubtful that cross-country speed requirements can be met using existing design approaches.

Salient Uncertainties

Overall, the risks associated with building highly mobile platforms may be less than that of developing perception sensors and software that will successfully guide a vehicle around every possible obstacle. We still need processing power several orders of magnitude greater than is currently available to reach the perception ability of the human brain. The alternative to a UGV with human-like perception is a highly mobile UGV that can negotiate any obstacle in its path. It may get knocked down, tumble, roll, and bounce off obstacles, but it will get to its destination due to superhuman (in terms of speed and strength) mobility. A highly mobile, minimal-perception UGV is also appealing in terms of survivability. After all, the first thing that the enemy is going to do is to shoot at and try to destroy the perception sensors on the vehicle. So the less we rely on these sensors, the more reliable the system will be. Finally, the U.S. Army has considerable experience in mobility design, as demonstrated with the design of the tank and other terrestrial and amphibious military vehicles. With this experience the path of minimal risk may be to spend more effort on developing highly mobile platforms.

Of course, developing highly mobile platforms that have less perception capability will alter the missions in which UGVs will first be used. Instead of scouting and stealthy reconnaissance missions UGVs would be used in the "red zone" in such brute-force missions as clearing fires, diversion, barricade bashing, and breaching minefields. These missions require great mobility, size, and speed. Obstacle avoidance sensing can be very simple with bumpers and local infrared (IR), radar, and acoustic proximity sensors.

Depending on requirements, UGV health maintenance technologies (discussed later in this chapter) for such things as diagnostics and self-maintenance will overlap with and simplify traditional supportability, maintainability, and reliability considerations. The scope of logistics concerns will be similar to those for other military mobility platforms, but the impact on logistics operations will depend on specific UGV capabilities.

Mobility requirements for ground vehicles continue to stress improved performance on off-road terrain. Smart active-suspension systems, new tire materials with controlled inflation, and high-performance traction with slip control for each wheel are examples of technologies in likely need of refinement to meet UGV requirements. An uncrewed vehicle has the advantage of not needing to be designed around human crew limitations but also has the disadvantage of needing mechanisms to replace human driving judgment. Thus, design requirements for UGV mobility platforms must be integrated with perception technologies to provide the capability to avoid obstacles, both positive and negative, that the platform is not hardened to overcome.

The foregoing discussion provides the basis for the answer to Task Statement Question 4.b as it pertains to mobility. See Box 5-2.

Areas of Research and Development

Research and development is needed in the following areas:

1. Many of the analysis tools and metrics developed for the design of military vehicles such as the tank and HMMWV use soil mechanics models that may not apply to smaller, lighter UGVs (Laughery et al., 2000). These models will need to be verified on smaller platforms and possibly altered based on theoretical analysis and experimental data.
2. The mobility platform is highly application dependent. The platforms developed for the Searcher, Donkey, Wingman, and Hunter-Killer applications

SUPPORTING TECHNOLOGIES

> **BOX 5-2**
> **Task Statement Question 4.b**
> **Mobility**
>
> **Question:** What are the salient uncertainties for the other main technology components of the UGV technology program (e.g., adaptive tactical behaviors, human–system interfaces, mobility, communications)?
>
> **Answer:** The mobility platform is highly application dependent. The platforms developed for different mission applications will need to be designed based on well-defined mission requirements.
>
> Mobility requirements for ground vehicles continue to stress improved performance on off-road terrain. While a UGV has the advantage of not needing to be designed around human crew limitations, it also has the disadvantage of needing mechanisms to replace human driving judgment. Thus, salient uncertainty surrounds how design requirements for UGV mobility platforms can be integrated with perception technologies to provide the capability to avoid obstacles, both positive and negative, that the platform is not hardened to overcome.
>
> Overall risks associated with building mobility platforms may be less than that of developing perception sensors and software for successful A-to-B mobility. An alternative might be a highly mobile UGV that can negotiate any obstacle in its path. It may get knocked down, tumble, roll, and bounce off obstacles, but it will get to its destination due to superhuman (in terms of speed and strength) mobility. A highly mobile, minimal perception UGV is also appealing in terms of combat survivability.

will need to be designed based on well-defined mission requirements. The current readiness of a 4 × 4 mobility platform such as the Demo III XUV vehicle is TRL 7. This platform has been successfully demonstrated in an operational environment. The new 6 × 6 mobility platform is at TRL 2. Building on past experience with designing and building the XUV, General Dynamics should be able to bring the 6 × 6 platform up to TRL 6 in two years. The proposed 8 × 8 platform mentioned in U.S. Army (1998) should also be investigated.

COMMUNICATIONS

Current military data links that are available to support UGV communications were developed to transport a specific set of information from one platform to another. Preplanning to arrange for using these communications channels is often logistically complex. Communications must generally be manually planned prior to an operation and controlled manually during the operation.

Over the last 15 years the commercial world has been the source of an enormous investment in data networking. Unfortunately, as the Internet has flourished in the commercial world, it has been largely tailored to the needs of a highly available static, wired network. Often these protocols that rely on a highly available static network do not meet the military requirements for mobility and redundancy.

In recent years, however, commercial Internet users have begun to desire support for a variety of mobile users with a certain level of horizontal and vertical data hand-off requirements. In some cases, the emerging protocols for IP mobility may be of some use to the military tactical mobility problem. Unfortunately, however, these protocols too often lack the necessary redundancy and independence from a fixed infrastructure that is required for a military network. When the commercial Internet speaks of mobility, it usually considers only mobility among users. A fixed infrastructure of wired routers is normally assumed.

The four example UGV systems defined in Chapter 2 support different military applications each having distinct requirements for communications. The Searcher would be a teleoperated UGV used to search urban environments and tunnels. The Donkey would be used as a logistical carrier to support timely delivery of supplies to the warfighter. The Wingman platform-centric autonomous ground vehicle would be used to provide combat support to a manned unit. The Hunter-Killer would be one of a collection of network-centric autonomous vehicles capable of ambushing and engaging enemy units.

The Searcher

Being a teleoperated vehicle, the Searcher requires a very-high-bandwidth, low-latency, and high-reliability communications system. Teleoperation places the communication channel directly into the control loop for the vehicle and its associated systems (sensors and weapons). Any delay in the communications system translates into a direct delay in the control loop for the vehicle. For example, fast-moving vehicles might hit an obstacle and be damaged if a 1- or 2-second delay is introduced in the time to transmit a video image back to a controller and send control back to the vehicle.

Teleoperation also implies that all of the sensor data from the unmanned vehicle that is necessary to control the vehicle must be sent back to an operator for consideration. Often this will imply real-time video or other high-bandwidth data streams. Teleoperation requires that the operator have a continuous line of communication with the vehicle being controlled. For a teleoperated vehicle a dead zone in the coverage could quickly become a black hole into which the vehicle is sent but never retrieved or operated.

Mitigating these severe requirements for bandwidth, latency, and reliability is the fact that the Searcher does not need to be operated over long distances and may even be able to drag a cable behind it. For operation with a cable, a state-of-the-art wire-line technology should be adequate to supply the required bandwidth, latency, and reliability. A number of options are available that can reliably deliver low-latency traffic at data rates sufficient to support compressed video.

Since cables may become entangled or cumbersome in buildings or cave complexes, short-range radio frequency (RF) communications may be necessary. The physical environment in which the Searcher operates limits the availability of direct-path communications. This makes communications at video data rates at 1 km slightly out of reach of current systems. A number of emerging systems, such as enhanced wireless local area network (EWLAN), JTRS Wideband Waveform, Surgical Strike, global mobile (GLOMO), and ultra wideband (UWB) systems, may be able to support operation up to 1 km at compressed video data rates. However, significant enhancements in seamless networking to support dynamic routing, QoS, and assured connectivity will be required if the operational range for the Searcher is increased significantly beyond 1 km. As the Searcher evolves and becomes capable of executing higher-level commands, such as "climb the stairs," the need for high-capacity, low-latency data communications will decrease. This will simplify the communications problem for the Searcher.

The Donkey

Because the Donkey is intended to carry supplies along a predefined path, communications is not a critical feature of the Donkey's mission. The primary required communication is to interact with humans located near the start and release points of the electronic path. With the addition of appropriate security there are a number of existing military and commercial systems that would be adequate for this low data rate, moderate latency, direct-path communications. As an augmentation to the basic concept for the Donkey, operators may want to redirect the Donkey en route or provide a level of situation awareness to enhance survival. Again, a number of currently available data links could perform this basic function; however, to provide this coverage reliably over long distances would require improvements in jamming protection, signal detectability, and advanced network-routing capability that is beyond the current state of the art.

The Wingman

Because the Wingman is intended to provide close support of a manned unit, it is important that communications between the Wingman and its controlling manned unit must be moderate to high bandwidth, moderate latency, and high reliability. These communications may include transmission of compressed video from the Wingman to its controlling unit, control back to the Wingman, and shared situational awareness.

Depending on the separation of the Wingman and the controlling unit, current data links might be able to serve this function. As distances become greater and communications become necessary around obstructions, state-of-the-art communications networks will not be able to support this function. As the Wingman becomes more advanced, enhanced networking that could allow the Wingman to share situational awareness information with other UGVs, UAVs, or manned systems would be beyond state-of-the-art systems.

The Hunter-Killer

The Hunter-Killer is intended to support a mission that is doctrinally quite straightforward for a small unit of soldiers. To accomplish this mission the individual UGVs that make up a Hunter-Killer team will need to be in close communication. Current state-of-the-art communications fall significantly short of the current requirements for the Hunter-Killer primarily because of its network-centric characteristics.

Since surprise is essential to the Hunter-Killer mission, low probability of intercept/low probability of detection (LPI/LPD) communications must be enhanced significantly to support this mission.

Dynamic, autonomous local area networking in a tactical environment is a technology area that is not currently mature enough, but it is beginning to emerge in advanced technology efforts. EWLAN, small unit operations (SUO), and the GLOMO efforts are some recent programs that have demonstrated a level of dynamic autonomous operation for tactical LANs.

In principle an autonomous UGV that utilizes situational awareness information (e.g., enemy, friendly, terrain, weather) distributed within the FCS network will be very dependent on high bandwidth and assured communications. An autonomous UGV with most of its situational awareness information obtained through organic sensors would have less dependency on high bandwidth and assured communications, but it would probably cost much more, might not be able to achieve the same level of situational awareness, and could make itself a much more expensive system. The latter may also have a higher technical risk in terms of being able to produce a sufficiently intelligent vehicle.

Technology Readiness Levels

For wire-line technologies to support the Searcher, a number of options are currently at TRL 6. Near-term wireless solutions for Searcher and Donkey are problematic. Network connectivity could easily be lost due to non-line-of-sight (NLOS) interference caused by terrain or obstacles (e.g., thick building wall). Directional communications systems using electronically steerable array antennas currently exist at TRL levels ranging from 2 to 4. Anti-jam and LPI/LPD communications systems with very wide spreading exist in the UWB and a C-band and above with TRL levels between 2 and 4. A number of prototype systems have been built up that have the potential to provide dynamic ad hoc networking for tactical users. Most of these systems exist only at the sub-network level, and some have achieved TRL levels of 4 and possibly 5; however, new versions of these sub-networks may be needed to support the anti-jam LPI/LPD waveforms necessary for UGV.

At the network level ad hoc technologies being developed under the CECOM MOSAIC program will be at TRL level 6 by 2004. The Air Force Multi-Sensor Command and Control Constellation (MC2C) program and the ACN (assign commercial network) program also seek to develop ad hoc networking technologies. Extensions to these technologies will probably be necessary to support UGV missions. Network security technologies are perhaps the furthest behind. While technologies have been developed and fielded for the Internet (TRL 6) these technologies are generally insufficient to support the requirements of the tactical military user of a UGV. Systems solutions that would provide multiple independent levels of security (MILS) data partitioning on a need-to-know basis across an entire network, authentication, key distribution, and intrusion detection and protection are currently largely just concepts and disjointed technologies at about TRL 2.

Capability Gaps

A number of gaps persist in the communications technology necessary to support the four example missions (DDR&E, 2002). These gaps exist in the areas of dynamic networking, security, LPI/LPD communications, and high-rate anti-jam communications. Specifically, technologies must be developed that will provide:

- Integration of all data over all channels
- Increased mobility, survivability, and flexibility
- Assured delivery of information from anywhere to everywhere
- Reduced logistics and manpower
- Unimpeded warfighter access to global information when and where needed
- Attainment of a seamless, integrated, strategic worldwide communications network to include joint, combined, and commercial, with interface to tactical communications systems
- Automated, self-healing, global network management
- Higher data rate—more bits per hertz.

Communications needs must also consider threat capabilities. The DARPA program manager (PM) for FCS communications is addressing this issue. For example, given the threat's capability to monitor and jam RF transmissions, the PM is assessing such features as directional antennas, burst transmissions, advanced frequency hopping, and frequency spectrum management (personal communication between James Freebersyser, DARPA/Advanced Technology Office, and Al Sciaretta, committee member, October 11, 2001).

Security issues that must be addressed include the need for a secure universal method of authenticating users, partitioning data, distributing keys, protection against intrusion, and detection of intrusion.

A robot without communications is a lost asset. A systems integration approach should consider losses of network connectivity due to terrain or feature (e.g., thick wall) masking or other associated problems. Information technology solutions should consider redundancy, relays, and changes in frequency, even if it means degradation in bandwidth. Another approach is the use of behavior adaptation design considerations, which could include tactical movement to positions to restore communications or dead-reckoning movements until communications are reestablished.

Feasibility and Risks

The four areas of highest risk for communications are these:

1. *Prevention of jamming or intercept.* Because UGVs may need to be controlled remotely, jamming of control signals to UGVs could render them useless. Intercept of required transmissions could make the system vulnerable to easy detection and destruction.
2. *Security.* Security attacks on a dispersed unmanned system could include denial of service, compromising of classified high-value tactical information, corruption of information, and, in the most unlikely but most dangerous case, usurpation of the system.
3. *Mobility management.* This involves the protocols necessary to be able to detect changes in network topology and reroute traffic. It also includes the tactical deployment necessary to ensure that there will always be network participants on station to provide relay when needed.
4. *Compatibility.* Because efforts are ongoing by many different companies involved in many different programs, it is important that these disparate endeavors be based on a common vision and conform as much as possible to common interface standards.

The foregoing discussion provides the basis for the answer to Task Statement Question 4.b as it pertains to communications. See Box 5-3.

Impact on Logistics

The impact that the communications system will have on UGV logistics will depend upon the specific UGV mission. With development of the proper technologies, however, there is no reason that the communications system has to represent a major portion of the overall UGV logistical cost.

For the Searcher mission, communication will be primarily between one Searcher and one controller, who is located relatively near the Searcher (within 1 km). Logisti-

> **BOX 5-3**
> **Task Statement Question 4.b**
> **Communications**
>
> **Question:** What are the salient uncertainties for the other main technology components of the UGV technology program (e.g., adaptive tactical behaviors, human–system interfaces, mobility, communications)?
>
> **Answer:** There is significant overlap with the uncertainties that apply to communications for manned combat systems, but the salient uncertainties in communications for UGVs are much more critical. These include prevention of jamming or intercept, security, and compatibility. Security attacks on dispersed unmanned systems could include denial of service, compromise of classified high-value tactical information, corruption of information, and in the extreme, usurpation of the system. Communications for mobility management, perhaps of increasing importance to network-centric operations, must be able to ensure that there will always be network participants on station to provide relay when needed.
>
> The efforts of disparate endeavors in communications for both manned and unmanned systems must be based on a common vision and conform as much as possible to common interface standards.

cally, coordination between the Searcher and the controller and generation and loading of cryptographic keys have to be locally coordinated only locally. Therefore, the communications system would not be expected to add significantly to UGV logistics.

For the Donkey, keys and authorization will have to be coordinated between all users who might need to communicate with the Donkey. This will be necessary to prevent unauthorized users from stealing, destroying, or diverting supplies that might be carried by the Donkey. Dynamic ad hoc networking and ad hoc cryptographic key distribution will be very important technologies to prevent the logistics for the Donkey's communications system from becoming a significant contributor to system operational cost.

For the Wingman as for the Searcher, the communication system is primarily between one Wingman and an associated controlling unit. As in the case of the Searcher the coordination of the communications system, including authorization, keys, and networking, can be performed locally. If information must be shared between a number of Wingman UGVs or between the larger tactical network and the Wingman, then the Wingman's communications system will require dynamic ad hoc networking and ad hoc cryptographic key distribution similar to the Donkey's.

Because of the rate of change in connectivities and individual links and because of the potential need to share information with other systems operating in the area, the Hunter-Killer will almost certainly require dynamic ad hoc network

and ad hoc cryptographic key distribution as described. Without these capabilities, the logistics required to design the networks and to assign and update keys could be so severe that the system would be impractical.

Recommended Areas of R&D

Research and development is necessary to fill the gaps in three areas: anti-jam LPI/LPD physical transmission waveform technologies, dynamic ad hoc networking for the tactical environment, and network security.

Anti-Jam LPI/LPD Physical Transmission Waveforms Technologies

Research and development is necessary along a number of fronts. Directionality shows significant promise in being able to supply covert robust communications at high data rates. Technologies need to be developed to control and steer directional communications in a mobile tactical environment. Improvements in electronically steerable antenna technology are necessary to facilitate directional communications in mobile environments. In addition to directionality, anti-jam and LPI/LPD performance can be improved by increasing the spreading factor of the transmission. Research and development is required for systems with very high degrees of spreading. These systems may include UWB systems or systems centered at very high frequencies, where signals can more easily be spread over 1 GHz or more.

Dynamic Ad Hoc Networking Technologies for the Tactical Environment

As data links proliferate in a mechanized UGV environment it will be more and more important that planning and logistics to support these data links be kept to a minimum. Traditional data links that require special data loads and significant network planning will not be possible in a fluid tactical environment that may contain thousands of nodes using several different physical waveforms. Dynamic ad hoc networking research and development is required at two different levels. First, research and development is necessary to develop and mature approaches to dynamic ad hoc networking at the sub-network level among network nodes that are using the same physical waveform. Second, research and development is necessary to develop and mature approaches to dynamic ad hoc networking at the network level, which will allow data to be seamlessly routed between sub-networks that may be using widely different physical waveforms.

Network Security

As data communications become more widespread and more tightly integrated into the tactical environment, protection of those communications from compromise or corruption is essential. Three areas of research are necessary in network security. First, we must develop technologies to allow networked data to be partitioned based on MILS so that information can freely flow through the network protected on a need-to-know basis. The second area for research and development is to develop a system to detect and protect against intrusion on the network. The third technology that must be developed to support a proliferation of communications in the UGV tactical environment encompasses secure methods for authenticating users within the network and for distributing appropriate keys. This third technology is essential to prevent the proliferation of communications nodes from overwhelming the logistical task of generating and distributing cryptographic keys.

POWER/ENERGY

This section defines the power/energy technology area as it relates to UGV systems. It assesses the state of the art in relevant technologies, estimates technology readiness levels, and identifies salient uncertainties.

Definition and Constraints

The energy source and the rate at which it can be utilized are key to robotic vehicles operating in the battlefield. At present, there are several options for energy sources, depending on the application. For small units the energy source can be a battery, rechargeable and nonrechargeable. For larger units the energy train can be fueled, allowing for motor-generator or hybrid-electric systems. The selection of the appropriate technology for use in any given robotic application must take into account all of the relevant factors that could influence mission success. These factors must be considered early in the development cycle.

The power train for robotic vehicles must support mobility, housekeeping, and mission package energy demands. The four example concepts, Searcher, Donkey, Wingman, and Hunter-Killer, can all be powered with existing technologies at some level. The difficulty comes when long, energetically demanding missions are required. For example, the battery technology to allow Searcher to go 1.5 km over modestly difficult terrain, perform a mission, and return to base would require more than 1 kWh of electrical energy. A rechargeable battery capable of delivering this amount of energy would be several kilograms in mass, leaving little room for any payload; therefore, it is imperative that small hybrid systems be available for Searcher to perform over the extended-duration mission envelope postulated. For Donkey, Wingman, and Hunter-Killer there are vehicle prototypes, both hybrid electrical and conventional diesel, that would suffice for many missions. If a stealth mode is required for long mission times, major developments in the power train will be necessary.

Mobility and Housekeeping

Depending on the mass, the power train may have to supply as much as 200 hp (~160 kW) to the drive mechanism for maximum speed and mobility. In addition, the onboard electronics (e.g., computer, communications) and sensor suites will require power/energy when in full-up operation. This could be reduced by half or less in quiescent or standby modes.

Mission Package

The UGV has two separate but integrated parts. The mobility package, which is the basic robot platform with the ability to navigate, to sense the environment, and so on, and the mission package, which must provide the mission function capabilities, such as weapons, logistics carrier, or reconnaissance scout hardware and software. Each mission package will have energy requirements of its own ranging from a few watts for long periods to kilowatts for short periods of time.

The most obvious factors impacting the energy supply are mission environment, mission time, vehicle mass, signature, cost, logistics support, size, and efficiency. These factors are not independent and they may be more severe and mission limiting for small robotic vehicles with the energy supplies that make up most of the mass and volume of the system.

The mission environment is taken to mean the local environment associated with any place in the world the military will conduct operations. Given that definition, the energy system must perform over a temperature range from approximately −65°F to well over 100°F in the desert. Further, sand, dust, salt fog and spray, and the possibility of chemical and biological environments are factors that could be superimposed on local environmental conditions.

Mission time can vary from days on station with little demand for energy for motive purposes or it can be full-up mobility and sensing at high rates of speed. Obviously the mission time is a strong function of the size, weight, and terrain through which the UGV is to operate; however, it is totally determined by the size of the energy store available for the vehicle. Due to the variable demand for energy it is better to express mission time in terms of kilowatt-hours needed for the mission. The mission time is also influenced by stealth considerations and the primary energy source available. Very probably stealth considerations will not be the same throughout a particular mission, allowing hybrid systems to be used when the secondary storage unit provides the energy for the stealth portion of the mission.

In general, consistent with the total energy requirement, weight must be minimized. The more mass devoted to energy the less there will be for payload. The mass factor is determined by mission profile, demand for stealth, and the energy demand for the payload. The mass factor is influenced by power requirements in terms of peak and continuous demands and by the mission duration in so far as it demands refueling.

Power/Energy Signatures

For survivability, robots in a combat mode must have minimal signature. The most obvious signatures are acoustic, electromagnetic, infrared, and visual. Acoustic noise generated by the power train can be detected by ear or by frequency-selective, amplified detectors specifically designed to receive particular frequencies that are characteristic of a particular device. This imposes the severe constraint, for stealth operations of having to mimic the natural background in any operation that demands mobility or trying to totally suppress any acoustic signature. Susceptibility to detection also depends on local environmental conditions and can vary by orders of magnitude depending on such factors as winds and weather.

Electromagnetic signature is more difficult to suppress. The robot will have to be in communication either continuously or periodically, depending on the degree of independence needed to carry out its mission. This signature may be minimized by using spread spectrum and/or pulsed mode to minimize the time necessary for autolocation by the enemy. In addition, many of the electrical systems generate noise that can be readily received and used as a homing signal. Motor noises and the low frequencies associated with conversion processes are examples.

Infrared signature is one of the most difficult to suppress. Fueled systems utilize the heat of combustion to drive such devices as motors in a well-defined thermodynamic cycle. The efficiency is limited to the Carnot cycle, and hence most of the heat of combustion must be ejected into the local environment. Present IR devices are sufficiently sensitive to detect objects that are less than a degree above or below ambient conditions. These signatures will be extremely difficult to suppress and will have great bearing on the survivability of UGV systems in the field. The general approach to date is to try to mimic the background environmental conditions.

Visual signature is a continual threat. It can be enhanced using electronic devices but is mitigated using typical camouflage color schemes and in the future with adapted camouflage that can be changed continuously to blend with the local environment.

Impact on Logistics

Since support personnel make up most of the armed forces, it is imperative that we seek to minimize the logistics requirements for the power train of UGVs. It may not be practical to eliminate one combat position and require three support personnel to service the robotic vehicle that replaced him. Within that context there would be requirements for training, spare parts, manuals, and special tools. A key factor is the use of any special fuel. As robotic vehicles mature, it will be imperative that the vehicle has the capability to refuel itself from prepositioned fuel or fuel that is air dropped in the vicinity. If it has smaller specialty robotic elements, it must be capable of fueling and programming them for specific tasks. At present, the military is moving to one battlefield fuel that limits many of the options for fueling UGVs in the near future and simultaneously eliminates several promising options.

Above all else, the energy source and its power train must be reliable, easily maintained, and available when needed. These factors are all interrelated with the logistical support and signature management. For high reliability, the system should be as simple as possible and be based on proven technology, preferably something with an enormous legacy within the civil or military sector. Redundancy in the items that represent single point failures will be necessary to ensure reliability in the field. For extended operation in the field, such items as spare parts and filters should be readily changeable; in some instances the UGV might be able to accomplish this task itself from prepositioned supplies or from a tactical logistics robot that could bring supplies when requested. In the absence of specific requirements there are no current values for reliability, maintainability, and availability.

To reduce logistics the volume and mass of the vehicle must be compatible with mission requirements and be locatable on the platform in a way that will provide an optimal center of gravity. Any given UGV will have volumetric constraints in addition to mass constraints that will need to be optimized.

The unmanned vehicles envisioned for the military run the gamut from microvehicles to miniature UGVs with short-range and highly specialized capabilities (Searcher) to multiton vehicles that can be used for convoy, tactical resupply, scout, weapons platforms, and resupply missions. In between are vehicles to serve as weapons platforms (Hunter-Killer), reconnaissance (Wingman), and rucksack-carrier (Donkey). The energy/power technologies appropriate for each mission must be formatted for that user. On the low end of the size range a small hybrid system or highly energetic primary batteries, even though costly, may be a solution; at the high end of the size range standard motor-generator technologies with the addition of hybrid concepts are applicable.

State of the Art

NRC (1997) contains a comprehensive description of the state of the art in advanced battery and fuel cell technology, both in the military and civil sectors. Table 5-3 shows various energy systems and conversion techniques and represents a concise statement of the state of the art in terms of technology readiness levels. Also included in Table 5-3 are estimates of the potential for improvements for each power

TABLE 5-3 Summary of Power/Energy Systems

Power System	State of the Art	Potential for Improvement	Key Issues	Scaling Laws	Potential for Unmanned Ground Vehicles	Hostile Signature	Suppression Potential	Fuel	Autonomy Time
Primary battery	Mature TRL 8-9	Moderate	Energy density Safety Power density Environmental impact, cost	Known	Existing inventory item Applicable to small UGVs Less weight Disposability	Minimal	Excellent	None	Hours/days
Secondary battery	Mature TRL 8-9	Moderate	Energy density Cycle life Power density	Known	Large industry investment Essential in hybrid power systems	Minimal	Excellent	None	Hours
Fuel cells (hydrogen)	Exploratory development TRL 6-8	Excellent	Fuel reformers Water management Safety	Known	New capability, large industry interest in automotive applications Less weight Fuel storage problems	Thermal	Excellent	Hydrogen	Days/weeks
Fuel cells (methanol)	Emerging TRL 4-6	Excellent	Fuel and fuel crossover Catalyst	Uncertain	New capability Requires new fuel	Thermal	Excellent	Methanol	Days/weeks
Nuclear isotope	Limited TRL 9 for space applications	Excellent	Safety Environmental impact Cost Public acceptance	Known	New capability, large NASA and DOE investment, fueled for lifetime of vehicle	Thermal Nuclear	Moderate	Special	Month/years
Internal combustion	Some versions mature TRL 9 for most applications	Moderate to excellent	Fuels Vibrations Life	Uncertain	Currently in extensive use Critical to hybrid power systems	Thermal Acoustic	Moderate	Multifuel (Some special)	Days/weeks
External combustion	Emerging, some mature versions TRL 4 for large sizes	Excellent	Low specific power	Known	New stealth capability Immature at high power levels, heavy	Thermal	Moderate	Multifuel	Days/weeks

SOURCE: Adapted from NRC (1997).

system technology with reference to its potential for UGV applications.

Hybrid Power/Energy Trains

In most cases the power train of choice should be one consisting of a fueled system that provides the primary energy store and an intermediate store that for stealth reasons will probably need to be a rechargeable battery. A small hybrid energy system would be sufficient for the Searcher, given the mission parameters postulated. Current development efforts focus on hybrid systems coupled with a secondary battery in the configuration shown in Figure 5-2.

The mission energy supply is contained within the fuel carried on the vehicle. In the near term a motor generator will be used to convert the energy to electrical energy usable by the UGV. In the far term, reformed logistic fuel coupled with a fuel cell may provide the initial electrical energy. Part of the energy is stored in the intermediate storage unit, usually a high-specific-energy, high-specific-power rechargeable battery. This unit is sized to give the UGV a predetermined amount of energy that will maintain power to the mission package and provide stealthy movement for some tactically significant time.

The controls package will have to be "intelligent" and may be part of the overall computational capability of the UGV. Clearly it must interface with whatever decision-making process determines the mode in which the vehicle as a whole is to operate. When the intermediate store has been expended to a point where it is necessary to recharge, the unit must determine if it is in danger and weigh the possibility of detection and destruction against running out of immediate energy and becoming disabled. The intermediate store and the prime energy store would be capable of operating simultaneously to give the UGV a sprint capability in an emergency. In a nonstealth mode, the motor gen-

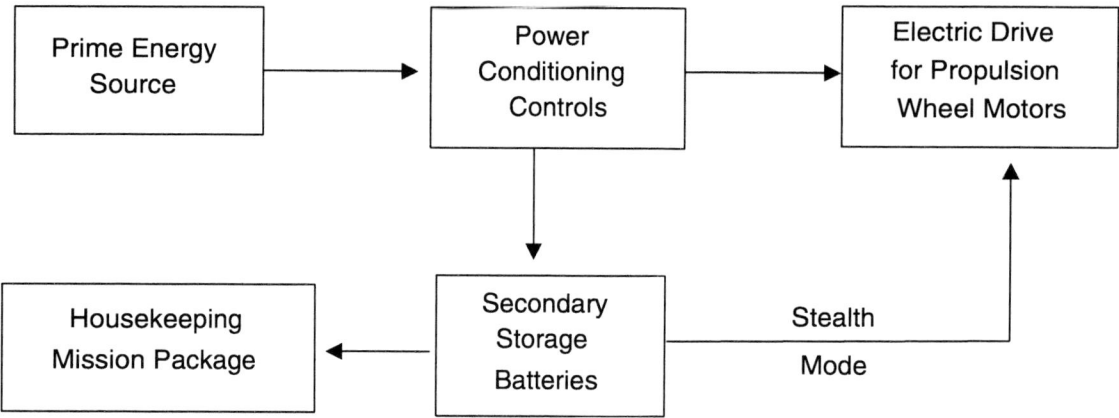

FIGURE 5-2 Schematic of typical hybrid electric power train for UGVs.

erator will provide energy for housekeeping and the mission packages while keeping the intermediate store fully charged. It is possible to extend the stealth mode to operation with the motor engaged. Tuned mufflers and other techniques can reduce motor noise to levels that cannot be detected beyond a few feet. This stealth technique only applies to the acoustic signature and is achieved at the expense of efficiency.

In the far term fuel cells offer the possibility of better fuel economy and inherently more stealthy operation. Due to the dynamics of reforming battlefield fuels and the operation of fuel cells in general, it will still be necessary to retain the battery-driven intermediate storage unit. The critical issue driving application of fuel cells in military systems is the problem of hydrogen generation and storage. Storage in hydrides is at most a few percent efficient by weight.

Reforming of battlefield fuels is hampered by the sulfur content of the fuels. Recent progress in microchannel reformers indicates that small efficient, poison-tolerant systems can be built that will enable the use of battlefield fuels in reformers for fuel cells (Irving et al., 2001).

For larger systems, Figure 5-3 shows approximate system mass for a hybrid electric power train as a function of mission duration measured in terms of kilowatt hours of energy. The hybrid systems chosen are 50 hp and 200 hp (0.76 kW/hp) with a motor generator with a high-specific-power, high-specific-energy intermediate storage battery for stealth mode, housekeeping, and mission package power for the near term, and a reformer fuel cell, intermediate storage unit that can be developed in the far term.

For small UGVs, such as those needed for tunnel investigation and building search and clearing, the energy requirements are much less than those for the large units described in Figure 5-3. NRC (1997) found that small hybrid systems would be the choice for dismounted soldier systems whose mission-time requirements exceeded a certain number of kWh. Figure 5-4 extrapolates this data from 50-W soldier systems to the 500-W systems more typical of a tactical mobile robot, such as the Searcher UGV. In Figure 5-4 a range of battery types are graphed along with two hybrid concepts. Hybrid devices excel above mission times of 4 kWh for all of the standard battery types investigated. Only the zinc-air system appears to be competitive on the basis of specific energy out to 8 kWh. Zinc-air in its current embodiment is, however, specific power limited, and some research would be needed to move it into a competitive position in this regime. Mission times in excess of approximately 8 kWh are clearly in the domain of fueled systems.

Technology Readiness

Miniature hybrid energy systems are estimated at TRL 4 but should achieve TRL 6 by 2006 if adequately funded to achieve Land Warrior program objectives (for soldier-portable power sources for individual soldier electronics). The high-specific-energy rechargeable battery is at TRL 4 and will not achieve TRL 6 until 2009. Compact logistic fuel reformers are at TRL 3 to 4.

Capability Gaps and Recommended R&D

The specific energy of rechargeable batteries is in some degree dependent on hazards to the humans who use them and their potential for abuse. For robotics, referring to Figures 5-3 and 5-4, improving the specific energy of the rechargeable battery by a factor of two would result in significant mass savings or more stealth time for the same mass. Similarly, fuel cells will not see widespread battlefield use until there is a compact, reliable fuel reformer capable of utilizing battlefield fuels. Both of these are high-impact areas and will need further work in order to extend mission times to acceptable levels.

Small robots are severely limited by the energy store. Current robots must rely on low-specific-energy recharge-

SUPPORTING TECHNOLOGIES 87

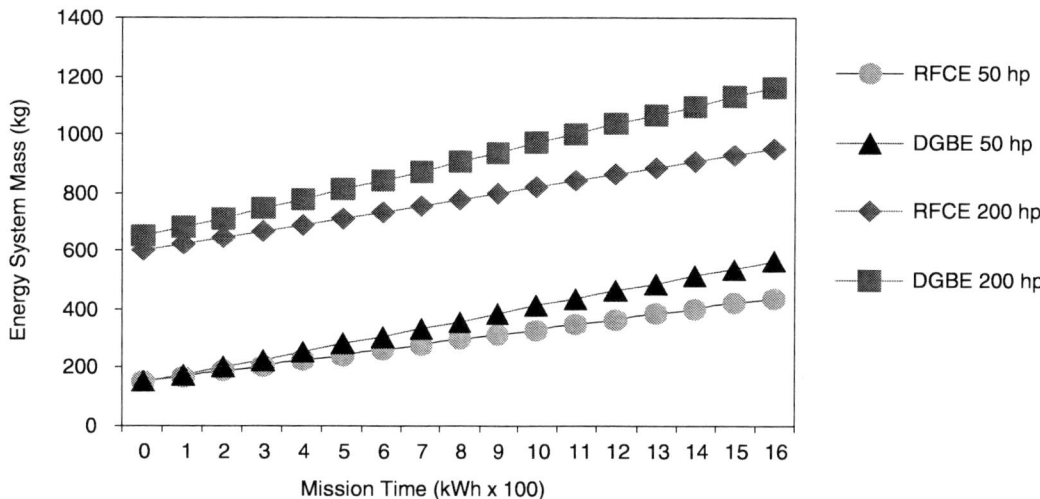

FIGURE 5-3 System mass as a function of mission energy requirements. RFCE = reformer fuel cell electric; DGBE = diesel generation battery electric + 20 min battery at 1/2 power. SOURCE: Bill (2001); DOE (1998b).

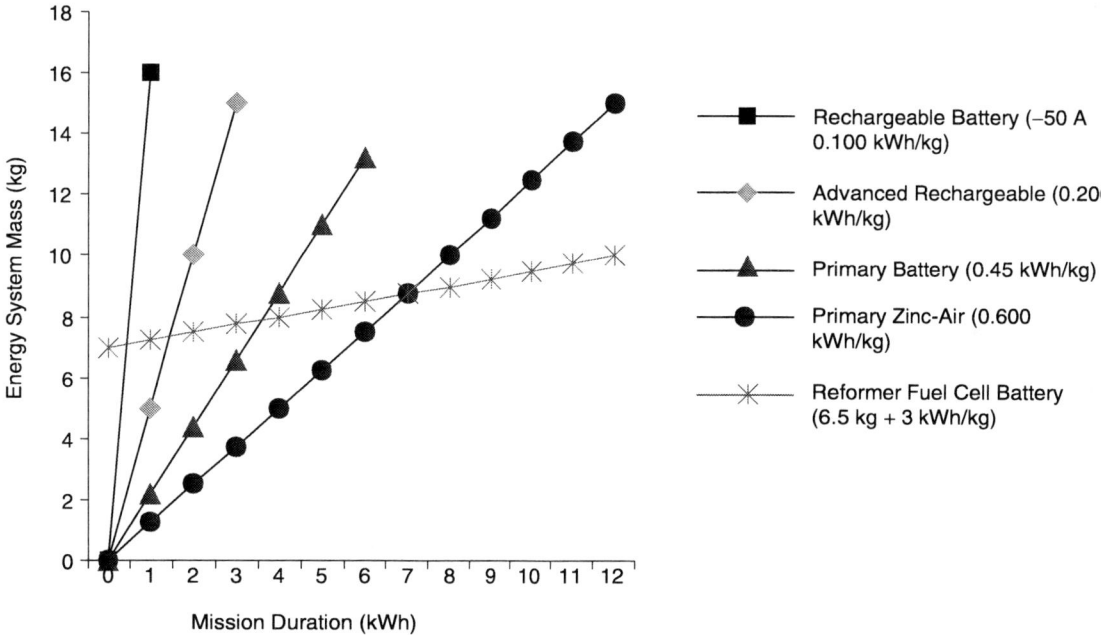

FIGURE 5-4 Hybrid UGV 50-watt to 500-watt systems. SOURCE: Data from NRC (1997).

able batteries or costly primary batteries that only extend the range by factors of two. Hybrid systems for small robots are a prime candidate for further investment.

The foregoing discussion provides the basis for the answer to Task Statement Question 4.b as it pertains to power/energy. See Box 5-4.

HEALTH MAINTENANCE

This section assesses health maintenance technologies for self-monitoring, diagnostics, and remediation of UGVs. It discusses the state of the art, technology readiness, and capability gaps.

> **BOX 5-4**
> **Task Statement Question 4.b**
> **Power/Energy**
>
> **Question:** What are the salient uncertainties for the other main technology components of the UGV technology program (e.g., adaptive tactical behaviors, human–system interfaces, mobility, communications)?
>
> **Answer:** Salient uncertainties for power/energy technology developments cannot be determined until one specifies a mission time in kWh. Short duration, low-mission-energy requirements can be met now. If the goal is to enable extended duration, high-energy UGV missions, the following issues must be addressed: catalysts for reforming fuel, thermal rejection processes, stealth, and energy storage and replenishment. These are all areas that are likely to be required for FCS systems.

Definition

Health maintenance has two distinct flavors. One is making the robot more physically robust; the other is detecting, diagnosing, and recovering from component failures (or from degradations in performance that may lead to mission failures). Of the two, making the robot more physically robust is well understood.

Such failures as engine overheating, loss of communications, or flat tires have solutions that are not unique to unmanned systems. UGV health maintenance technologies resolve aspects of failures that are either unique to UGVs or require machine awareness. This specifically includes technologies targeted at preventing or mitigating failures of sensors or electronics for robot vehicles.

UGVs operating as part of the FCS must be exceptionally robust for many reasons. They are likely to be limited in number and there may not be many backup or replacement vehicles. In addition, the short duration of the mission may not permit a replacement; consider a UGV that takes 8 hours to position itself, fails, then another 8 hours is spent positioning a replacement for a total of 16 hours of unavailable asset at a desired location.

Vehicle health monitoring and maintenance are intended to ensure that the robot performs its current set of tasks reliably and within acceptable parameters, as well as to project those capacities for the future (e.g., a robot just a few hours away from needing an overhaul should not be tasked for a new mission). Subtle sensor failures could lead to false positives or false negatives on targets, leading to an overall mission failure without the operators ever suspecting a fault. Likewise, UGVs should have monitoring systems that aid the maintenance technicians and reduce the time out of service.

Vehicle health monitoring is expected to be accomplished using machine intelligence for several reasons. If a vehicle cannot self-detect problems, it may actually make them worse while waiting for a human operator to notice the dysfunction, for example, spinning itself deeper into the mud, oblivious to being stuck. The number of sensors and the impact on performance may be too subtle for a human operator to discern and respond to in a timely fashion. The time lag may be too large for critical situations to wait for human involvement. For example, suppose a navigational sensor is damaged by a sniper, the robot cannot just sit there and wait for the human to notice and take charge. Instead, it should either swap to a safety behavior using another sensor or continue with its current behavior using either a physically or logically redundant sensor.

Health monitoring and maintenance are similar to fault tolerance. Fault tolerance connotes the ability of a system to compensate for failure conditions. Many engineers now refer to fault-tolerant methods with the acronym "FDIR" (fault detection, identification, and recovery) to emphasize the steps involved in accomplishing fault tolerance. However, both fault tolerance and FDIR may not adequately capture the breadth of issues in vehicle health monitoring and maintenance or the source of errors. In the manufacturing and aerospace domains failures are synonymous with hardware, obscuring the role of control and mission-planning software in intelligent vehicles. The fault-tolerance community is concerned with what may be termed catastrophic failures, complete failure of a critical component. This concern ignores the potential impact on performance due to gradual degradations in such components as actuators and sensors over time or transient failures. It also overlooks the contribution of the environment to the dysfunction of a component (e.g., mud causing wheels to slip, fog covering a sensor lens). In intelligent vehicles each of the local components can be behaving correctly yet still show a failed or incorrect performance due to errors in the mission plan (e.g., instructing the robot to do the wrong thing at the wrong time).

The breadth of issues involved in vehicle health monitoring and maintenance include:

- Identification of both degradation and catastrophic failure of components, either singly or cumulatively, that result (or will soon result) in an unacceptable level of task performance.
- Identification of the cause with sufficient granularity as to effect a recovery (e.g., the vehicle does not need to know what component on the camera failed, just that the camera cannot be used). As noted in Murphy and Hershberger (1999) and Basu and Paramasivam (2000), faults originating from one source may appear to stem from another, such as a software failure may appear to be a hardware failure.
- Identification and implementation of a recovery procedure to restore performance, either by adapting or recalibrating processes, pre-empting and reallocating hardware and/or software, changing to different behaviors, or replanning.

State of the Art

Fault tolerance and machine health have long been a focus area in manufacturing. Manufacturing facilities often use statistical process control to track deviations in the output quality of machines, adapt the production process, and determine when to pull a machine for maintenance. Statistical process control (SPC) is not well suited for industrial robots or robot vehicles because it only points to a defective machine, not what makes the machine produce defective parts. There is also usually a lag time in SPC before the problem is discovered. Real-time fault tolerance is needed for vehicle systems. It is commonplace, in computing, where critical computers often work in mirrored pairs so that if one fails the other takes over; this explicit redundancy is undesirable with limited numbers of expensive vehicles.

The need for fault tolerance in robot systems has been noted since, at least the mid-1980s (Moore, 1985). While the need for fault tolerance has been discussed for unmanned underwater vehicles (UUVs) (Rae and Dunn, 1994) and UGVs (Yan et al., 2000), the UAV community appears to have a lead in actual implementation of such systems. This lead may stem from its aerospace roots, where flight critical systems are explicitly designed for fault tolerance with triple physical redundancy on many components. The UAV community has a decided advantage over UGVs; UGVs must operate in much more complex and adversarial terrain and carry far more sensing software.

In general, fault-tolerant methods for mobile robots in the literature concentrate on detection and diagnosis and assume that recovery is straightforward. The majority of work focuses on catastrophic failures, ignoring normal degradations with only a few exceptions (Murphy and Hershberger, 1999; Liu, 2001). Some recent efforts claim to be fault tolerant; they bypass the FDIR cycle by simply imposing limits or thresholds on what is an acceptable signal from an actuator or sensor (Schreiner, 2001; Okina et al., 2000). Clearly these approaches are too limited for FCS-like domains.

The dominant methods for fault detection and diagnosis are offshoots of control theory or artificial intelligence research in diagnosis. The earliest work on robot diagnostics developed toolsets (Carnes and Misra, 1996) and expert systems and other interactive tools (Krishnamurthi and Phillips, 1992; Nikam and Hall, 1997), including ROBODOC (Patel and Kamrani, 1996). Model-based methods are popular (Jackson, 1997; Clements et al., 2000), particularly variants of Kalman filtering (Washington, 2000; Roumeliotis et al., 1998; Vos and Motazed, 1999). Actual demonstrations of model-based methods have been shown on ground robots for navigational errors in (Goel et al., 2000; Soika, 1997) and with the Bell Helicopter Eagle-Eye UAV (Rago et al., 1998).

Despite their popularity, model-based methods have many disadvantages. The models are difficult to construct and fall apart if the model is not correct and complete. Reasoning over the model is often computer and time intensive, preventing it from being used in real time (though ever increasing chip speeds may provide the ultimate solution). They tend to be monolithic in nature, though some systems explicitly reflect the hardware, software, environment, and planning errors by creating a hierarchy of layers (Visinsky et al., 1995). Murphy and Hershberger (1999) developed a system that relied on only partial causal models, avoiding these drawbacks, and demonstrated it on two different mobile robot systems. Other detection and diagnostic methods proposed for robot fault tolerance include the use of rules (Djath et al., 2000), automata theory (Perraju et al., 1996), and fault trees (Madden and Nolan, 1999).

Of the fault-tolerant activities, detection is possibly the most commonly targeted topic. Techniques for the detection of hardware and software failures include learning with neural-networks, UUVs (Deuker et al., 1998), pattern matching in complex streams of information (Oates and Cohen, 1996), and using kinematic and dynamic models of wheels (Dixon et al., 2001) besides the more traditional approach of hardcoding monitoring routines. Sheldon et al. (1993) offer a method of selecting the best software for a simplified UGV. Techniques for the detection of navigational errors, particularly not reaching expected locations within expected time or other modeling constraints, have been studied by Bikfalvi and Lorant (1999) and Stuck (1995). Detection of planning errors or failures has been demonstrated for space vehicles (Rasmussen, 2001; Aghabarari and Varney, 1995) and in mobile robots (Lamine and Kabanza, 2000; Bergeon et al., 1994; Lam et al., 1989). The detection of adverse changes in the environment has not been considered, except cursorily by Murphy and Hershberger (1999).

Unfortunately, each of the above methods is vulnerable to the same drawbacks. Each ignores transient errors (e.g., loose connection jolted off, then back on). Multiple simultaneous failures are ignored or assumed to be statistically independent, but battle damage is not statistically independent; for example, all the components on one side of the vehicle may be affected by an impact.

Isolated diagnostic methods include the use of logics from AI (Portinale and Torasso, 1999), while explorations in recovery appear limited to the use of genetic algorithms (Baydar and Saitou, 2001). The research in these areas is theoretical and has not been applied outside of simulation.

There is no reason to suspect a breakthrough in fault tolerance from foreign technology. Academic research seems to be split between Europe and the United States, with some participation by the Japanese. There seems to be no clear lead, and it is expected that the relative levels of academic research represent the relative levels of implementation in industry.

Current Army Capabilities

The current state of the art in academia is quite limited in regards to vehicle health monitoring and maintenance. Current capabilities do not include even the basic interactive expert systems available since the mid-1990s. This is not surprising given the emphasis on component technologies and demonstrations in the Army UGV projects. Fault tolerance and system health are currently not active considerations in the Army UGV program.

Technology Readiness

The current level of readiness for UGV health maintenance and health-monitoring technologies is at TRL 0. Much of this work is theoretical and may not be readily transferred to Army systems. Assuming that data on failure modes is available, it is reasonable to expect that within 5 years technology should be able to provide an interactive expert system that will at least aid a teleoperator in diagnosis.

Although not part of health maintenance technology per se, physical reliability and maintenance characteristics for the Searcher and Donkey examples are at TRL 5, with TRL 6 achievable within the near term. Cooperative diagnostics and monitoring to link with the tactical skill of calling for help has not been demonstrated for robotic platforms and is estimated at TRL 3 or less.

If algorithms for self-diagnostics could be adapted from those for existing platforms, this technology could be at TRL 3. Other platform-dependent technologies, such as intentional design for combat survivability as would be required for the Wingman and Hunter-Killer, is at TRL 1 or less. Similarly, self-repair technologies for Hunter-Killer are at TRL 1.

Capability Gaps

In the long term, work in reconfigurable robots is expected to be of use in having robots identify damaged components and adapt accordingly. An often overlooked aspect of health maintenance is how the robot responds to handling by unauthorized personnel. In this case it is expected that the robots will have non-lethal responses to unauthorized handling.

An important aspect of individual vehicle health monitoring and maintenance is the application of what is learned about one vehicle to other vehicles. An FCS system that could share, propagate, and exploit information gained from one vehicle to another would be much more useful. Imagine a vehicle having a problem with a FLIR sensor due to unexpected interaction of fog and foliage. Once identified, all vehicles could be alerted to this possible failure mode and proactively attempt to recognize it before the FLIR reported suspect readings that might lead to a wrong decision or take valuable time and bandwidth to resolve.

Fault detection, diagnosis, and recovery is generally acknowledged to be a difficult problem, especially in systems without the triple redundancy and a relatively static environment found in aerial vehicles. The first level of response, required by all four of the example systems, will be for the robot to help monitor itself so that the operator can respond appropriately and timely. The levels escalate from there.

Higher-order health maintenance capabilities, such as self-diagnostics and self-repair, depend to a great degree upon advances in similar technologies for manned systems.

Salient Uncertainties

One fortunate aspect of vehicle health monitoring and maintenance is that it does not have to be perfect or autonomous to improve performance. Any system that can reduce the number of errors or the time it takes to discover that an error is occurring is a win, even if the final diagnosis and recovery plan are generated manually. The only other options are to have a human constantly scrutinize the vehicle for signs of failures (a poor use of manpower and questionable as to the amount of attention that can be maintained) or to just assume that everything is working correctly.

The literature offers promising avenues to explore in health monitoring but there is no clear solution. One interpretation is that development of a health-monitoring system is not feasible. The interpretation we chose is that health monitoring for UGVs is probably more difficult than for manufacturing or aerospace, since UGV military operations represent a more demanding domain for real-time fault tolerance than any of the test domains reported in the literature, but it can be done if the demand for it is made clear. The amount of research spent on fault tolerance for mobile robot systems is almost negligible compared with navigation, path planning, localization, and mapping. Likewise, vehicle platforms remain sensor impoverished, focusing only on the ideal sensor needed to navigate, not on adding expensive physical redundancy or identifying logical redundancy, or adding internal sensing and processing algorithms.

The real question of feasibility depends on the FCS deployment profile: a trade-off between the tempo of operations and the vehicle redundancy. If extra vehicles are available to be predeployed to rapidly replace a failed robot, then allowing a UGV to die in place may be acceptable. Note that this only eliminates the diagnostic and recovery functions of fault tolerance; the robot or operator still must be able to detect failures. If the number of vehicles is limited, every asset must remain as functional as possible.

The foregoing discussion provides the basis for the answer to Task Statement Question 4.b as it pertains to UGV health maintenance. See Box 5-5.

Recommended Research and Development

In order to have fault-tolerant vehicles, the Army has to incorporate the expectation of fault tolerance into its development programs. The Army may be able to use "black box" technologies to record what prototypical robots are sensing and doing; this is critical to begin the rigorous data collection needed to serve as the basis for any health-monitoring system or even what it means to be survivable. Steps that should be taken now are as follows:

SUPPORTING TECHNOLOGIES

> **BOX 5-5**
> **Task Statement Question 4.b**
> **Health Maintenance**
>
> **Question:** What are the salient uncertainties for the other main technology components of the UGV technology program (e.g., adaptive tactical behaviors, human–system interfaces, mobility, communications)?
>
> **Answer:** Toolsets consisting of sensors, diagnostics, and recovery methods will be based on FCS operational requirements. Coming up with a calculus of diagnosis and recovery similar to the triple-redundancy development systems and design logic used for NASA systems will be a major challenge. Like human–robot interaction technologies, health maintenance developments will require extensive experimentation and testing.

- Begin a program of rigorous data collection on failure modes in UGVs, including software and AI functions. The majority of approaches to FDIR require models and/or frequency of failure data; this data is usually collected from empirical studies, particularly in the presence of hard-to-model environmental effects. The data can be used by any probabilistic, model-based, or expert system FDIR scheme, and so a premature commitment to any particular approach is not required.
- Develop a toolset of common sensors, diagnostics, and recovery methods.
- Explore using fault-tolerant techniques developed for the aerospace community for ground vehicles.
- Explore mechanisms for explicitly designing in physical and logical redundancy in mobility and sensing subsystems.

SUMMARY OF TECHNOLOGY READINESS

Table 5-4 summarizes the technology readiness assessments made in each of the preceding sections as relates to the example systems defined in Chapter 2. Technology developments in UGV supporting technology areas of power/energy, communications, and health maintenance depend heavily on system-level requirements for UGVs that have not yet been defined. These may well prove to be pacing items for the overall FCS UGV system development program. This provides basis for the answer to Task Statement Question 3.b in Box 5-6.

Capability Gaps

The capability gaps discussed in the previous sections that must be filled by the Army to support development of the four example systems are summarized in Table 5-5. For each gap the committee estimated a degree of difficulty and risk (indicated by shading in Table 5-5) according to the following criteria:

- *Low difficulty/low risk*—Single short-duration technological approach needed to be assured of a high probability of success
- *Medium difficulty/medium risk*—Optimum technical approach not clearly defined; one or more technical approaches possible must be explored to be assured of a high probability of success
- *High difficulty/high risk*—Multiple approaches possible with difficult engineering challenges; some basic research may be necessary to define an approach that will lead to a high probability of success.

Tables 5-4 and 5-5 provide the basis for answers to Task Statement Question 3.d. See Box 5-7.

> **BOX 5-6**
> **Task Statement Question 3.b**
>
> **Question:** Are all the necessary technical components of a UGV technology program identified and in place, or if not, what is missing?
>
> **Answer:** Missing technical components include supporting developments in human–robot interaction, power/energy, communications, and vehicle health maintenance that will depend heavily on system-level requirements.

TABLE 5-4 Estimates for When TRL 6 Will Be Reached in UGV Supporting Technology Areas

Technology Areas	Searcher	Donkey	Wingman	Hunter-Killer
Human–robot interaction	Near-term	Near-term	Mid-term	Mid-term
Mobility	Near-term	Near-term	Near-term	Near-term
Communications	Mid-term	Near-term	Mid-term	Far-term
Power/energy	Mid-term	Mid-term	Far-term	Far-term
Health maintenance	Mid-term	Near-term	Mid-term	Mid-term

Near-term
Mid-term (2006–2015)
Far-term (2016–2025)

TABLE 5-5 Capability Gaps in Supporting Technology Areas

Degree of Difficulty/Risk

- Low
- Medium
- High

Technology Areas	Capability Gaps			
	Searcher	Donkey	Wingman	Hunter-Killer
Human–robot interaction (HRI)	Telesystem HRI algorithms that support 1 operator per robot.	Semiautonomous HRI algorithms that support 1 operator per 5 homogeneous robots.	Natural user interfaces.	Natural user interfaces.
	Multimodal interfaces (nlp, gesture).	Natural user interfaces.	Diagrammatic and multimodality interfaces.	Methods for interacting and intervention under stress.
			Semiautonomous HRI algorithms that support multiple operators.	Near-autonomous HRI algorithms that support multiple operators and robots.
Mobility	Ability to right itself in restrictive passages/areas.	Platform capable of handling 40 km/h on smooth terrain with sensitive payload.	Platform capable of handling 100 km/h on smooth terrain with sensitive payload.	Heterogeneous marsupialism to transport specialized robots and sensors.
		Platform capable of handling 40 km/h on rough terrain with sensitive payload.	Platform capable of handling 100 km/h on rough terrain with sensitive payload.	Platform capable of handling 120 km/h.
Communications	High bandwidth for secure video; local to group.	Low bandwidth for "breadcrumbs," local to group.	Medium bandwidth for mobile network, local to group.	High bandwidth for secure and reliable network-centric communications.
	Wireless backup for line-of-sight communications.			Large amounts of distributed information.
Power/energy	High energy density rechargeable battery.	High energy density rechargeable battery.	Highly efficient stealth energy system.	Long standby (30 days).
	Small, hybrid energy system.			Highly efficient stealth energy system.
				High-speed mobility enablers.
Health maintenance	High physical reliability, low maintenance.	High physical reliability, low maintenance.	Design for combat survivability.	Self-repair by reconfiguring components.
		Cooperative diagnostics for remote operator.	Algorithms for self-diagnosis.	Self-repair by self-reprogramming.
		Ability to know when to call for help.		

> **BOX 5-7**
> **Task Statement Question 3.d**
> **Supporting Technology Areas**
>
> **Question:** What technology areas merit further investigation by the Army in the application of UGV technologies in 2015 or beyond?
>
> **Answer:** The committee postulated operational requirements for four example UGV systems and determined critical capability gaps in multiple UGV technology areas. Technology areas meriting further investigation by the Army are listed in Table 5-5.

6

Technology Integration

It will take more than technology development to field operationally capable, supportable, and affordable unmanned ground vehicles (UGVs) for the Army. The Army will also need to pursue a system engineering approach to component and subsystem design and integration. Key considerations for integration of UGV technologies are life-cycle support, software engineering and computational hardware, assessment methodology, and modeling and simulation.

System engineering is defined as an integrated design approach, from requirements definition, through design and test, to life-cycle support, that optimizes the synergistic performance of a system, or system of systems, so that assigned tasks can be accomplished in the most efficient and effective manner possible. Each component of each system, and each system within a system of systems, is designed to function as part of a single entity (the platform) and network (network-centric environment). Overall performance of a system is enhanced with the inclusion of manpower, reliability, maintainability, supportability, preplanned product improvement, and safety considerations.

A successful UGV technology integration strategy must seek to meaningfully evolve the overall UGV architectures (technical, systems, and operational) within the Future Combat Systems (FCS) architectures, the UGV interfaces with humans and organizations, and UGV relations with external environments (e.g., maintenance and supply). Without such a strategy, it will be difficult, if not impossible, to integrate and transition UGV technologies.

STATUS OF UNMANNED GROUND VEHICLE SYSTEM DEVELOPMENT

As the committee reviewed the UGV activities in the Army and elsewhere, it was apparent that the projects and programs are pursuing independent objectives and are not adequately coordinated to lead to higher-level developments. Top-level planning for UGV systems is not happening (at least, the committee was not made privy to any top-level planning) and the efforts are technology driven, rather than requirement or system driven.

The history of Army and Defense Advanced Research Project Agency (DARPA) UGV programs has been one of developing new UGV platforms in almost every program, which consumes a sizable fraction of the money and time available to each program without leaving substantial legacy to following programs, in terms of reusable development platforms. The goal of advancing perception technology, for example, would be furthered more effectively by making reasonably standardized, low-cost, low-maintenance test-bed vehicles available. The DARPA Perceptor program has taken an approach along these lines, but even that program required about $1 million and 1 year per vehicle for design, fabrication, software development, and integration to produce test-bed vehicle systems ready for new perception research, even though the program took commercial all terrain vehicles (ATVs) at under $10,000 each as the point of departure. The Demo III program spent a major share of its money on adapting and outfitting the experimental unmanned vehicle (XUV) platform. As the FCS architecture slowly evolves, the multiple efforts to develop UGV capabilities will require an effective degree of overall or enterprise management to make UGVs a viable part of the FCS and Objective Force concepts. Thus, there is an immediate need for a disciplined system engineering approach to determine UGV performance and platform design. This, in turn, will ensure an adequate flow-down of functional performance design requirements and a subsequent definition of technology requirements.

Missing Ingredients

Activity to define a UGV architecture that would provide an effective and economic path to modernization for technology insertion over time is missing from the Army UGV program. System structures that ensure operational robustness and economic manufacturability are also absent. As implied above, the important role and fit of a UGV platform into the higher-level FCS system of systems structure (operationally, functionally, and synergistically) is missing.

Engineering process discipline can be a major force to ensure that the Army does the right thing as well as does the thing right. The committee applied such discipline to the conduct of its study by postulating UGV scenarios and related example UGV systems to guide its technology assessments. Regardless of whether the examples coincide with Army requirements for FCS, they provide a rational basis for identifying needed elements in the Army's program.

From a technical viewpoint, important system engineering and design processes need attention. Besides a system engineering approach, technology integration considerations include designation of a lead for system engineering and life-cycle cost management, software engineering, development of an effective assessment methodology, and use of modeling and simulation assessment tools. These considerations are discussed further in this chapter.

To optimize system engineering efforts the Army should consider:

- Assigning all technology integration and system engineering responsibilities to a single person (office) with resources and ability to influence changes in design and development
- Identifying, developing, and integrating technologies early in UGV design that can reduce life-cycle support requirements
- Developing an effective and efficient software engineering effort
- Developing an integrated assessment methodology that includes experimental analysis throughout the design and development of the UGV, as well as early user evaluation and test of UGV components and systems
- Developing metrics to support the above assessment and
- Utilizing modeling and simulation, where appropriate, to support experimentation and test.

Integration and Advocacy

In the absence of clear requirements to drive UGV development efforts the Army UGV program must be one of developing capabilities, but without focus and advocacy a capabilities-driven process is likely to suffer from diffusion and incoherence. Although the existing technology demonstration and science and technology objective (STO) programs have explicit objectives (see Chapter 3), these efforts do not add up to a coherent, coordinated UGV program.

The Army needs to pursue an integrated science and technology (S&T) user approach to UGV technology developments. This integrated approach should consider all relevant S&T programs, identify gaps in capabilities and technology, support FCS planning and programming, and create a system engineering environment for development of UGV technologies and systems. Just as special high-level emphasis was needed for successful planning and implementation of digitization initiatives for the Army, the committee believes that it will take a strong advocate or office high in the Army chain of command to advance UGV technologies and systems. While the Army Digitization Office is an excellent example of a successful high-level integration office, the Army's UGV programs are not mature enough nor funded enough to warrant such an office at this time.

There are multiple UGV technology programs being pursued across the Army, Department of Defense (DOD), and other services and agencies. There are multiple spokespersons for the development of UGVs for FCS, including Army Research Laboratory (ARL), Army Aviation and Missile Command (AMCOM), and Tank-Automotive and Armaments Command (TACOM) within the Army; the DOD Joint Program Office; DARPA; and FCS program managers. These multiple efforts to produce UGV systems are not focused and require integration and visibility for overall program success.

Past attempts at integration have involved various approaches in the Army S&T and user communities, including technology roadmaps, integrated product teams, advanced technology demonstrations, and warfighting experiments. However, development of supporting technologies, such as mobility, communications, and power, are associated with separate functional branches of the Army research and development (R&D) community, each an S&T advocate in its own right. Additionally, the numerous potential functions for UGV systems make it difficult to identify a single user advocate.

Using system engineering principles, this advocate could influence the development and assessment of UGV operational concepts, as well as the direction and level of effort for UGV S&T programs. This person could capitalize on robotics technology development, both military and commercial, and maximize technology integration and transition efforts for prospective FCS UGV systems.

The immediacy of the essential system-level considerations, which are discussed in the remainder of this chapter, make the designation of a board-selected Army program manager (PM) for UGV technologies, serving as an agent for the ASA (ALT), a desirable and recommended course of action. The PM could serve as principal advocate and focus on the development and integration of several experimental UGV prototypes to expedite the evolution of viable systems. Working closely with the user community, this PM could significantly influence the requirements and the acceptance of UGV systems. This position would contrast with the present program managers (for FCS and Objective Force), who are focused on objectives that can be achieved with or without a dollar of investment in UGV technology.

LIFE-CYCLE SUPPORT

Support requirements that may have technology solutions (hardware or software) must be considered and designed early into UGV development programs, because integration late in system design may become very costly or even impossible. Determining support requirements for systems in development is usually based on reviewing historical support data of similar fielded predecessor systems and adjusting support requirements for new technology benefits or disadvantages.

The difficulty with determining future support infrastructure needs for a UGV system is that there is little to no historical information from operational uses of UGVs. There is indirectly related information from unmanned aerial vehicles (UAVs) and manned vehicles. The committee considered this information as well as its subject matter expertise to identify potential support infrastructure needs.

Support requirements should first consider manpower, skill level, and training needs of the operators, and then maintenance and transportation needs.

Manpower

A large UAV may have as many as 20 people involved with its operations. UGVs may need more than one operator to share tasks like maneuvering, sensing, and engagement. Optimally, one person should be able to operate at least one and probably a few UGVs. The human to UGV ratio needed for UGV operations in a 24-hr period will be dependent on the autonomy of the UGV, the complexity of the interface, the number of operators, and the cognitive loads on the operator(s). The cognitive loads should take into account operational mission tasks not associated with the UGV (e.g., an observer or loader in a manned attack vehicle may have an additional duty of being the operator for a UGV Wingman).

Skill Levels

Current Army UGVs require highly skilled personnel to operate them. The skill level of a UGV operator might have to be that of a combat vehicle commander (an E-6). Operation of multiple UGVs may require the skill levels of armored vehicle team leaders (an E-7 for two UGVs) and platoon leaders (an O-1 for three to four UGVs). Simulation experiments should be conducted to assess these needs. There will have to be separate UGV skill identifiers.

Training Needs

UGV trainers will be needed for operational users (at all TRADOC schools that utilize UGVs within their perspective branches). Training levels will have to be maintained with system (e.g., software) upgrades. Training will vary from that required for teleoperation to training needed to manage network centric UGVs.

To begin to understand the requirements for UGV personnel (operators and trainers), the Army should conduct technical and simulation assessments of manpower, skill levels, and training. The human–robot interaction technologies will significantly impact these personnel needs. Trade-off analyses need to be conducted to assist in determining levels of effort and design requirements for human–robot interaction technology.

Maintenance Needs

Part of the problem with current UGVs is that they need more people to operate and maintain them than the general military population realizes. By today's standards, when a UGV returns to its maintenance location, two levels of maintenance capability may have to be available immediately: Level I (organic) for checking fluid levels, tire pressures, and so on, and Level II for UGV hardware (organic and automotive), vetronics, and software and the mission package (numbers and types relative to the mission package).

Cooperative diagnostic systems (for an integrated assessment of UGV hardware, vetronics, software, and mission package) must be designed into the UGV to assist maintenance personnel in identifying UGV and mission package faults early enough to prevent catastrophic failures. These diagnostic systems must maintain historical records of faults that occur during operations, just as a human would relay problems during operations. The less capable the cooperative diagnostics, the more highly skilled and cross-trained the Level II maintenance personnel will have to be. For a UGV with learning software the software technician may need to have the capability and skills to recognize that unwanted behavior has been learned and be able to modify the UGV's behavior to erase that fault. Additional personnel may be needed as UGV maintenance instructors at maintenance schools. Cooperative diagnostic technology should be designed and integrated into the Army UGV program now. Without it preventative maintenance and repairs may become too resource intensive.

Postulated maintenance requirements for the Wingman are at a level that it could self-diagnose its maintenance problems and allow that information to be downloaded for action by human maintainers. When receiving alerts from its diagnostic software, the Wingman would also notify the section leader of serious problems that could impact current operations. The Wingman concept was predicated on humans doing the maintenance.

The Hunter-Killer team would be a step beyond that in maintenance capability—each robot would not only have self-diagnostic capabilities but also be able to do some self-repair. NASA (National Aeronautics and Space Administration) is initiating research on robotic planetary explorers that can perform self-repair or that can repair other robots, thus extending the lives of robots.[1] Using such a concept, the Hunter-Killer marsupial UGV could actually do some field repairs on other Hunter-Killers or on themselves. Another concept is that maintenance could be conducted between missions, in safe areas, by other highly specialized robots.

This self-repair capability would significantly reduce the requirement for human maintenance, thus enabling a very small ratio of humans to UGVs—the idea being that the Hunter-Killer team would have evolved to a state where "unmanned" means "almost no people." Whether the committee's number is too big or too small is obviously a matter of judgment and should be investigated further. The committee was attempting to think "out of the box" not only about UGV operational capabilities but also about maintenance support.

Transport Needs

For rapid, precise movements in small areas (e.g., roll on, roll off ship) or movements over large distances (moving in convoys on roads over large distances in a short amount of time), large UGVs may need to have an override driver's position for humans to operate the vehicle for those times when a human can move the vehicle more efficiently than the UGV (tethered, semiautonomous, or even autonomous). This may require standby drivers for the UGVs.

Figure 6-1 illustrates how most of the life-cycle cost decisions are made early in the development of a system (top line) when relatively small amounts of money are available for the program (bottom line). By the end of the concept design phase, for which S&T technologies have a major impact, decisions have already been made that impact 70 percent of a system's life-cycle costs. Thus, the use of system engineering processes in UGV S&T programs will significantly influence the life-cycle costs of UGV systems. Additionally, it is important for the Army's UGV S&T program to define and assess technologies and concepts that when integrated into a UGV system, may significantly reduce life-cycle costs.

The foregoing provides the basis for the answer to Task Statement Question 5.c in Box 6-1.

SOFTWARE ENGINEERING

The effectiveness of the software engineering effort for a program the size and scope of the UGV will be determined by key programmatic decisions. Among these decisions are the architecture of the hardware and software infrastructure, the general development methodology, and software development environment and tool sets.

Software Architecture Considerations

Well-defined software architecture is a necessary antecedent to efficient software engineering. The architecture must be defined at several levels of abstraction, encompassing the basic computing infrastructure (e.g., processor and operating system), the inter-application communications infrastructure and services (communication profiles and middleware), and ancillary support infrastructure (e.g., system health, fault monitoring, and recovery; software loading). To enable efficient long-term evolution of the UGV computing infrastructure and software, an open system architecture approach is mandatory. Open systems leverage industry-standard application programming interfaces and enable large systems to be constructed quickly from existing components. Further, open-system architectures enable the use of commercial off-the-shelf (COTS) and/or open source hardware and software for major infrastructure components such as the processor, operating systems, communication stacks, and middleware. Substantial non-recurring cost savings will be realized during initial development as well as during maintenance and evolution.

The use of COTS (and/or open source) hardware, operating system, and middleware ensures a degree of hardware and software independence. Of these major parts of the software infrastructure, the middleware layer is perhaps the most critical, as it enables definition of hardware-independent services and inter-application interfaces. In addition, middleware allows cooperating applications to be located as needed on the communications network in order to meet system requirements for responsiveness or redundancy. It also allows applications to be developed and deployed without concern for the processor or network topology and technology. Changes in the processing hardware or network design may impact the open source or middleware implementations, but revising the infrastructure software would only need to be done once for a large system. Software applications may simply need to be recompiled.

Software Development Methods and Technologies

By taking the open-system approach to UGV system and software architecture, numerous opportunities are afforded to software developers. Most significant among these is the ability to develop and test applications and systems of applications on ordinary microcomputers and workstations. Transferring applications from the development hosts to lab equipment for final testing can be as simple as copying the binary if the development environment accurately replicates the embedded hardware. This allows for rapid, iterative software development and integration without dependence on embedded hardware development.

---

[1] See, for example, Mitra, S., and E.J. McCluskey. 2002. Dependable reconfigurable computing design diversity and self repair, presented at the 2002 NASA/DoD Conference on Evolvable Hardware, Alexandria, Va., July 15–18, and other papers presented at this conference on evolvable hardware.

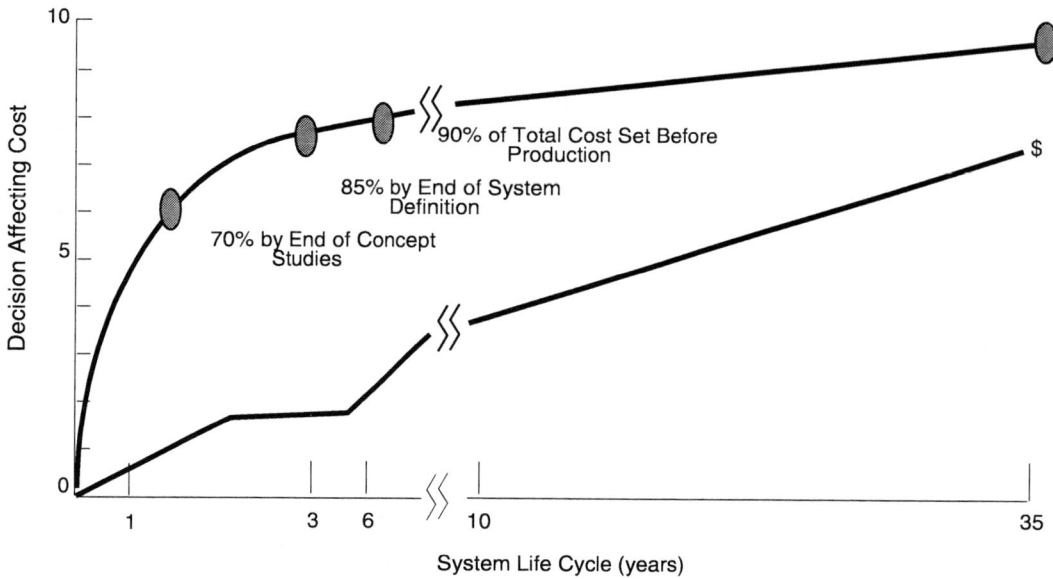

FIGURE 6-1 Life-cycle cost decisions. SOURCE: Adapted from Jones (1998).

Industry-standard middleware technology enables an object-oriented component-based application development approach. Use of object-oriented technology allows the use of existing COTS tools and methods in the specification of the software engineering environment. Note, however, that use of object-oriented technology in high-assurance systems can be problematic if certain restrictions are not enforced. Recent research in the application of object-oriented technology to high-assurance systems promises to improve the predictability of such systems, while retaining the benefits of the object-oriented approach: abstraction, encapsulation, inheritance, and polymorphism.

To increase the efficiency of software development, a model development approach should be used. In the same way that high-level languages obviated the need for software developers to be concerned with how processor registers are used, model development eliminates the concern about how a source module is coded. Automatic code generation and automatic test case generation based on detailed requirements and design models substantially reduce development effort. Examples of commercially available model development approaches include the Object Management Group's (OMG) Model Driven Architecture and Rational's Unified Process.

Where necessary, formal methods can be applied to prove the correctness of software requirements for high-assurance applications through the use of formal modeling languages and model checkers.

Given well-defined software architecture and prescriptive application construction guidelines based on an open architecture (i.e., software building codes), applications can be efficiently created, integrated, and maintained.

Software Technology Gaps

To put the capabilities of today's software technology in perspective, weaknesses must be considered. For example, current software technology focuses on specification and implementation of individual applications or individual software components. Software technology is weak in the specification and testing of systems of real-time software components (software systems of systems). Object-oriented and component-based software development approaches provide the potential to solve these problems, but more work is needed in methods and tools aimed at the inter-application level. One effort focusing on this level of abstraction is the Society of Automotive Engineers' (SAE) emerging standard AS-5C, which defines an Avionics Architecture Definition Language (AADL). The AADL has the potential to evolve into a general-purpose-architecture description language for hard real-time embedded systems.

Specification of complex systems is only half of the battle. Typically, validation approaches for key technical system-level performance measures (e.g., whether deadlines can be met, whether application interactions meet requirements) are ad hoc. There is no standardized approach to validating or verifying inter-application performance requirements. Once an inter-application specification language (like AADL) is in place, automated approaches to validating and verifying the interfaces can be developed.

End user configuration management of software systems of systems can be problematic due to evolving application functionality. Certain applications in the UGV application domain will evolve more rapidly than others. Without making a concerted effort to maintain backward compatibil-

> **BOX 6-1**
> **Task Statement Question 5.c**
>
> **Question:** Are there implications for Army support infrastructure for a UGV system? For example, will other technologies need to be developed in parallel to support a UGV system, and are those likely to pose significant barriers to eventual success in demonstrating the UGV concept or in fielding a viable system?
>
> **Answer:** Cooperative diagnostic systems (capable of providing an integrated assessment of UGV hardware, vetronics, software, and mission-function equipment) must be designed into the UGV to assist maintenance personnel in identifying UGV/mission-package faults early enough to prevent catastrophic failures. These must be developed in parallel to support UGV systems. Other support requirements that may have technology solutions (hardware or software) must be considered and designed early into UGV development programs, because integration late in system design may become very costly or even impossible.
>
> Support requirements are usually based on reviewing historical data of similar fielded systems and adjusting as appropriate. There is little historical information on UGV systems; the committee had to resort to similar information on UAV system and manned vehicle systems operations and experiences to identify likely support infrastructure needs and weaknesses. These include requirements for operators (number, skill level, and training), vehicle maintenance, and deployment transportation.

ity, software upgrades will need to be performed in a block (i.e., multiple software components would need to be upgraded simultaneously). This complicates vehicle maintenance and logistics significantly. For example, inter-application compatibility would need to be determined and compatibility data consulted whenever a software upgrade was planned or performed. This problem is similar to that of installing new or updated software on a microcomputer and discovering that device drivers also need to be updated at the same time.

Among the most problematic software technologies to specify and verify is that of adaptive algorithms. Validation of nonadaptive high-assurance systems is well understood and can be efficiently performed through the application of test automation tools and coverage analyzers. The challenge is how to validate systems that change their behavior in response to external conditions. Research into these methods is being performed, but to date no standardized approach exists.

While the software technology principles that have been discussed may be common to all system developments, the area of adaptive algorithms and intelligent software may prove to be critical to the timeline for development of acceptable UGV systems. As discussed in Chapter 4, significant uncertainty exists regarding the degree to which methods of machine learning will apply to UGV systems. The Army must determine and invest in decision making and control algorithms that will be adequate to support required levels of performance, because the impact on time and effort to complete software developments could be severe.

For over two decades, formal methods have held out the promise of mathematically precise specifications and automated verification of system and software specifications. Industrial adoption has been slow for a variety of reasons, including notations that engineers find difficult to use, limited offerings in U.S. universities, and lack of automated tools; however, this is changing rapidly. In recent years, several companies, mostly in Europe, have successfully used notations such as State Charts, Lustre, Esterel, and B on large commercial products. These initiatives routinely report savings of 25 to 50 percent in development costs and an order-of-magnitude reduction in errors. These gains are achieved through early identification of errors, better communication between team members, automated verification of safety properties, and automatic generation of code and test cases.

COMPUTATIONAL HARDWARE

An ideal UGV would be capable of replacing an equivalent manned vehicle. Unfortunately, present-day limits on computing power, among other factors, make this expectation highly unrealistic. It is clear that a brute-force approach to mimicking the computational power of the brain cannot be supported by semiconductor technology. While software can do much to enable machines to mimic human behaviors, the upper limit on the computing power has always been determined by hardware, and this will continue to be the case for the future. Thus it is useful to examine present and projected hardware performance not only to estimate how much room exists for improvement but also to ensure that the Army aims for reasonable targets and avoids unrealistic expectations.

Human Versus Semiconductor-Based Technology

The human brain is capable of performing about 10^{14} computations/second (c/s) (Moravec, 1999; Albus and Meystel, 2001), has the capacity to store some 10^{15} bits of information, and achieves this level of performance with a relatively low power input of 25 watts. In its major conscious task of visual perception it is aided by the fact that 90 percent of the image processing of the output of the approximately 10^8 pixels of the eye is actually done in the retina, which is connected to the brain by approximately 10^6 nerve fibers that operate in a pulse-code-modulation mode at approximately 50 Hz (Werblin et al., 1996).

In UGVs the equivalent capacity is supplied by semiconductor technology. One characteristic of semiconductor technology that is often taken for granted is its continuing rapid improvement of capabilities. Over the last 30 years the various measures of performance have faithfully followed

Moore's law, which states that capabilities double approximately every 18 months (Intel, 2002). This essentially self-fulfilling prediction is perhaps better understood as a guideline used by the industry to ensure that semiconductor technology advances uniformly across an enormously complex front. The questions that should be asked in the context of UGVs are where are we today and where are we likely to be in the future?

Considering first computing power, laptop computers currently operate at about 10^9 c/s with 100 W of power. High-end, reasonably portable systems operate at about 10^{10} c/s and 1,000 W. Thus computational power is presently about four orders of magnitude short of human performance and has considerably greater power requirements.

Future performance capabilities can be estimated in several ways. One possibility is to count the number of functions (transistors) per chip, which for high-end computers is predicted to rise from its present value of about 2×10^8 to about 2×10^{10} in 2014 (SIA, 2001). Another, but less accurate measure is to extrapolate the performance gains made over the last 30 years. Using the 1995 trend, Albus and Meystel project a human-equivalent performance level of 10^{14} c/s at 2021 (Albus and Meystel, 2001); using the more conservative average trend from 1975 to 1985, Moravec predicts this level at 2030 (Moravec, 1999). Again following Albus and Meystel, the 10^{11} and 10^{12} c/s performance levels needed for good driving skills and average human driving performance, respectively, are expected to be reached in 2006 and 2011, respectively. Of course, "driving" is not the only activity that will require computational resources for a sophisticated Wingman or Hunter-Killer UGV.

Impending Limits

Aside from the brute-force nature of their analyses, the problem with the Moravec and especially the Albus and Meystel estimates is that they do not take into account the impending limits of silicon technology. Up to now, speed has been increased by reducing the sizes of the individual devices (scaling), but scaling has reached the point where the $Si-SiO_2$ materials system on which the last 30 years of progress has been based is approaching its theoretical limits. Some improvements are expected with the replacement of SiO_2 with so-called high-k dielectrics. However, Intel terminates its Moore's law estimates at 2011 at a point still about one or two orders of magnitude short of human performance (Intel, 2002). Advances of the scale envisioned by Albus and Meystel and by Moravec will almost certainly require a new hardware technology that is not presently developed. These constraints are "hard" in the sense that given the scale of the semiconductor industry (approximately U.S. $150 billion in worldwide sales in 2001), it would be effectively impossible for any single source to provide enough financing to influence these trends.

To provide a more definitive illustration of the shortfall in computing power, a 640×480 pixel array (somewhat over 300,000 pixels total) operating at a typical readout rate of 30 Hz and an 8-bit gray scale with three colors delivers information at a rate of about 200 MHz. Comparing pixel ratios and data rates, one sees that the human visual-perception system is presently superior by nearly three orders of magnitude. Coupling this with the four-orders-of-magnitude lower performance of computers, the relatively primitive perception capabilities of UGVs are easily understood.

It should be noted that this argument presupposes exclusive use of general-purpose microprocessors. Digital signal processors (DSPs) may achieve a MIPS/watt (million instructions per second/watt) advantage of an order of magnitude over general-purpose machines for many UGV applications; also, field programmable gate arrays (FPGAs) have been said to achieve a two-orders-of-magnitude advantage for specific applications. This suggests that once algorithms are mature enough to be stable, implementing them in dedicated digital hardware (i.e., programmable logic) may provide a key path to enable small, low-power systems with limited degrees of semiautonomy. These solutions would thus forestall the overall shortcoming in computational power needed for near-human levels of autonomy and offer a way around the impending limits in silicon technology.

In addition to such breakthroughs in computational power, the most direct approach toward achieving autonomy goals for UGVs will be to augment visual perception systems with other sensors. The multimodal approach will likely need to be combined with analog or optical processing techniques to overcome any future deficits in semiconductor processing power. While existing or forthcoming processors should be adequate for meeting the anticipated computational loads for this approach, a major system engineering problem is to optimize the perception system hardware and software architecture, including sensors, embedded processors, coded algorithms, and communication buses.

The use of hardware and software in-the-loop simulation, appropriately instrumented, would move architecture design and optimization in this direction and would also support algorithm benchmarking. As noted in the previous section, the quality of software is also a major issue, adding three or more years for software engineering, re-implementation, and performance testing before system fielding to the UGV system development time line.

ASSESSMENT METHODOLOGY

Demonstrations alone do not provide statistically significant data to assess the maturity, capabilities, and benefits of a particular technology both as an individual technology and as part of a larger system or system-of-systems concept. For example, while it may be reasonable to assume that the Demo III program has advanced the state of the art over the

Demo II program, there is no way to know in the absence of statistically valid test data. To make progress emphasis should be placed particularly on data collection in environments designed to stress and break the system, e.g., unknown terrain, urban environments, night, and bad weather.

No quantitative standards, metrics, or procedures exist for evaluating autonomous mobility performance, particularly off-road. It is difficult to know where deficiencies may exist and where to focus research. Similarly, there is no way to determine if the algorithms being used are the best available. In the "A-to-B" mobility context, for example, no system-level process exists for benchmarking algorithms described in the literature and evaluating them for incorporation in UGVs. This must represent a major lost opportunity. All off-road perception research is handicapped because of a lack of ground truth, lack of consistent data packages for researchers to use, and a lack of quantitative standards for evaluation.

The assessment methodology should be designed to assess technology issues in the operational context of FCS and the Objective Force. This assessment methodology should describe the objectives, issues, and analytical methodology required to address the key issues for the UGV within the FCS and Objective Force architectures. The methodology should also identify input and support requirements, key assumptions, and time lines for the assessments.

The assessment methodology should include a series of experiments that initially begin with analyses of concepts and technologies and mature to technology-integration warfighting experiments that approach levels of assessment similar to operational test and evaluation. The level of detail of the assessments should grow with the level of maturity of the UGV technologies and their ability to be integrated into an FCS.

The assessment methodology should be iterative, or as some call this approach in Army acquisition, a spiral development approach. Experimentation, defined as "an iterative process of collecting, developing and exploring concepts to identify and recommend the best value-added solutions for changes to DOTML-P (doctrine, organization, training, materiel, leadership and people), is required to achieve significant advances in future joint operational capabilities" (VCDS, 2002).

Metrics

The assessment methodology must have appropriate metrics for both operational and technical issues. Metrics are needed at all evaluation levels: technical, tactical, and operational. For example, the evaluation of a Hunter-Killer UGV will require metrics on at least three levels:

- *Level 1: technical performance metrics.* Examples of technical performance include such things as detection range, probability of detection (P_D), probability of false alarm (P_{FA}), time needed to process and send critical C2 data, and ability to communicate with other FCS elements. These metrics can be called measures of performance (MOPs).
- *Level 2: tactical effectiveness metrics.* Examples of tactical effectiveness are the impact of the UGV on the mean time required by an intelligence officer to acquire a target accurately, the percentage of accurate acquisitions, target selections by a commander, sensor-to-shooter times, and probability of kill (P_K). These metrics can be called measures of effectiveness (MOEs).
- *Level 3: operational utility metrics.* The impact of the UGV on battle outcomes (e.g., force exchange ratios, percentage and numbers of indirect fire kills) is an example of operational utility. These metrics can be called measures of value (MOVs).

It is important to realize that the determination of a single valid technical performance metric, such as the appropriate measure for obstacle negotiation, is nontrivial and may by itself be the basis for extensive research. To ensure that the UGV metrics are relevant to the warfighter it would seem prudent to follow a process that identifies a warfighter's goals; derives performance objectives and criteria that relate to these goals (performance objectives are usually expressed in terms of key issues or, as in the Army, essential elements of analysis [EEAs]); and develops appropriate MOPs, MOEs, and MOVs.

A recommended approach for developing UGV metrics should include:

1. *Development of analytical requirements.*
2. *Determination of subjective metrics.* Subjective metrics of importance to the warfighter could relate to warfighter utility or operational issues. Some examples of non-quantifiable metrics may be
   - Usefulness. User assessment of value added, completeness of information, and accuracy of information are examples of usefulness metrics.
   - Usability. Human factors, interoperability, accessibility, and consistency of information are examples of usability metrics.
   - User assessment of performance. Standards compliance, overall capability, bandwidth requirements, and system availability are examples of metrics that may reflect the user's personal opinion.
3. *Determination of quantifiable metrics.* Quantifiable metrics for technologies being evaluated in an experiment should
   - Be relevant to the warfighter's needs (which include system requirements and system specifications).

- Provide flexibility for identifying new UGV requirements.
- Be clearly aligned with an objective.
- Be clearly defined, including the data that is to be collected for each metric.
- Identify a clear cause and effect or audit trail for the data being collected and the technology being evaluated.
- Minimize the number of variables being measured, with a process identified for deconflicting data collected from and perhaps impacted by more than one variable.
- Identify nonquantifiable effects (e.g., leadership, training) and impacts of system wash-out (i.e., a technology's individual performance is lost in the host system performance), and control (or reduce) them as much as possible.

4. *Documentation of each measure.*

Some examples of UGV-specific metric-generating issues may include:

- Example operational issues
  —Command and control issues may include the impact of UGVs on command efficiency (e.g., timeliness of orders, understanding of the enemy situation and intentions) of a small unit leader or unit commander and his staff
    - What is the cognitive workload (e.g., all critical events observed, accuracy of orders) of a small unit leader or unit commander and staff?
    - What can an array of UGVs do that a single UGV cannot?
    - What information needs to be shared to support collaboration among systems (manned and unmanned)?
- Example technical issues
  —Mobility issues such as those enumerated in the High Mobility Robotic Platform study (U.S. Army, 1998) will provide a basis for UGV mobility metrics. UGV-specific mobility issues, such as the mobility metrics used in the DARPA Tactical Mobile Robot program should be considered. The measures should be based on specific applications being evaluated (e.g., logistics follower on structured roads, soldier robot over complex terrain).
  —Other supporting technology issues
    - How do data compression techniques impact UGV performance?
    - What is the optimal trade-off between local processing and bandwidth?
    - How much of a role should ATR play in perception?

There are many more issues and metrics that need to be defined for Army UGV assessments. The process must be to identify UGV objectives first, then the issues generated by each objective, then the hypotheses for each issue, and finally the measures needed to prove or disprove the hypotheses.

A major goal going forward must be a science-based program for the collection of data sufficient to develop predictive models for UGV performance. These models would have immediate payoff in support of system engineering and would additionally provide a sound basis for developing concepts of operation (reflecting real vehicle capabilities) and establishing requirements for human operators, e.g., how many might be required in a given situation. Uncertainties with regard to these last represent major impediments to eventual operational use.

MODELING AND SIMULATION

Modeling and simulation (M&S) is an essential tool for analyzing and designing the complex technologies needed for UGVs. Much has been written on the use of simulations to aid in system design, analysis, and testing. The DOD has also developed a process for simulation-based acquisition. However, little work has been done to integrate models and simulations into the system engineering process to assess the impact of various technologies on system performance and life-cycle costs.

To fully realize the benefits of M&S the use of M&S tools must begin in the conceptual design phase, where S&T initiatives have the most impact (Butler, 2002). For example, early in the conceptual design phase M&S can be used to evaluate a technology's impact on the effectiveness of a UGV concept, determine whether all the functional design specifications are met, and improve the manufacturability of a UGV. By using simulations in this fashion, S&T programs can support significant reductions in design cycle time and the overall lifetime cost of future UGVs. Simulations provide a capability to perform experiments that cannot be realized in the real world due to physical, environmental, economic, or ethical restrictions. Because of this, they should play a very important role in implementing the assessment methodology used to assess UGV concepts and designs.

Just as with training, UGV system experimentation could be supported with any one or mix of the following types of simulations:

- Live simulations—real people operating real systems.
- Virtual simulations—real people operating simulated systems.
- Constructive simulations—simulated people operating simulated systems (note: real people stimulate,

or make input, to these simulations, but they are not involved with determining the outcomes).

Simulations, however, are meaningful only if the underlying models are adequately accurate and if the models are evaluated using the proper simulation algorithms. Technical experiments will be difficult to conduct, since data is lacking for detailed engineering models of UGVs and detailed multispectral environment representations. The multispectral environment includes detailed terrain elevation (<1-meter resolution) data, feature (natural and manmade) data, and effects of weather (including temperature).

Both the use of UGVs and the FCS in an operational environment are relatively new concepts, so little data has been accumulated that could be used to develop verified and validated models. The Army has made good strides toward overcoming similar deficits in this area by extrapolating the results of laboratory experiments, using information from similar fielded systems and applying subject matter expertise from the Joint Virtual Battlespace at the Joint Precision Strike Demonstration Project Office (DMSO, 2001).

While existing M&S tools are adequate for the near term, complex UGV systems in the far term are likely to require M&S tools designed specifically to address system engineering issues. In the future, for example, material and structural systems will have sensors and actuators embedded so that the material serves multiple functions simultaneously (e.g., solving problems to achieve particular thermal properties, electromagnetic properties, sensing properties, antenna functions, mechanical and strength functions, and control functions). The kinds of mathematical and numerical tools that would be required to jointly optimize these disciplines, or to develop mathematical models that are appropriate for such system design efforts, are not currently being investigated. These tools would be invaluable aids to determine performance-limiting factors and to integrate technologies into multiple disciplines.

7

Roadmaps to the Future

This chapter provides roadmaps for the development of unmanned ground vehicle (UGV) systems that would be similar to the examples described in Chapter 2. It also provides a science and technology (S&T) roadmap for the Army that takes into account technologies of immediate importance to the Future Combat Systems (FCS) program as well as investigations that will be required for longer-term upgrades (2015 and beyond) of UGV capabilities for the Objective Force.

MILESTONES FOR SYSTEM DEVELOPMENT

Trends in robotics developments and autonomous intelligence support an evolution of UGV systems. Figure 7-1 illustrates that the Army envisions UGVs as playing an evolving role in combat missions and operating with increasingly higher levels of autonomy.

In Chapter 2 the committee postulated four examples of unmanned ground vehicles with a progression of capabilities: Searcher, Donkey, Wingman, and Hunter-Killer. It was necessary to do this to provide a basis for estimating technology readiness levels (TRLs) for each of the technology development areas in Chapters 4 and 5. The technical requirements for the example systems provide for increasing capabilities consistent with the evolution depicted in Figure 7-1. Further, given a sufficiently funded and dedicated effort, each of the postulated systems could be developed within predictable time frames. To be successful, however, it will be necessary to capitalize on vigorous evolutionary and spiral development processes, and to effectively integrate the technologies using modeling and simulation, assessment, and software engineering methods as described in Chapter 6.

The level of autonomy required is clearly higher for more complex applications and missions. As complexity advances from a basic Searcher UGV to a sophisticated Hunter-Killer, the degree of trust in and the independence of the robotic vehicle system increase to a level of "responsible" autonomy, in which only minimum-acceptable controls over the robot have been retained. But total autonomy in robotic vehicles is unlikely to be achieved even in the far term, and it should not be the goal in developing UGV systems for the Army.

The examples were all postulated to operate as part of human–robotic teams with varying requirements for autonomy that will depend upon such things as military doctrine, rules of engagement, and local dictates of a field commander. Army UGVs should be designed to function as part of a soldier-robot team, and much of the requirement for new technology development will depend on the degree to which the UGV is expected to function on its own versus in a team environment.

Figure 7-2 depicts developmental relationships between system capabilities of the four example UGV systems. Each system builds on one or more of the capabilities demonstrated by a predecessor system so that the technology developments have a cumulative effect, providing a path for the evolutionary development of multiple UGV capabilities for the Army.

The chronology at the bottom of Figure 7-2 results from the earlier TRL estimates of technological maturity for the fundamental UGV technologies required for the example systems. Between successive systems the figure lists the mission capabilities that would also need to be pursued with the enabling UGV technologies to accomplish the applications envisioned for the postulated example systems. Many of these capabilities are now being pursued in separate programs by the Army.

FCS Program Planning

The notional program plan for the Future Combat Systems, as briefed to the committee in 2001, is shown in Figure 7-3 (Johnson, 2001). The committee determined that the

ROADMAPS TO THE FUTURE

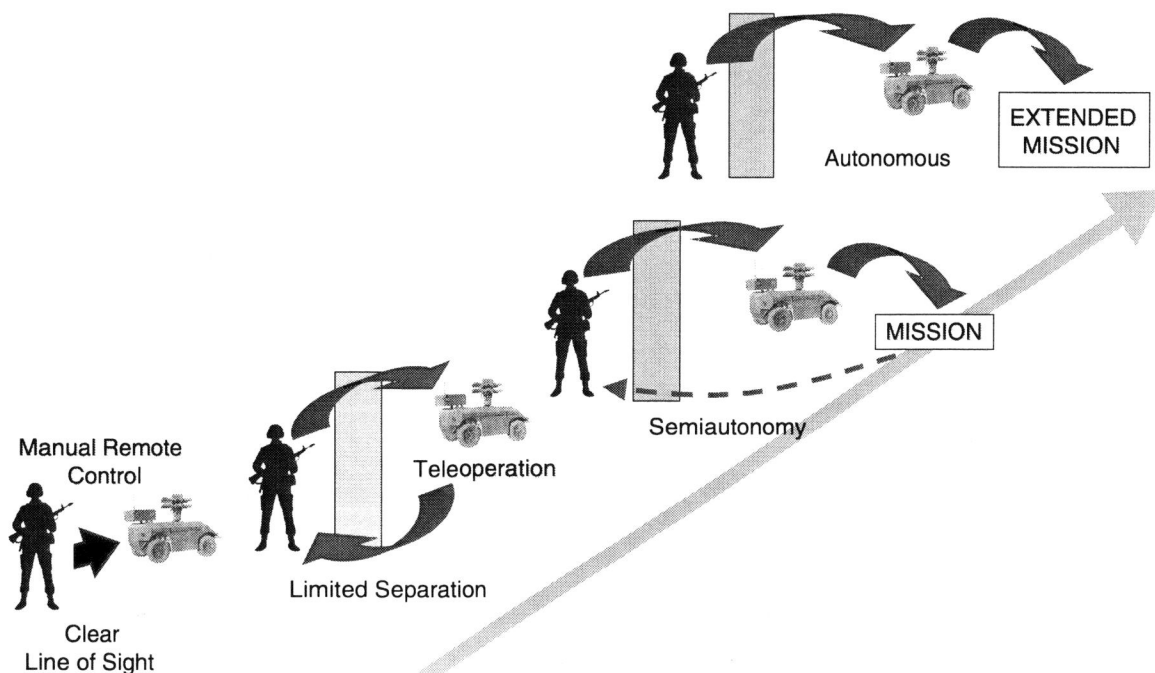

FIGURE 7-1 Evolution of UGV systems. SOURCE: AMCOM (2001).

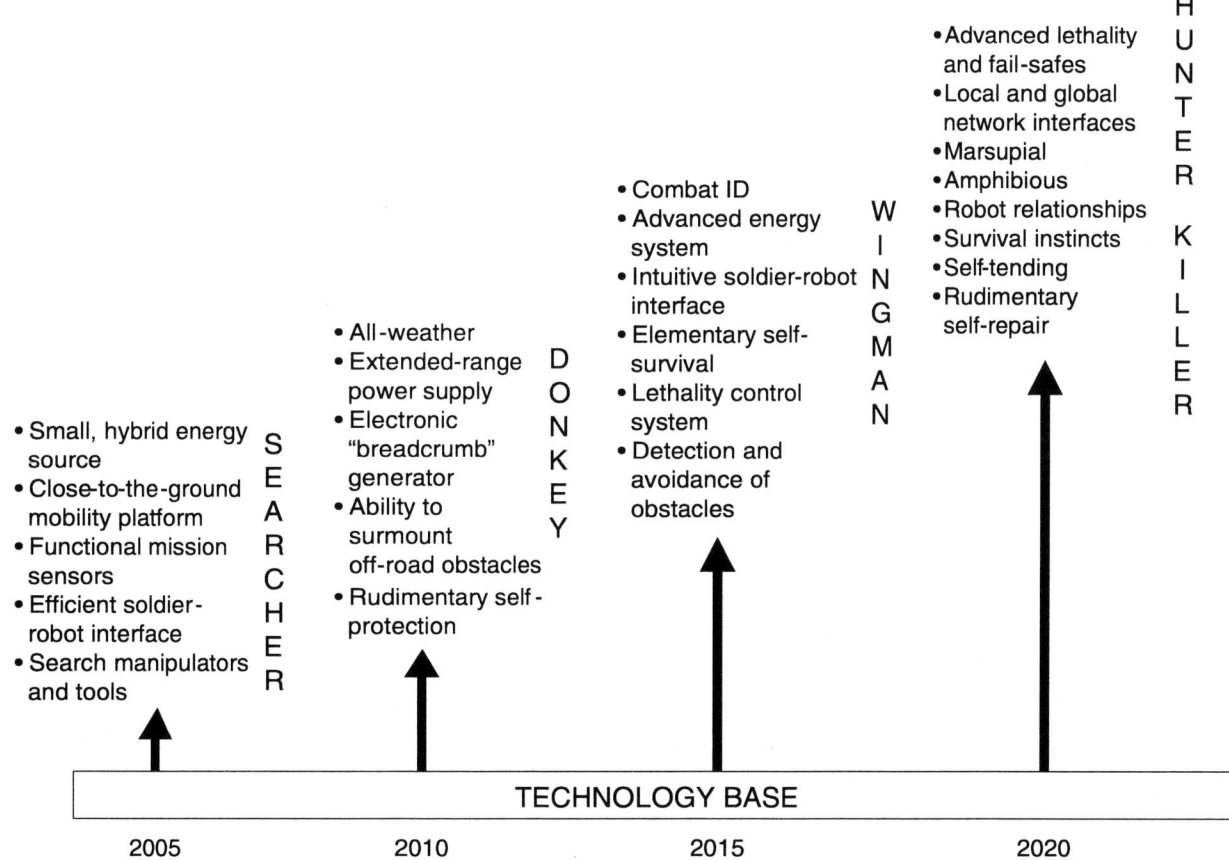

FIGURE 7-2 Possible evolution of UGV system capabilities.

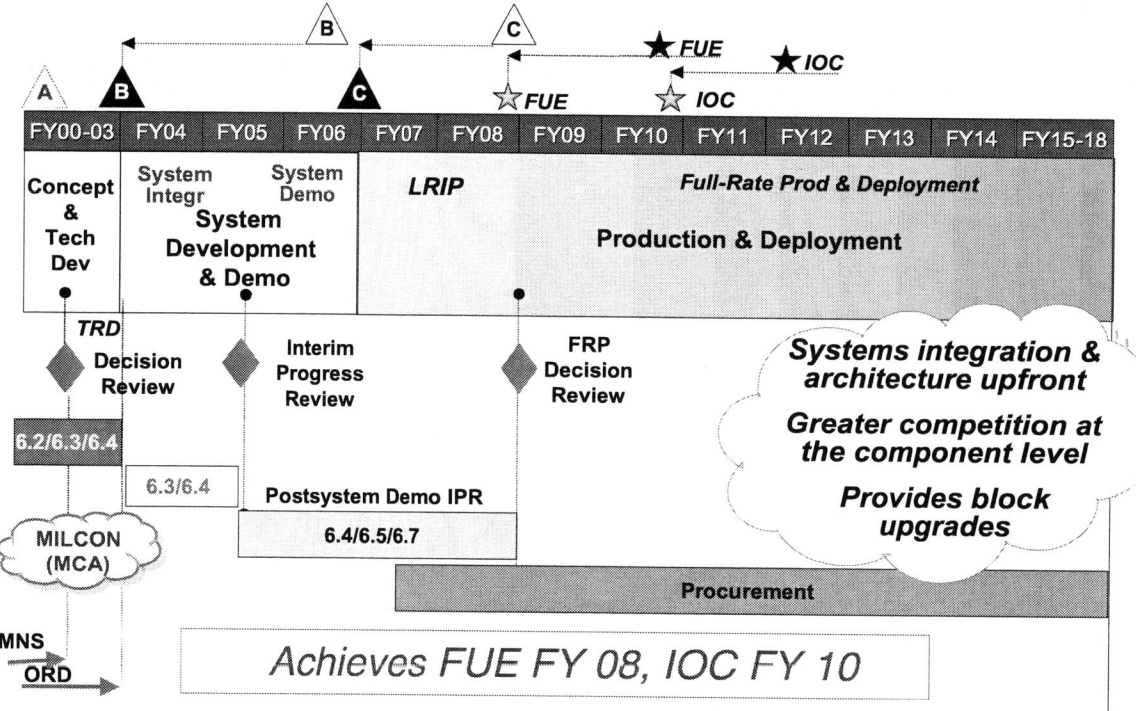

FIGURE 7-3 Notional FCS acquisition program. SOURCE: Johnson (2001).

overall Army UGV development program does not currently provide a basis for including UGVs with a high degree of autonomy in the initial FCS. Depending on the FCS requirement for UGVs, however, a special task force for the Assistant Secretary of the Army for Acquisitions, Logistics, and Technology (ASA [ALT]) concluded in early 2002 that a system-level insertion of a UGV system in the semiautonomous preceder/follower class should be possible as a block upgrade to FCS by 2009 and a UGV system in the network-centric autonomous capability class by 2025.[1]

Figure 7-4 depicts the most optimistic milestone dates for development of the example applications from a pure engineering perspective. It shows that the particular progression of example UGV system developments postulated by the committee could lead to insertion of a Hunter-Killer UGV in FCS in 2025. But the milestone dates assume that all of the capability gaps, even those identified as difficult and risky, will be filled in a timely fashion (see Tables 4-6 and 5-5 in Chapters 4 and 5, respectively). It is important to note that these milestones depend on knowing the objective capabilities desired for Hunter-Killer at the outset and that they do not take into account the many parallel develop-

ments not unique to UGVs, such as the mission-function equipment, high-performance engines, and network-centric communications, that undoubtedly will also be needed for FCS systems.

The examples used in the study resulted in lower TRLs (and more extended time lines) in large part because the committee believes that requirements for FCS UGVs will be much more demanding than indicated in field demonstrations of component technologies. Virtually all of the research and development work is being conducted under conditions that are much less harsh than the battlefield conditions under which UGVs will have to operate to be useful to the Army. In particular, the committee found that technologies that may appear to be relatively advanced when tested in good weather, on known terrain, with difficult obstacles removed, and with no enemy countermeasures are at a significantly lower TRL for military applications that are likely to be conducted in adverse weather over unmapped terrain in the presence of obstacles, obscurants, and electronic countermeasures.

Roadmaps for Technology Development

Figures 7-5 through 7-8 are technology development roadmaps for the specific examples described in Chapter 2. As with Figure 7-4, the roadmaps assume that all capability

---

[1]The chair and three members of the NRC Committee on Army Unmanned Ground Vehicle Technology served on the task force.

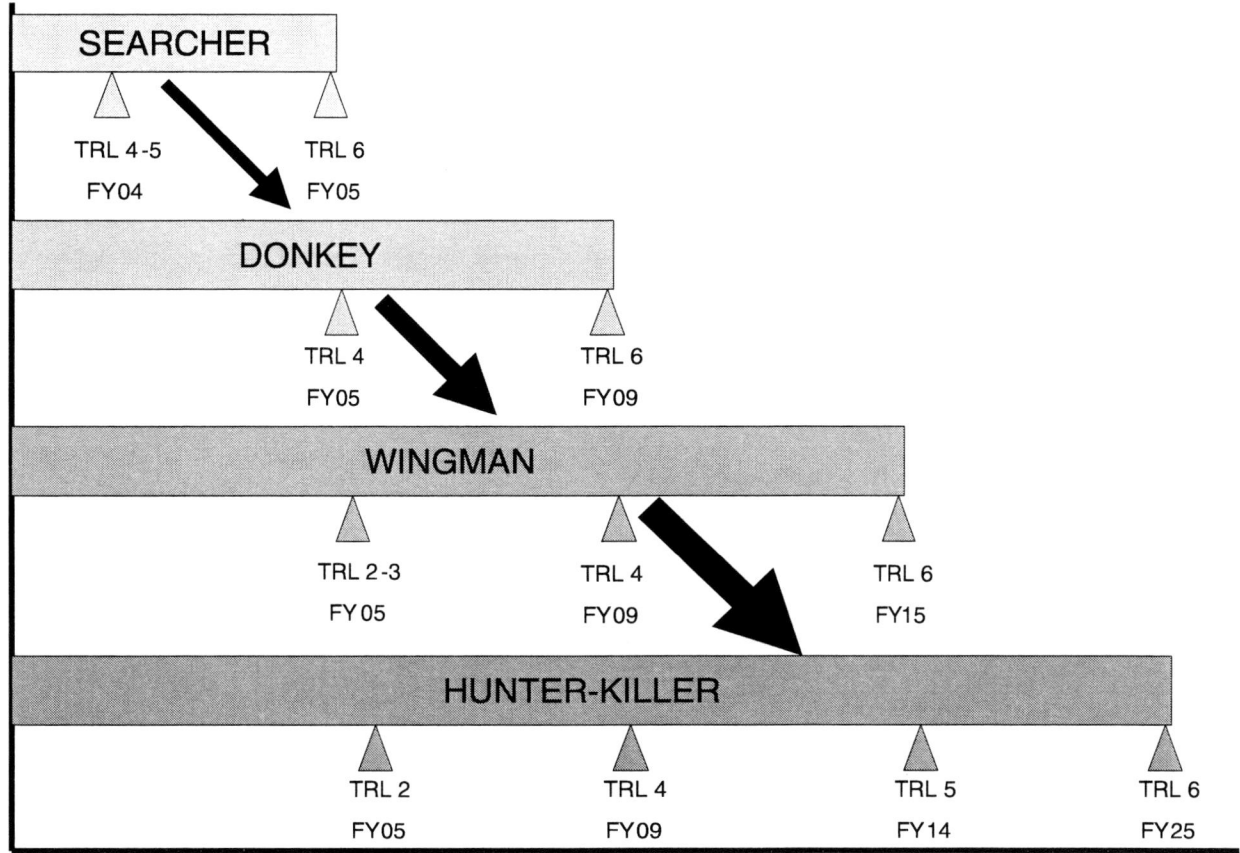

FIGURE 7-4 Time lines for development of sample UGV systems, assuming progressive capability developments.

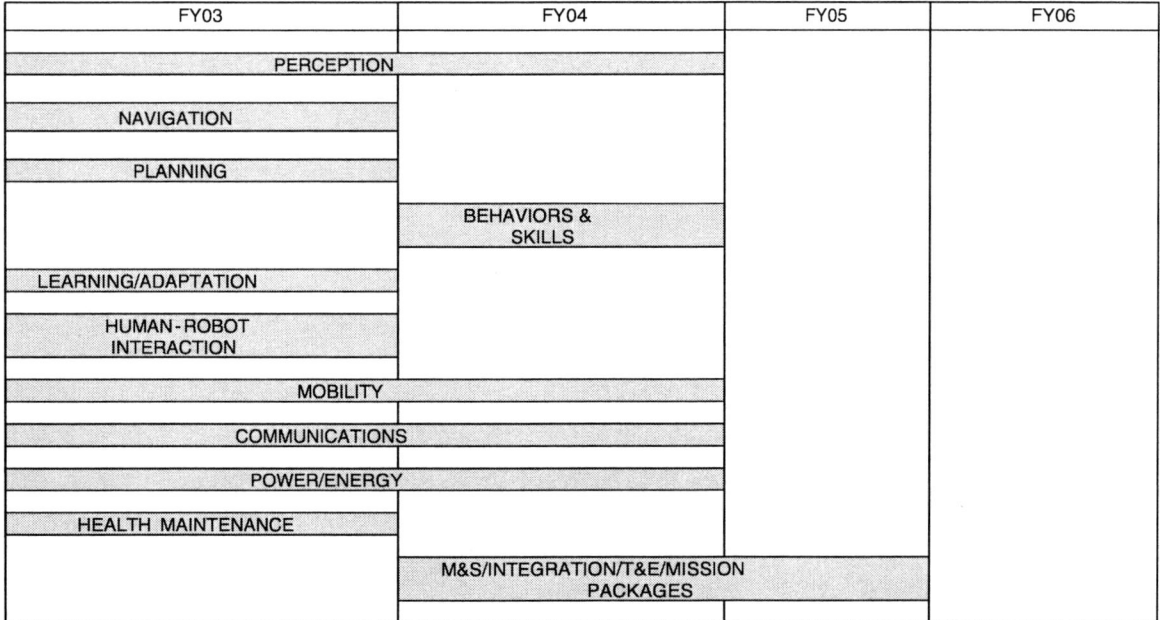

FIGURE 7-5 Technology development roadmap for the Searcher indicating the year TRL 6 will be reached for the system.

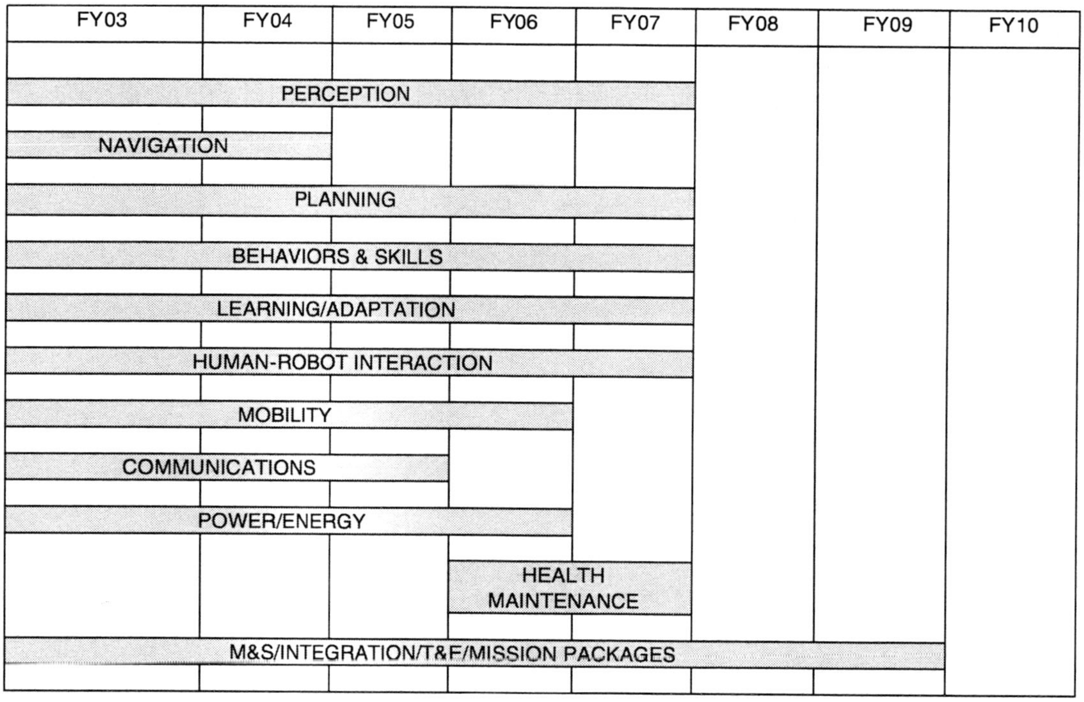

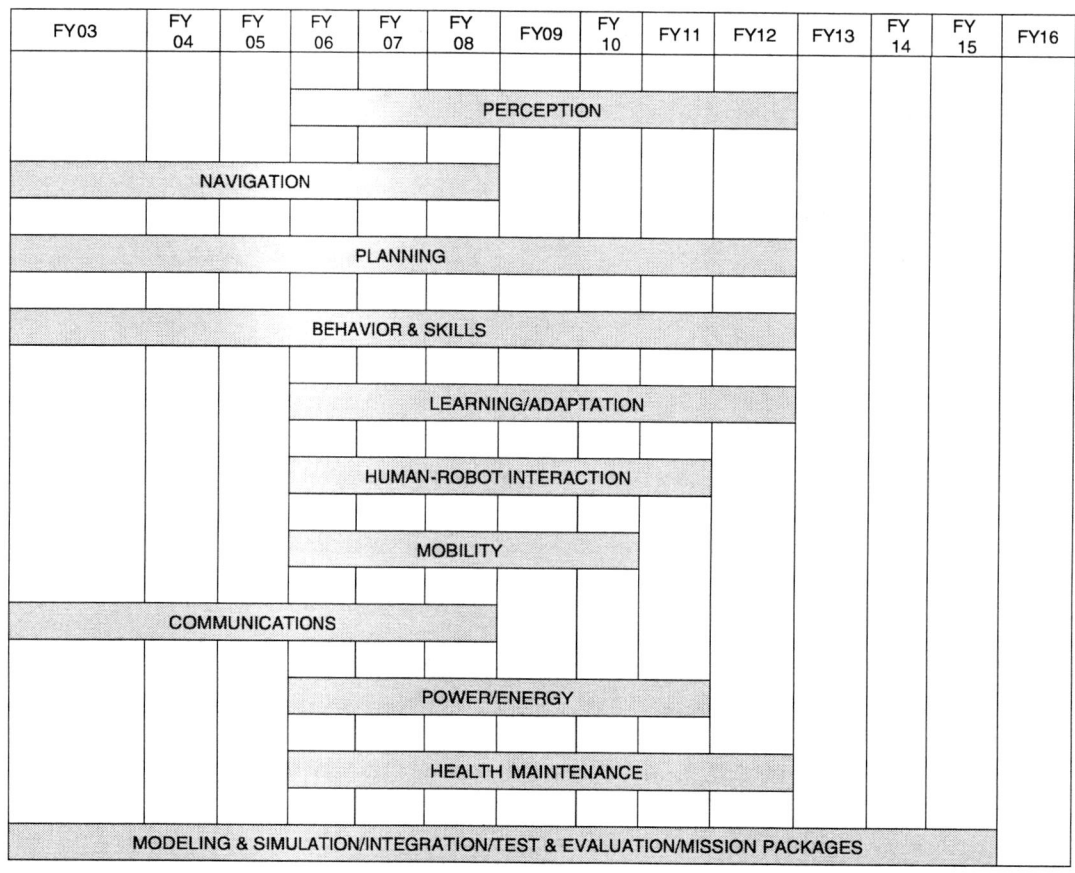

FIGURE 7-6 Technology development roadmap for the Donkey indicating the year TRL 6 will be reached for the system.

FIGURE 7-7 Technology development roadmap for the Wingman indicating the year TRL 6 will be reached for the system.

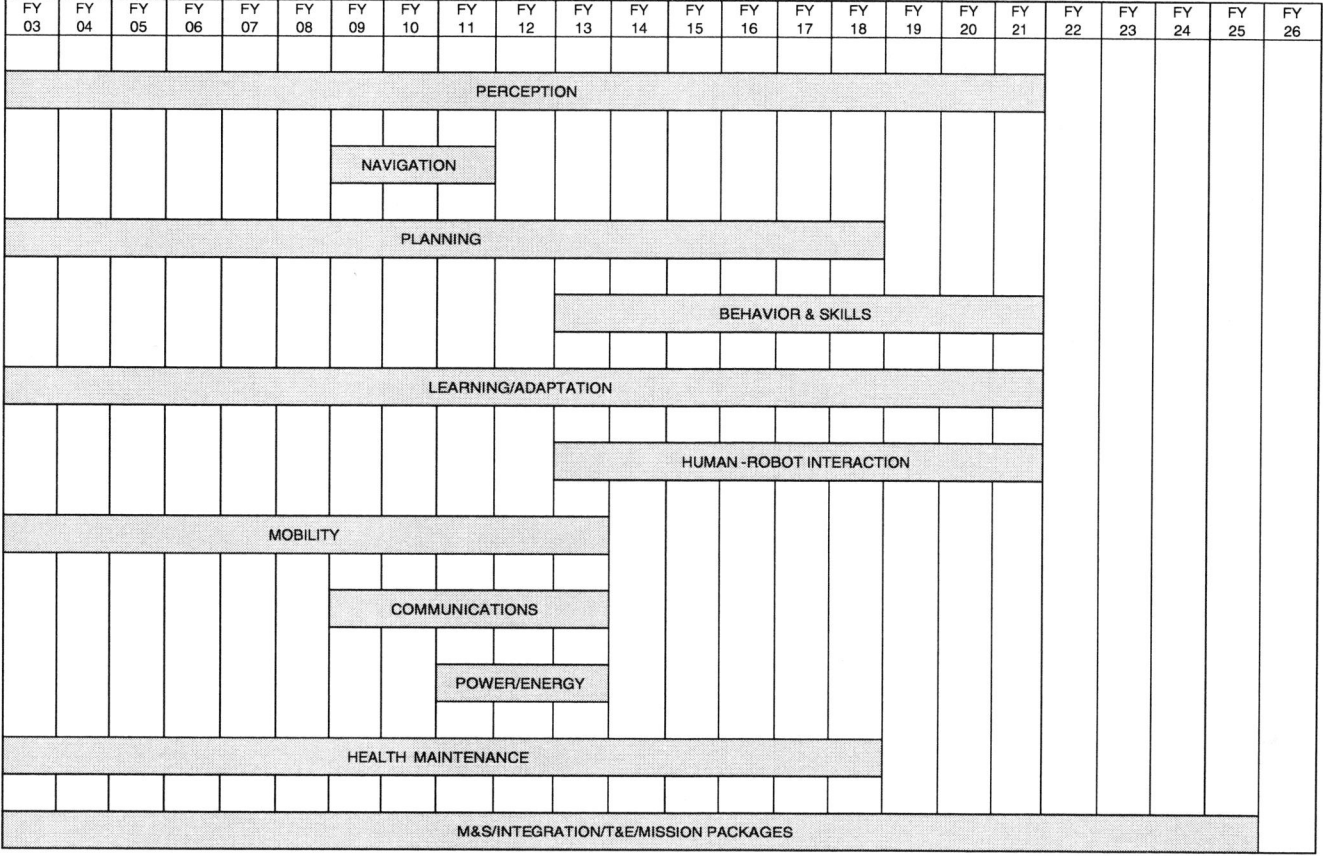

FIGURE 7-8 Technology development roadmap for the Hunter-Killer indicating the year TRL 6 will be reached for the system.

gaps identified by the committee are filled in a timely fashion. The roadmaps are interdependent, each building on research and development (R&D) accomplished to achieve capabilities needed in a preceding system. While the TRL 6 estimates in Chapters 4 and 5 were based on readiness to support individual example systems, the roadmaps in Figures 7-5 through 7-8 are based on the expectation that the advanced capabilities needed by the Hunter-Killer would be identified as goals from the beginning.

TIME LINES FOR GENERIC UGV SYSTEMS

Having evaluated the readiness of various UGV technologies to support development of specific UGV systems postulated for each of four broad UGV capability classes, the committee was able to estimate time lines for the development of tactically significant, generic, "entry-level" systems. In doing so, it assumed that "entry-level" teleoperated ground vehicles (TGVs) now exist. Figure 7-9 takes the technology development roadmaps for specific examples and generalizes to estimated time lines for the Army to field an "entry-level" semiautonomous preceder/follower (SAP/F), platform-centric autonomous vehicle (PC-AGV), and network-centric autonomous ground vehicle (NC-AGV).

The committee's estimates are based in part on the TRL assessments made in Chapters 4 and 5 and include an additional 2 years for system engineering, technology integration, and test and evaluation for each system. These time lines take into account that technology integration for common capabilities can begin earlier when preceded by prior system developments. They do not consider parallel developments that might be required for technology developments to support mission-function packages.

Figure 7-9 also illustrates how the efforts will vary in each of the UGV technology areas. Actual differences will reflect the times needed to fill critical capability gaps in the current UGV development programs. The time line estimates in Figure 7-9, combined with Figure 7-4 and the discussion of the FCS program plan, provide the basis for the answer to Task Statement Question 5.a in Box 7-1.

Priorities for UGV Development

Current Army developments address near-term capabilities, and the study showed that the achievement of technology objectives in pursuit of defined systems would facilitate the development of multiple capabilities, both near- and far-term. This should be an important consideration in establishing Army priorities. While the Army S&T plan should concentrate on difficult technological challenges, advances are needed in all areas if UGV systems are to be fielded with FCS. Specific recommendations for the priority of UGV technology developments in the Army's S&T plan are given in Recommendation 1 of Chapter 8.

> **BOX 7-1**
> **Task Statement Question 5.a**
>
> **Question:** From an engineering perspective, what are reasonable milestone dates for a UGV system development program leading to production? For example, does the current FCS program have a coherent plan and roadmap to build UGVs for FCS and the Objective Force?
>
> **Answer:** UGV requirements for FCS have not yet been defined, so the current FCS program does not have a coherent plan and roadmap to build UGVs. Figure 7-4 provides reasonable milestone dates from an engineering perspective for development of four specific example systems with many of the capabilities that may be needed for FCS. It should be emphasized that the milestones are optimistic estimates for UGV systems of unquestioned utility on the battlefield and not for entry-level systems or prototypes. Figure 7-9 provides reasonable estimates for "entry-level" UGV system developments in generic capability classes.

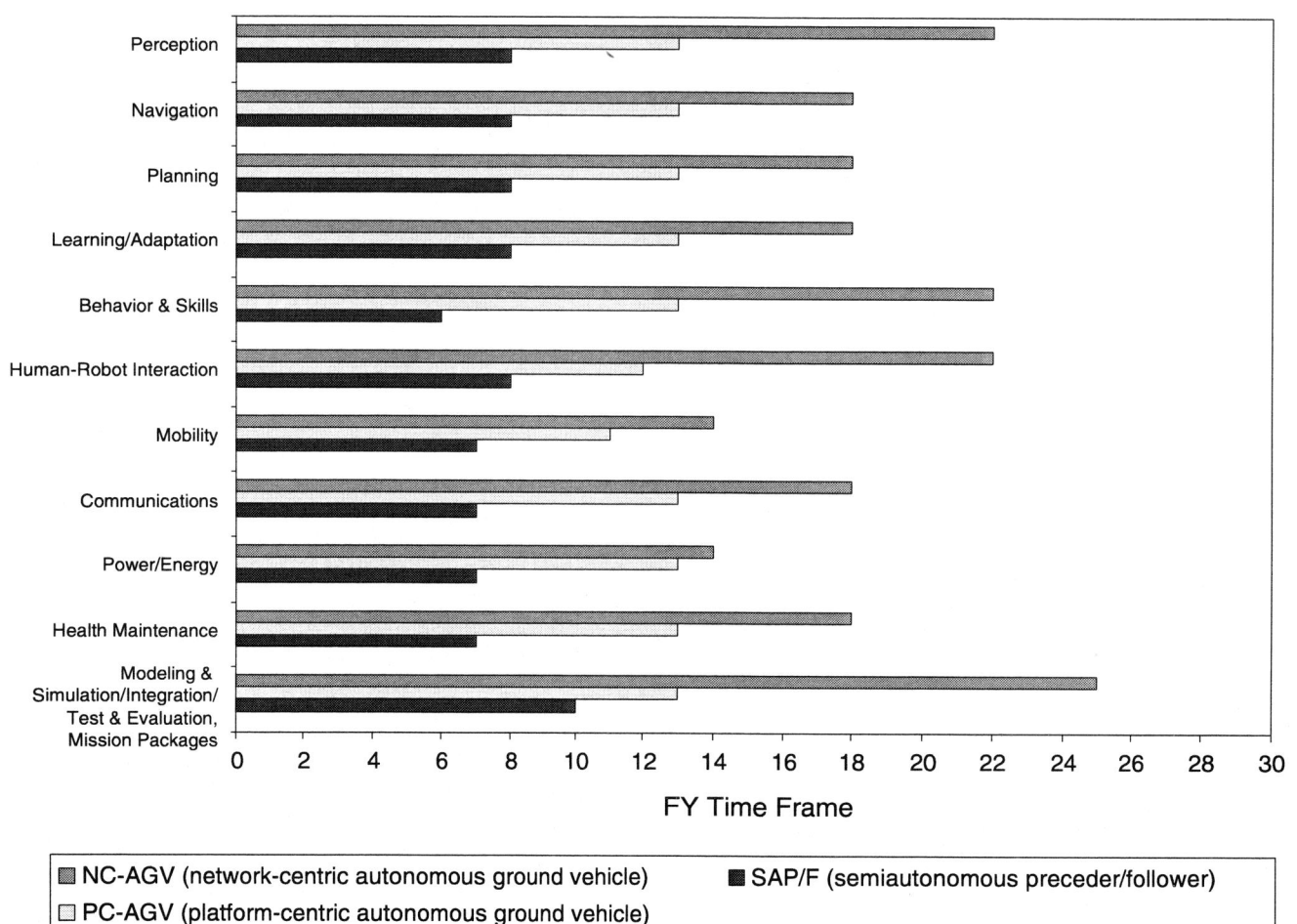

FIGURE 7-9 Technology roadmap for development of generic "entry-level" systems in capability classes. The milestones are based on achievement of TRL 6 in each technology area.

8

Findings and Recommendations

This chapter summarizes recommendations for the Army UGV Technology program emphasizing, in particular, where the Army should focus its attention to make unmanned ground vehicles (UGVs) a reality for the Future Combat Systems (FCS) program and the Objective Force.

TECHNOLOGY DEVELOPMENT PRIORITIES

Findings

The statement of task for the study challenged the committee to identify technology areas that merit further investigation by the Army. Tables 4-6 and 5-5 summarize the capability gaps that remain to be filled to implement the four application-oriented example systems postulated for the analysis. In general, items in the dark-shaded cells in the tables represent the most difficult challenges and should be accorded high priority for resolution if the Army desires to achieve the particular, or similar, systems postulated in the report. The committee assumed that the Army would fill all of the gaps identified to develop the example systems it postulated. Within this context, the following technologies stand out for their significance as potential "showstoppers" for developing UGV systems for the FCS.

Technology Areas Meriting Further Investigation

Clearly the highest priority for the Army should continue to be the development of perception technologies for autonomous mobility. Basic shortcomings in this area have been highlighted in numerous past studies and provide impetus for the current Army Research Laboratory (ARL) science and technology objective (STO).

On-road and off-road A-to-B mobility is fundamental to the acceptability of three of the four systems postulated by the study. On-road capability is immature and has not been emphasized. Although a start was made with small platforms in the Defense Advanced Research Project Agency (DARPA) Tactical Mobile Robot (TMR) program, little or no attention has been paid to A-to-B mobility in urban settings. Essentially no capability exists for mobility on unstructured roads. Off-road perception capability for mobility is extremely limited and has not been evaluated in unknown terrain, at night, in bad weather, or in the presence of obscurants. There is no evidence that the current level of perception capability can support an autonomous cross-country traverse of tactical significance, at tactical speeds, under combat conditions. Detection of obstacles, especially negative obstacles, cannot be done reliably, and there is essentially no capability to detect tactical features or to conduct situation assessments, yet the use of perception technology to extend situational awareness is essential to both the Wingman and Hunter-Killer example systems.

Perception technologies, including the sensors, algorithms (particularly for data fusion and for active vision in multiple modalities), and processing capabilities, must be perfected or UGV systems will prove a liability on the battlefield. The Army, Joint Program Office (JPO), and DARPA have made some progress, but much more must be done.

Improvements in individual sensor capabilities and algorithms are needed, but a big problem, largely unacknowledged, is optimizing the perception system hardware and software architecture: sensors, embedded processors, coded algorithms, and communications buses. There is currently no way to know how perception performance is reduced by suboptimized architecture or where improvements might be made. This is a very complicated systems engineering problem that is exacerbated by having work carried out by separate organizations in separate programs.

While perception technologies for A-to-B mobility and situation awareness are clearly the top priority, the committee found that other priorities for attention are dependent on capability class.

Teleoperated Ground Vehicles

For teleoperated ground vehicles, human–robot interaction, health maintenance, communications, and power/energy technologies assume major prominence. Current robots rely on teleoperation using microcomputer user interfaces, such as keyboards, mice, joysticks, and touch screens. Most are demanding to operate and have not been validated by human-factors personnel. Techniques to augment external navigation controls with algorithms for real-time mapping and localization would reduce the stress on operators. Such algorithms, which have been developed for indoor urban missions and have not transferred well to irregular outdoor environments, would help to reduce the number of operators required per robot.

Current unmanned systems, both ground and air, require many more technicians for repair and preventive maintenance than are required by manned systems. Future teleoperated ground vehicles (TGVs) (and UGVs in all classes) must be able to self-monitor and to provide information to remote locations for diagnosis and possible recovery. Such vehicles should be designed with behaviors and characteristics that facilitate their own survivability. These issues are not currently being addressed and constitute a major gap that must be filled to produce vehicles that will work in the field, be accepted by the user, and eventually reduce the logistics footprint compared to manned systems.

Semiautonomous Preceder/Follower Unmanned Ground Vehicle (SAP/F-UGV)

For the SAP/F-UGV class of vehicles, mobility, navigation, tactical behaviors, and health-maintenance technologies are all high priorities. Successful integration of navigation technologies with an all-terrain mobility platform could enable preceder/follower UGVs to serve not only as logistics carriers but also as lead elements for small-unit patrols or soldier-portable robot vehicles on security outposts. These UGVs must be operationally reliable to a degree greater than manned vehicles.

Depending on application, basic tactical behaviors will be required to ensure that SAP/F-UGVs can perform missions without becoming a burden on the battlefield. Developers must have clear operational guidance before these behaviors, many peculiar to Army field operations, can be programmed and tested.

Platform-Centric Autonomous Ground Vehicles (PC-AGV)

Priority technologies include those for the SAP/F-UGV class (mobility, navigation, tactical behaviors, and health maintenance) as described above plus learning/adaptation technologies. Tactical behaviors are central to the utility of PC-AGVs, and developers must have operational guidance to focus and direct implementation software. To be useful for extended durations as part of the FCS, PC-AGVs must be capable of adapting embedded tactical behaviors to changing situations without requiring reprogramming in the field. Ideally, lessons learned would be cumulative and could be transferred to other AGV systems.

Network-Centric Autonomous Ground Vehicles (NC-AGV)

Communications, including mobile self-configuring networks and distributed knowledge bases, become all-important for this class of UGV. To respond to multiple demands NC-AGVs must be tightly networked with other FCS elements and information systems on the battlefield. Other priority technologies include mobility (to provide versatile, multifunction platforms), human–robot interaction (to ensure proper task allocation between soldier–robot and robot–robot), and learning/adaptation (to expand the range of autonomous behaviors).

Recommendation 1. The Army should give top priority to the development of perception technologies to achieve autonomous mobility. In addition, it should focus on specific technologies depending on UGV capability class.

Recommendation 1 provides the basis for the answer to Task Statement Question 3.a in Box 8-1.

FOCUS ON COMPELLING ARMY APPLICATIONS

Findings

The compelling reason for incorporating UGVs in FCS and the Objective Force is that they can save soldiers' lives by taking on some of the most life-threatening missions soldiers now perform. The committee found no compelling arguments that UGVs will reduce force structure requirements for combat operations in the time frames envisioned for FCS or the Objective Force. Neither will they reduce force structure in the theater of operations. On the contrary, without progress in key UGV technology areas, such as perception and tactical skills, the ratio of personnel required for operating and maintaining each vehicle is likely to approach the 4:1 and higher ratios needed to support small UAV systems. Only in the far term (2025 and beyond) are UGVs likely to operate in field conditions at ratios approaching 1:1 with the soldiers handling them.

The capabilities of UGVs should complement what humans can do better, with the aim of maximizing the additional benefits that can be gained by introducing UGVs. Each UGV class should be specified and designed to do what robots can do better (or at lower risk) than humans, rather than trying to imitate what humans already do very well.

The potential user (warfighter) community, represented by Training and Doctrine Command (TRADOC), remains skeptical about the promise of benefits from UGVs. There is

> **BOX 8-1**
> **Task Statement Question 3.a**
>
> **Question:** What technologies should next be pursued, and in what priority, to achieve a UGV capability exceeding that envisioned in the ARL STO?
>
> **Answer:** The major technology thrust beyond the ARL STO should be in the area of perception technologies to support increasing levels of autonomous mobility and situation awareness. A particular focus is needed on the fusion of a balanced suite of active and passive onboard sensory, contextual, and external data. Other priorities depend upon the capability class of the UGV system to be developed. For the semiautonomous preceder/follower UGV envisioned by the ARL STO, these include technologies for the mobility platform, integrated navigation, human–robot interaction, tactical behaviors, and vehicle health maintenance.

good reason for this skepticism when the benefits are stated in terms of replacing soldiers in the force structure, rather than aiding soldiers in performing their missions. Another basis for skepticism is the survivability of UGVs on the battlefield. In the committee's judgment, only if basic utility and robustness can be demonstrated with experimental, application-driven vehicles will the user community begin to accept UGV missions requiring higher levels of autonomy.

Until requirements are validated in the Army user community there can be no commitment to UGV systems and applications. The existing statements of requirements are insufficient to guide and stimulate technological evolution. This void has forced UGV development into a technology push mode, rather than a balance between technology push and requirements pull modes.

For UGV systems to be included with FCS and the Objective Force, a process of spiral development involving the user will be necessary, including successive iterations of application and capability refinements. Technologies that merit special development attention can then be identified and developed using a "skunk-works" approach that achieves the focus and centralized leadership necessary to reach goals set by the user community. Prototypes resulting from targeted technology development and integration can be used for higher-level developments or for experiments involving particular mission-package applications by the user. Such technology-integrating experiments will help users determine which concepts have the most value.

The committee believes that the Army UGV program is best served by developing a small number of experimental vehicle types capable of applications with compelling value for FCS and the Objective Force. The objective would be to develop and integrate the technologies required for several classes of vehicle capability. In this sense the program would be capabilities driven, rather than requirements driven. A requirements development process could evolve later, when the user community is comfortable with making system commitments based on demonstrated UGV capabilities.

A "skunk-works" approach to develop autonomous A-to-B mobility would consolidate and focus the development of technologies essential to FCS UGVs under a single manager, eliminate duplication of effort, and provide the basis for standardized research platforms to be used in the spiral development of UGV systems. The process would also be best served by systematic testing and refinement under severe operating conditions.

The committee believes that Army mission needs and operational requirements for UGV systems can evolve in a spiral development process as a technology integration program advances, provided the program is focused on maturing the underlying technologies and achieving system integration of those technologies at several useful levels of vehicle capability. As documented in Chapter 2, the committee found that current operational requirements or mission-needs statements are inadequate to focus a capabilities-driven development program for experimental prototypes. In that chapter the committee defines four UGV capability classes and describes example military applications for each. Although the examples are essential to the committee's assessment of technology readiness levels, they are not intended to suggest what the Army requirements should be, or even which applications the Army should undertake for technology integration experiments.

Focusing on a few specific applications for the experimental prototypes, some of which may be simulated, is essential to maturing the needed technologies and resolving the significant issues of system integration. The focus on applications would organize the capabilities development effort into manageable components, each with a clear operational outcome to be achieved. While capabilities may mature at different rates, the program as a whole would address technical challenges of all applications concurrently. The application prototypes should be selected to develop capabilities needed for FCS and the Objective Force. The roadmaps developed in Chapter 7 were built around four such applications, but they illustrate only one of many possible combinations for evolutionary development.

Recommendation 2. The Army should adopt a "skunk-works" approach to develop technologies necessary for autonomous A-to-B mobility, so that such capabilities can be fielded with a small number of unmanned ground vehicle (UGV) classes, each of which is an experimental prototype for a compelling military application. TRADOC and the research and development community should commit to a spiral development process for refining and evolving concept-based requirements for UGVs, depending on what is learned from these technology-integrating prototypes.

SYSTEMS ENGINEERING CHALLENGE

Findings

Even when all underlying technologies for a UGV application have reached TRL 6, a great deal of work will be required for integrating specific technologies into one or more UGV systems capable of accomplishing FCS missions. In fact, the committee concluded that the greatest technical challenge for fielding UGVs of significant value to FCS and the Objective Force is likely to be technology integration and systemization as described in Chapter 6.

Adequate time must be allowed for the technologies that are developed to be put together and tested in the field in ways that give the developer and the user community feedback on how to improve a given concept. The user and developer communities must work together to provide direction for the technology integration to implement vehicle experiments. These directions should feed into the spiral development process from experimental prototypes to requirements-based systems following the established development process. For example, application parameters must be formulated to address the integration of the mission package technologies, mobility technologies, and communications technologies that are necessary for each experimental prototype.

The UGV program must adopt a systems development approach. Performance metrics and other assessment methodologies must be established that provide objective feedback to developers and users on how well an application-oriented experimental prototype is performing as an integrated system. Such an approach is needed to ensure integration across the presently stovepiped programs for individual contributing technologies.

Systems engineering discipline could be introduced by emphasizing hardware and software in-the-loop simulations. When appropriately instrumented, such simulations could aid in accomplishing architecture design and optimization and, most importantly, algorithm benchmarking. The goals would be to establish a focused UGV technology base, encourage rapid experimental prototyping, and enable near-real-time performance assessment.

Several supporting technologies, while not part of the autonomous behavior architecture, will nonetheless be critical to UGV system developments. Technologies needed for human–robot interaction (HRI), mobility, power, communications, and vehicle health maintenance will be different from those developed for manned systems. Research in these areas is heavily dependent on systems engineering to identify requirements.

Software quality is also an important issue, requiring extensive software engineering, re-implementation, and performance assessments to field a given system. The impact software quality has on the performance of current UGVs is unknown.

Systems engineering is also important for the spiral process of defining requirements for UGV vehicles as integral components of an FCS unit of operation. The FCS is a system of systems, and UGV systems must operate within the broader FCS system architecture. The overall architectural decisions for the system of systems should determine many of the requirements that are imposed on individual elements such as the UGVs, rather than vice versa. Technology development in all areas will be heavily dependent on system prototypes.

Recommendation 3. The Army should begin planning for unmanned ground vehicle UGV system development now. Systems-engineering processes should be used to inform and guide the development of UGV operational concepts and technology.

ADVOCATE FOR UGV DEVELOPMENT

Findings

In the absence of clear requirements to drive UGV development efforts, the UGV technology program must be one of developing capabilities. Without focus and advocacy, a capabilities-driven process is likely to suffer from diffusion and incoherence. Although the existing STO programs may have specific capability objectives (see Chapter 3), the efforts do not automatically add up to a coherent, coordinated program for UGV technology development.

The Army's UGV program must cut across existing program stovepipes and increase resources. If the objective is to field a network-centric autonomous ground vehicle for the Objective Force, then the Army must dedicate resources now to 6.2–6.4 developments with a common focus on achieving this end. A strong central advocate is needed.

Experience has shown that the Army responds well to challenges that are represented by high-level positions or organizations dedicated to a single purpose. The Army Digitization Office, established in the 1990s by the Army Chief of Staff, provides a good example of how such focus can be used to move a project forward that might otherwise become lost in the bureaucracy. Similarly, special Army selection boards exist to select highly qualified personnel for designation as program managers (PMs) for technology and system developments of high-level importance.

Although UGV system concepts and requirements are not sufficiently advanced to merit the same approach at this time, extraordinary measures analogous to the Digitization Office initiative should be considered as the UGV program matures beyond the science and technology (S&T) stage. In the interim, a board-selected PM for UGV technology and system developments would be able both to serve as an advocate for autonomous systems and to focus development effort on achieving A-to-B mobility capabilities and developing experimental prototypes, thereby advancing the experimentation and acceptance of UGV systems. This new position would contrast with the present PM positions (for FCS and Objective Force), which are focused on objectives that can be achieved with or without a dollar of investment in underlying UGV technology. The new position would not duplicate the functions of the DOD UGV PM position, which is focused on integrating UGV systems using existing technologies in response to specific DOD-endorsed requirements.

Recommendation 4a. The Army should designate a Program Manager for Unmanned Ground Vehicles (PM-UGV) to coordinate research, development, and acquisition of Army UGV systems. The PM-UGV would act for the Assistant Secretary of the Army (Acquisition, Logistics, and Technology) to manage Army UGV technology developments, approve technology base planning, provide acquisition guidance, and oversee resource allocation. The PM would be the Army's principal advocate for unmanned ground systems and single point of contact for UGV developments with the Joint Program Office, the Defense Advanced Research Projects Agency, and other agencies.

Recommendation 4b. As the unmanned ground vehicle (UGV) program matures beyond the S&T stage, the Army should consider additional extraordinary measures, analogous to the successful Army digitization initiatives, to ensure sufficient focus on developing and fielding UGV systems for the Future Combat Systems and the Objective Force.

BOX 8-2
Task Statement Question 5.b

Question: What can be recommended on the technical content, time lines and milestones based on these assessments?

Answer: Recommendations 1–4 address the technical content, time lines, and milestones of UGV efforts for FCS. First, the Army should focus S&T efforts on the perception technology area, with other priority areas dependent upon the particular capability class that is determined for UGV systems in FCS. Second, the Army should adopt a "skunk-works" approach to develop the essential perception technologies to enable autonomous A-to-B mobility capabilities that can be fielded with multiple UGV systems as experimental prototypes for possible insertion in the FCS program. Third, the Army must begin immediately to fill a void in systems engineering by defining system requirements, planning for life-cycle support, establishing milestones for development of assessment methods and metrics for UGV systems, and taking advantage of modeling and simulation tools. Finally, the Army should designate a high-level advocate to accelerate S&T time lines and take the lead in integrating UGV technologies into prototypical systems.

The study's four recommendations provide the basis for the answer to Task Statement Question 5.b in Box 8-2.

References

AFRL (U.S. Air Force Research Laboratory). 2002. Data provided to the Committee on Army Unmanned Ground Vehicle Technology. Tyndall Air Force Base, Fla.: U.S. Air Force Research Laboratory.

Aghabarari, E., and J. Varney. 1995. Electrical power system failure detection, isolation and recovery on the International Space Station Alpha. Pp. A49–A54 in Proceedings of the 30th Intersociety Energy Conversion Engineering Conference. D.Y. Goswami, L.D. Kannberg, T.R. Mancini, and S. Somasundaram, eds. New York, N.Y.: American Society of Mechanical Engineers.

Albus, J.S., and A.M. Meystel. 2001. Engineering of Mind: An Introduction to the Science of Intelligent Systems. New York, N.Y.: Wiley.

AMCOM (U.S. Army Aviation and Missile Command). 2001. Data provided to the Committee on Army Unmanned Ground Vehicle Technology. Redstone Arsenal, Ala.: U.S. Army Aviation and Missile Command.

Andrews, A.M. 2001. Accelerating the pace of transformation. Briefing by A. Michael Andrews, Deputy Assistant Secretary of the Army for Research and Technology/Chief Scientist, to the Board on Army Science and Technology, National Research Council, Washington, D.C., December 10.

Arkin, R.C. 1992. Cooperation without communication—Multiagent schema-based robot navigation. Journal of Robotic Systems 9(3): 351–364.

ASA (ALT) (Assistant Secretary of the Army (Acquisitions, Logistics, and Technology)). 2001a. Army Science and Technology Master Plan. Washington, D.C.: Assistant Secretary of the Army (Acquisitions, Logistics, and Technology).

ASA (ALT). 2001b. UGV Technology Roadmap. Washington, D.C.: Assistant Secretary of the Army (Acquisitions, Logistics, and Technology).

ASB (Army Science Board). 2001. Army Science Board FY 2000 Summer Study Final Report, Technical and Tactical Opportunities for Revolutionary Advances in Rapidly Deployable Joint Ground Forces in the 2015–2025 Era, April. Arlington, Va.: Army Science Board.

Astrom, K.J., and B. Wittenmark. 1989. Adaptive Control. Reading, Mass.: Addison-Wesley.

Balch, T., and R.C. Arkin. 1998. Behavior-based formation control for multirobot teams. IEEE Transactions on Robotics and Automation 14(6): 926–939.

Baluja, S. 1996. Evolution of an artificial neural network based autonomous land vehicle controller. IEEE Transactions on Systems, Man and Cybernetics, Part B 26(3): 450–463.

Basu, D., and R. Paramasivam. 2000. An approach to software assisted recovery from hardware transient faults for real time systems. Pp. 264–274 in Proceedings of the 19th International Conference on Computer Safety, Reliability and Security (SAFECOMP 2000). F. Koornneef, and M. van der Meulen, eds. Berlin, Germany: Springer-Verlag.

Baydar, C.M., and K. Saitou. 2001. Automated generation of robust error recovery logic in assembly systems using genetic programming. Journal of Manufacturing Systems 20(1): 55–68.

Beni, G., and P. Liang. 1996. Pattern reconfiguration in swarms—convergence of a distributed asynchronous and bounded iterative algorithm. IEEE Transactions on Robotics and Automation 12(3): 485–490.

Bergeon, B., A. Zolghadri, Z. Benzian, J.L. Ermine, and M. Monsion. 1994. Specification of a real-time knowledge based supervision system. Pp. 571–576 in Automatic Control—World Congress 1993, Volume 2—Robust Control, Design and Software, Proceedings of the 12th Triennial World Congress of the International Federation of Automatic Control. G.C. Goodwin and R.J. Evans, eds. Oxford, United Kingdom: Pergamon Press, Ltd.

Biktalvi, P., and I. Lorant. 1999. Improvements in navigation and fault diagnosis of an autonomous mobile robot. Pp. 215–221 in Proceedings of the Workshop on European Scientific and Industrial Collaboration, WESIC '99, Promoting: Advanced Technologies in Manufacturing. G.N. Roberts and C.A.J. Tubb, eds. Newport, United Kingdom: University of Wales College.

Bill, R. 2001. Fuel Efficiency on the Battlefield. Presentation by Robert Bill, Army Research Laboratory Vehicle Technology Directorate, to the Hybrid Electric Vehicles Workshop, University of Texas at Austin, Austin, Texas, September 6, 2001.

Bozorg, M., E.M. Nebot, and H.F. Durrant-Whyte. 1998. A decentralised navigation architecture. Pp. 3413–3418 in Proceedings of the 1998 IEEE International Conference on Robotics and Automation. New York, N.Y.: Institute of Electrical and Electronics Engineers, Inc.

Brooks, R.A. 1986. A robust layered control system for a mobile robot. IEEE Journal of Robotics and Automation 2(1): 14–23.

Brooks, R.A., and A.M. Flynn. 1989. Fast, cheap and out of control: A robot invasion of the Solar System. Journal of the British Interplanetary Society 42(10): 478–485.

Brumitt, B., and M. Hebert. 1998. Experiments in autonomous driving with concurrent goals and multiple vehicles. Pp. 1895–1902 in Proceedings of the 1998 IEEE International Conference on Robotics and Automation. New York, N.Y.: Institute of Electrical and Electronics Engineers, Inc.

Brumitt, B.L., and A. Stentz. 1998. GRAMMPS: A generalized mission planner for multiple mobile robots in unstructured environments. Pp. 1564–1571 in Proceedings of the 1998 IEEE International Conference on Robotics and Automation. New York, N.Y.: Institute of Electrical and Electronics Engineers, Inc.

Butler, D.C. 2002. The integration of models and simulations: A life cycle approach. Presentation by David C. Butler, George Mason University, to the INCOSE Mid-Atlantic Regional Conference, Sheraton Reston Hotel, Reston, Va., April 5–8.

Cao, Y.U., A.S. Fukunaga, and A.B. Kahng. 1995. Cooperative mobile robotics: Antecedents and directions. Pp. 226–234 in Proceedings of IROS '95, The 1995 IEEE/RSJ International Conference on Intelligent Robots and Systems: Human Robot Interaction and Cooperative Robots. Los Alamitos, Calif.: Institute of Electrical and Electronics Engineers, Inc. Computer Society Press.

Carnes, J.R., and A. Misra. 1996. Model-integrated toolset for fault detection, isolation and recovery (FDIR). Pp. 356–363 in Proceedings IEEE Symposium and Workshop on Engineering of Computer-Based Systems. Los Alamitos, Calif.: Institute of Electrical and Electronics Engineers, Inc. Computer Society Press.

Chen, C.H. 1996. Fuzzy Logic and Neural Network Handbook. New York, N.Y.: McGraw-Hill.

Chen, Q., and J.Y.S. Luh. 1994. Coordination and control of a group of small mobile robots. Pp. 2315–2320 in Proceedings of the 1994 IEEE International Conference on Robotics and Automation. Los Alamitos, Calif.: Institute of Electrical and Electronics Engineers, Inc. Computer Society Press.

Choset H., and K. Nagatani. 2001. Topological simultaneous localization and mapping (SLAM): Towards exact localization without explicit localization. IEEE Transactions on Robotics and Automation 17(2): 125–137.

Clements, N.S., B.S. Heck, and G. Vachrsevanos. 2000. Component based modeling and fault tolerant control of complex systems. Pp. 6F4/1–6F4/4 in Proceedings of the 19th Digital Avionics Systems Conference. Piscataway, N.J.: Institute of Electrical and Electronics Engineers, Inc.

Congress. 2000. Enactment of Provisions of H.R. 5408, The Floyd D. Spence National Defense Authorization Act for Fiscal Year 2001: Conference Report to Accompany H.R. 4205. Washington, D.C.: United States Congress.

Covey, S.R., A.R. Merrill, and R.R. Merrill. 1994. First Things First: To Live, To Love, To Learn, To Leave a Legacy. New York, N.Y.: Simon & Schuster.

DARPA (Defense Advanced Research Projects Agency). 2001. Next Generation Explosive Ordnance Disposal Remote Controlled Vehicle (NGEODRCV): Common Systems Architecture, Draft version 5, October. Arlington, Va.: Defense Advanced Research Projects Agency.

DDR&E (Director, Defense Research and Engineering). 2002. Information Systems Technology (IST) Defense Technology Area Plan, February. Washington, D.C.: Department of Defense.

Desai, J.P., J. Ostrowski, and V. Kumar. 1998. Controlling formations of multiple mobile robots. Pp. 2864–2869 in Proceedings of the 1998 IEEE International Conference on Robotics and Automation. New York, N.Y.: Institute of Electrical and Electronics Engineers, Inc.

Desai, J.P., J.P. Ostrowski, and V. Kumar. 2001. Modeling and control of formations of nonholonomic mobile robots. IEEE Transactions on Robotics and Automation 17(6): 905–908.

Deuker, B., M. Perrier, and B. Amy. 1998. Fault-diagnosis of subsea robots using neuro-symbolic hybrid systems. Pp. 830–834 in Proceedings of the IEEE OCEANS Conference and Exhibition on Engineering for Sustainable Use of the Oceans (OCEANS 98). New York, N.Y.: Institute of Electronics and Electrical Engineers, Inc.

Dissanayake, M.W.M.G., P. Newman, S. Clark, H.F. Durrant-Whyte, and M. Csorba. 2001. A solution to the simultaneous localization and map building (SLAM) problem. IEEE Transactions on Robotics and Automation 17(3): 229–241.

Dixon, W.E., I.D. Walker, and D.M. Dawson. 2001. Fault detection for wheeled mobile robots with parametric uncertainty. Pp. 1245–1250 in Proceedings of the 2001 IEEE/ASME International Conference on Advanced Intelligent Mechatronics. New York, N.Y.: Institute of Electrical and Electronics Engineers, Inc.

Djath, K., M. Dufaut, and D. Wolf. 2000. Mobile robot multisensor reconfiguration. Pp. 110–115 in Proceedings of the IEEE Intelligent Vehicles Symposium 2000. New York, N.Y.: Institute of Electrical and Electronics Engineers, Inc.

DMSO (Defense Modeling and Simulation Office). 2001. Joint Virtual Battlespace (JVB)-Defense Modeling and Simulation Office (DMSO) Technical Exchange Meeting, May 17.

DOD (Department of Defense). 2001. Department of Defense Report on the Joint Robotics Program Master Plan FY2002: Out Front in Harm's Way. Washington, D.C.: Department of Defense.

DOE (Department of Energy). 1998a. Robotics and Intelligent Systems Roadmap. Albuquerque, N. Mex.: Sandia National Laboratories.

DOE. 1998b. Cost analysis of Fuel Cell Transportation, Final Report Ref 49739, SFAA No. DE-SC02-98EE50526. Washington, D.C.: Department of Energy.

Dohrmann, C.R., G.R. Eisler, and R.D. Robinett. 1996. Dynamic programming approach for burnout-to-apogee guidance of precision munitions. Journal of Guidance, Control, and Dynamics 19(2): 340–346.

Dubois, L., T. Fukuda, F. Delmotte, and P. Borne. 1997. Approximate reasoning for the control of a robot in an uncertain environment: A multi-model approach. Pp. 823–828 in Proceedings of the 1997 IEEE International Conference on Robotics and Automation. New York, N.Y.: Institute of Electrical and Electronics Engineers, Inc.

Eicker, P. 2001. Army Unmanned Ground Vehicle Technology Roadmap. Briefing by Patrick Eicker, Director, Intelligent Systems and Robotics Center, Sandia National Laboratories, to the Committee on Army Unmanned Ground Vehicle Technology, National Academy of Sciences, Washington, D.C., August 28.

Feddema J., and D. Schoenwald. 2001. Decentralized control of cooperative robotic vehicles. Pp. 136–146 in Unmanned Ground Vehicle Technology III, Proceedings Of SPIE Volume 4364. G.R. Gerhart and C.M. Shoemaker, eds. Bellingham, Wash.: The International Society for Optical Engineering.

Feddema, J.T., C. Lewis, P. Klarer, G.R. Eisler, and R. Caprihan. 1999. Cooperative robotic sentry vehicles. Pp. 44–54 in Sensor Fusion and Decentralized Control in Robotic Systems II, Proceedings of SPIE Volume 3839. G.T. McKee and P.S. Schenker, eds. Bellingham, Wash.: The International Society for Optical Engineering.

Feddema, J., C. Lewis, and D. Schoenwald. 2002 (in press). Decentralized control of cooperative robotic vehicles: Theory and application. IEEE Transactions on Robotics and Automation.

Fukuda, T., D. Funato, K. Sekiyama, and F. Arai. 1998. Evaluation on flexibility of Swarm Intelligent System. Pp. 3210–3215 in Proceedings of the 1998 IEEE International Conference on Robotics and Automation. New York, N.Y.: Institute of Electrical and Electronics Engineers, Inc.

Fukuda, T., H. Mizoguchi, K. Sekiyama, and F. Arai. 1999. Group behavior control for MARS (micro autonomous robotic system). Pp. 1550–1555 in ICRA '99: IEEE International Conference on Robotics and Automation. New York, N.Y.: Institute of Electrical and Electronics Engineers, Inc.

Gad-el-Hak, M. 2001. The MEMS Handbook. Boca Raton, FL: CRC Press.

Goel, P., G. Dedeoglu, S.I. Roumeliotis, and G.S. Sukhatme. 2000. Fault detection and identification in a mobile robot using multiple model estimation and neural network. Pp. 2302–2309 in Proceedings of the 2000 IEEE International Conference on Robotics and Automation. Piscataway, N.J.: Institute of Electrical and Electronics Engineers, Inc.

Goldberg, D.E. 1989. Genetic Algorithms in Search, Optimization, and Machine Learning. Reading, Mass.: Addison-Wesley.

Goldsmith, S.Y., J.T. Feddema, and R.D. Robinett. 1998. Analysis of decentralized variable structure control for collective search by mobile robots. Pp. 40–47 in Sensor Fusion and Decentralized Control in Robotic Systems, Proceedings of SPIE, Volume 3523. P.S. Schenker and G.T. McKee, eds. Bellingham, Wash.: The International Society for Optical Engineering.

Hancock, J., and C. Thorpe. 1995. ELVIS: Eigenvectors for land vehicle image system. Pp. 35–40 in Proceedings of IROS '95, The 1995 IEEE/

RSJ International Conference on Intelligent Robots and Systems: Human Robot Interaction and Cooperative Robots. K. Ikeuchi and P. Khosla, eds. Los Alamitos, Calif.: Institute of Electrical and Electronics Engineers Computer, Inc. Society Press.

Haykin, S. 1999. Neural Networks, A Comprehensive Foundation, 2nd Ed. Upper Saddle River, N.J.: Prentice-Hall.

Hougen, D.F., M.D. Erickson, P.E. Rybski, S.A. Stoeter, M. Gini, and N. Papanikolopoulos. 2000. Autonomous mobile robots and distributed exploratory missions. Pp. 221–230 in Distributed Autonomous Robotic Systems 4. L.E. Parker, G. Bekey, and J. Barhen, eds. New York, N.Y.: Springer-Verlag.

Howard, A., H. Seraji, and E. Tunstel. 2001. A rule-based fuzzy transversability index for mobile robot navigation. Pp. 3067–3071 in Proceedings of the 2001 IEEE International Conference on Robotics and Automation. New York, N.Y.: Institute of Electrical and Electronics Engineers, Inc.

Hsu, A., S.R. Sachs, F. Eskafi, and P. Varaiya. 1991. The design of platoon maneuvers for IVHS. Pp. 2545–2550 in Proceedings of 1991 American Control Conference. Piscataway, N.J.: Institute of Electrical and Electronics Engineers, Inc.

Hurtado, J.E., R.D. Robinett, C.R. Dohrmann, and S.Y. Goldsmith. 1998. Distributed sensing and cooperating control for swarms of robotic vehicles. Pp. 175–178 in Proceedings of the 1st IASTED International Conference on Control and Applications. M.H. Hamza, ed. Calgary, Canada: ACTA Press.

Hutchinson, H. 2001. Foreign Robotics Initiatives. Briefing presented by Harold Hutchinson, U.S. Army Tank-Automotive and Armaments Command, to the Committee on Army Unmanned Ground Vehicle Technology, U.S. Army Tank-Automotive and Armaments Command, Warren, Mich., September 20.

INEEL (Idaho National Engineering and Environmental Laboratory). 2002. INEEL Capability: Remote, Robotics, and Automated Systems. Available online at <http://www.inel.gov/capabilities/robotics> [August 19, 2002].

Intel. 2002. Moore's Law. Available online at <http://www.intel.com/research/silicon/mooreslaw.htm> [August 16, 2002].

Irving, P., W.L. Allan, Q. Ming, and T. Healy. 2001. Novel catalytic fuel cell reforming with advanced membrane technology. Available online at <http://www.eren.doe.gov/hydrogen/pdfs/30535r.pdf> [August 15, 2002].

Jackson, E. 1997. Real-time model-based fault detection and diagnosis for automated systems. Pp. 26–28 in Proceedings of the IEEE Industry Applications Society 1997 Dynamic Modeling Control Applications for Industry Workshop. New York, N.Y.: Institute of Electrical and Electronics Engineers, Inc.

Jennings, J.S., G. Whelan, and W.F. Evans. 1997. Cooperative search and rescue with a team of mobile robots. Pp. 193–200 in Proceedings of the 8th International Conference on Advanced Robotics (ICAR 97). New York, N.Y.: Institute of Electrical and Electronics Engineers, Inc.

Jochem, T. 2001. Safe-TRAC Technical Brief. Available online at <http://www.assistware.com/Tech_Brief.PDF> [August 8, 2002].

Jochem, T., D. Pomerleau, and C. Thorpe. 1995a. Vision guided lane transition. Pp. 30–35 in Proceedings of the Intelligent Vehicles '95 Symposium. Piscataway, N.J.: Institute of Electrical and Electronics Engineers, Inc.

Jochem, T.M., D.A. Pomerleau, and C.E. Thorpe. 1995b. Vision-based neural network road and intersection detection and traversal. Pp. 344–349 in Proceedings of IROS '95, The 1995 IEEE/RSJ International Conference on Intelligent Robots and Systems: Human Robot Interaction and Cooperative Robots. K. Ikeuchi and P. Khosla, eds. Los Alamitos, Calif.: Institute of Electrical and Electronics Engineers, Inc. Computer Society Press.

Johnson, W. 2001. Information submitted by William Johnson, Program Manager, Future Combat System, to the Committee on Army Unmanned Ground Vehicle Technology, September 16, 2001.

Jones, J.V. 1998. Integrated Logistics Support Handbook, Special Reprint Edition. New York, N.Y.: McGraw Hill.

Kadmiry, B., P. Bergsten, and D. Driankov. 2001. Autonomous helicopter control using fuzzy gain scheduling. Pp. 2980–2985 in Proceedings of the 2001 IEEE International Conference on Robotics and Automation. New York, N.Y.: Institute of Electrical and Electronics Engineers, Inc.

Kaga, T., J. Starke, P. Molnar, M. Schanz, and T. Fukuda. 2000. Dynamic robot-target assignment—dependence of recovering from breakdowns on the speed of the selection process. Pp. 325–334 in Distributed Autonomous Robotic Systems 4. L.E. Parker, G. Bekey, and J. Barhen, eds. New York, N.Y.: Springer-Verlag.

Kaufman, H., I. Bar-Kana, and K. Sobel. 1994. Direct Adaptive Control Algorithms: Theory and Applications. New York, N.Y.: Springer-Verlag.

Kim, T.W., and J. Yuh. 2001. A novel neuro-fuzzy controller for autonomous underwater vehicles. Pp. 2350–2355 in Proceedings of the 2001 IEEE International Conference on Robotics and Automation. New York, N.Y.: Institute of Electrical and Electronics Engineers, Inc.

Koren, Y., and J. Borenstein. 1991. Potential field methods and their inherent limitations for mobile robot navigation. Pp. 1398–1404 in Proceedings of the 1991 IEEE International Conference on Robotics and Automation. Los Alamitos, Calif.: Institute of Electrical and Electronics Engineers, Inc. Computer Society Press.

Kosuge, K., T. Oosumi, M. Satou, K. Chiba, K. Takeo. 1998. Transportation of a single object by two decentralized-controlled nonholonomic mobile robots. Pp. 2989–2994 in Proceedings of the 1998 IEEE International Conference on Robotics and Automation. New York, N.Y.: Institute of Electrical and Electronics Engineers, Inc.

Krishnamurthi, M., and D.T. Phillips. 1992. An expert system framework for machine fault-diagnosis. Computers and Industrial Engineering 22(1): 67–84.

Kube, R.C., and H. Zhang. 1994. Collective robotics: From social insects to robots. Adaptive Behavior 2(2): 189–218.

Lam, R.K., R.S. Doshi, D.J. Atkinson, and D.M. Lawson. 1989. Diagnosing faults in autonomous robot plan execution. Pp. 399–408 in Sensor Fusion: Spatial Reasoning and Scene Interpretation, Proceedings of SPIE Volume 1003. P.S. Schenker, ed. Bellingham, Wash.: The International Society for Optical Engineering.

Lamine, K.B., and F. Kabanza. 2000. Using fuzzy temporal logic for monitoring behavior-based mobile robots. Pp. 116–122 in Proceedings of the IASTED International Conference Robotics and Applications. M.H. Hamza, ed. Anaheim, Calif.: IASTED/ACTA Press.

Landau, I.D., R. Lozano, and M. M'saad. 1998. Adaptive Control. New York, N.Y.: Springer.

Langer, D., and C.E. Thorpe. 1997. Sonar-based outdoor vehicle navigation. Pp. 159–186 in Intelligent Unmanned Ground Vehicles: Autonomous Navigation Research at Carnegie Mellon. M. Hebert, C.E Thorpe, and A. Stentz, eds. Boston, Mass.: Kluwer Academic Publishers.

Latombe, J.C. 1991. Robot Motion Planning. Boston, Mass.: Kluwer Academic Publishers.

Laughery, S., G. Gerhart, and P. Muench. 2000. Evaluating mobility using Bekker's equations. Pp. 53–61 in Proceedings of the 11th Annual Ground Target Modeling and Validation Conference. W.R. Reynolds and T.T. Macki, eds. Calumet, Mich.: Signature Research, Inc.

Liu, G. 2001. Control of robot manipulators with consideration of actuator performance degradation and failures. Pp. 2566–2571 in Proceedings of the 2001 IEEE International Conference on Robotics and Automation. New York, N.Y.: Institute of Electrical and Electronics Engineers, Inc.

Liu, Y., K.M. Passino, and M. Polycarpou. 2001. Stability analysis of one-dimensional asynchronous swarms. Pp. 716–721 in Proceedings of the 2001 American Control Conference. New York, N.Y.: Institute of Electrical and Electronics Engineers, Inc.

Lozano-Perez, T. 1983. Spatial planning—A configuration space approach. IEEE Transactions on Computers 32(2): 108–120.

Madden, M.G.M., and P.J. Nolan. 1999. Monitoring and diagnosis of multiple incipient faults using fault tree induction. IEE Proceedings-Control Theory and Applications 146(2): 204–212.

Malhiot, M. 2002. Briefing by Marshall Malhiot, Head Force Protection Branch (AFRL/MLQF), to Frank Rose, Committee on Army Unmanned Ground Vehicle Technology, Tyndall AFB, Fla., March 15.

Man, K.F., K.S. Tang, and S. Kwong. 1999. Genetic Algorithms: Concepts and designs. New York, N.Y.: Springer.

Meyrowitz, A., and A. Schultz. 2002. Navy Center for Applied Research in Artificial Intelligence. Briefing by Alan Meyrowitz and Alan Schultz, Naval Research Laboratory to the Committee on Army Unmanned Ground Vehicle Technology, Naval Research Laboratory, Washington, D.C., March 27.

Michalewicz, Z. 1992. Genetic Algorithms + Data Structures = Evolution Programs. New York, N.Y.: Springer-Verlag.

Mitchell, M. 1997. An Introduction to Genetic Algorithms. Cambridge, Mass.: MIT Press.

Molnar, P., and J. Starke. 2000. Communication fault tolerance in distributed robotic systems. Pp. 99–108 in Distributed Autonomous Robotic Systems 4. L.E. Parker, G. Bekey, and J. Barhen, eds. New York, N.Y.: Springer-Verlag.

Moore, T.N. 1985. A health monitoring system for robots. Pp. 30–34 in Proceedings of the IASTED International Symposium Advances in Robotics. M.H. Hamza, ed. Anaheim, Calif.: Acta Press.

Moravec, H.P. 1999. Robot: Mere Machine to Transcendent Mind. New York, N.Y.: Oxford University Press.

Murphy, R.R. 2000. Introduction to AI Robotics. Cambridge, Mass.: MIT Press.

Murphy, R.R., and D. Hershberger. 1999. Handling sensing failures in autonomous mobile robots. International Journal of Robotics Research 18(4): 382–400.

Murphy, R.R., and E. Rogers. 2001. Human–Robot Interaction, Final Report for DARPA/NSF Study on Human–Robot Interaction. Available online at <http://www.csc.calpoly.edu/~erogers/HRI/HRI-report-final.html> [August 12, 2002].

Nikam, U., and E.L. Hall. 1997. A fault diagnostic system for a mobile robot. Pp. 186–196 in Intelligent Robots and Computer Vision XVI: Algorithms, Techniques, Active Vision, and Materials Handling, Proceedings of SPIE Volume 3208. D.P. Casasent, ed. Bellingham, Wash.: The International Society for Optical Engineering.

Noreils, F.R. 1992. Multi-robot coordination for battlefield strategies. Pp. 1777–1784 in Proceedings of the 1992 IEEE/RSJ International Conference on Intelligent Robots and Systems. New York, N.Y.: Institute of Electrical and Electronics Engineers, Inc.

Noreils, F.R. 1993. Toward a robot architecture integrating cooperation between mobile robots—Application to indoor environment. International Journal of Robotics Research 12(1): 79–98.

NRC (National Research Council). 1997. Energy-Efficient Technologies for the Dismounted Soldier. Washington, D.C.: National Academy Press.

Numrich, S.M. 2002. Briefing on Modeling and Simulation Technology Overview. Briefing by S.M. Numrich, DMSO, to the Information Systems Technology (IST) Technology Area Review and Assessment (TARA), Fort Monmouth, N.J., April 16.

Oates, T., and P.R. Cohen. 1996. Searching for structure in multiple streams of data. Pp. 346–354 in Machine Learning, Proceedings of the Thirteenth International Conference on Machine Learning (ICML '96). L. Saitta, ed. San Francisco, Calif.: Morgan Kaufmann Publishers.

Okina, S., K. Kawabata, T. Fujii, Y. Kunii, H. Asama, and I. Endo. 2000. Study of a self-diagnosis system for an autonomous mobile robot. Advanced Robotics 14(5): 339–341.

ORNL (Oak Ridge National Laboratory). 2002. CESAR. Available online at <http://www.cesar.ornl.gov/robotics-index.html> [August 19, 2002].

Parker, L.E. 2000. Current state of the art in distributed autonomous mobile robotics. Pp. 3–12 in Distributed Autonomous Robotic Systems 4. L.E. Parker, G. Bekey, and J. Barhen, eds. New York, N.Y.: Springer-Verlag.

Patel, S.A., and A.K. Kamrani. 1996. Intelligent maintenance system for robots. Pp. 74–79 in Proceedings of the Fifth Industrial Engineering Research Conference. R.G. Askin, B. Bidanda, and S. Jagdale, eds. Norcross, Ga.: Institute of Industrial Engineers.

Perraju, T.S., S.P. Rana, and S.P. Sarkar. 1996. Specifying fault tolerance in mission critical systems. Pp. 24–31 in Proceedings of the IEEE High-Assurance Systems Engineering Workshop. Los Alamitos, Calif.: Institute of Electrical and Electronics Engineers, Inc. Computer Society Press.

Pomerleau, D.A. 1992. Progress in neural network-based vision for autonomous robot driving. Pp. 391–396 in Proceedings of the Intelligent Vehicles '92 Symposium. Pittsburgh, Pa.: Carnegie Mellon University.

Portinale, L., and P. Torasso. 1999. Diagnosis as a variable assignment problem: A case study in space robot fault diagnosis. Pp. 1087–1095 in Proceedings of the Sixteenth International Joint Conference on Artificial Intelligence. T. Dean, ed. San Francisco, Calif.: Morgan Kaufmann Publishers.

Rae, G.J.S., and S.E. Dunn. 1994. On-line damage detection for autonomous underwater vehicles. Pp. 383–392 in Proceedings of the 1994 Symposium on Autonomous Underwater Vehicle Technology. New York, N.Y.: Institute of Electrical and Electronics Engineers, Inc.

Rago, C., R. Prasanth, R.K. Mehra, and R. Fortenbaugh. 1998. Failure detection and identification and fault tolerant control using the IMM-KF with applications to the Eagle-Eye UAV. Pp. 4208–4213 in Proceedings of the 37th IEEE Conference on Decision and Control. New York, N.Y.: Institute of Electrical and Electronics Engineers, Inc.

Rajamani, R., and S.E. Shladover. 2001. An experimental comparative study of autonomous and cooperative vehicle-follower control systems. Journal of Transportation Research Part C 9(1): 15–31.

Rasmussen, R.D. 2001. Goal-based fault tolerance for space systems using the mission data system. Pp. 2401–2410 in 2001 IEEE Aerospace Conference Proceedings. New York, N.Y.: Institute of Electrical and Electronics Engineers, Inc.

Roumeliotis, S.I., and G.A. Bekey. 2000. Distributed multi-robot localization. Pp. 179–188 in Distributed Autonomous Robotic Systems 4. L.E. Parker, G. Bekey, and J. Barhen, eds. New York, N.Y.: Springer-Verlag.

Roumeliotis, S.I., G.S. Sukhatme, and G.A. Bekey. 1998. Sensor fault detection and identification in a mobile robot. Pp. 1383–1388 in Proceedings of the 1998 IEEE/RSJ International Conference on Intelligent Robots and Systems—Innovations in Theory, Practice and Applications. New York, N.Y.: Institute of Electrical and Electronics Engineers, Inc.

Schalkoff, R.J. 1997. Artificial Neural Networks. New York, N.Y.: McGraw-Hill.

Schneider, F.E., D. Wildermuth, and H.-L. Wolf. 2000. Motion coordination in formations of multiple robots using a potential field approach. Pp. 305–314 in Distributed Autonomous Robotic Systems 4. L.E. Parker, G. Bekey, and J. Barhen, eds. New York, N.Y.: Springer-Verlag.

Schreiner, K. 2001. Hyperion project follows sun. IEEE Intelligent Systems 16(5): 4–6.

Schulze, M. 1997. CHAUFFEUR—The European way towards an automated highway system. Paper No. 2311 in Proceedings of Fourth World Congress on Intelligent Transport Systems. Berkeley, Calif.: ITS America at the University of California, Berkeley.

SciAm (Scientific American Newsletters). 2001a. High customer demand for Cadillac night vision. Inside ITS, Scientific American Newsletters 11(15): 8.

SciAm. 2001b. DaimlerChrysler demos laser headlights in "Active Night Vision" system. Inside ITS, Scientific American Newsletters 11(15): 8–9.

Sheldon, F.T., H. Mei, and S.M. Yang. 1993. Reliability prediction of distributed embedded fault-tolerant systems. Pp. 92–102 in Proceedings of the Fourth International Symposium on Software Reliability Engineering. Los Alamitos, Calif.: Institute of Electrical and Electronics Engineers, Inc. Computer Society Press.

Shephard's. 2001. Unmanned Vehicles Handbook 2001. Bucks, England: The Shephard Press.

Shoemaker, C. 2001. ARL Robotics. Briefing by Charles Shoemaker, Robotics Program Office, Army Research Laboratory to the Committee on

Army Unmanned Ground Vehicle Technology, National Research Council, Washington, D.C., October 26.

SIA (Semiconductor Industry Association). 2001. ITRS Roadmap. Available online at <http://public.itrs.net> [August 16, 2002].

Simonin, O., A. Liegeois, and P. Rongier. 2000. An architecture for reactive cooperation of mobile distributed robots. Pp. 35–44 in Distributed Autonomous Robotic Systems 4. L.E. Parker, G. Bekey, and J. Barhen, eds. New York, N.Y.: Springer-Verlag.

SNL (Sandia National Laboratories). 2001. The Intelligent Systems and Robotics Center. Available online at <http://www.sandia.gov/isrc/Roboticvehicles.html> [August 19, 2002].

Soika, M. 1997. A sensor failure detection framework for autonomous mobile robots. Pp. 1735–1740 in IROS '97, Proceedings of the 1997 IEEE/RSJ International Conference on Intelligent Robot and Systems: Innovative Robotics for Real-World Applications. New York, N.Y.: Institute of Electrical and Electronics Engineers, Inc.

Steinvorth, R. 1991. Model reference adaptive control of robots. Springfield, Va.: National Technical Information Service.

Stuck, E.R. 1995. Detecting and diagnosing navigational mistakes. Pp. 41–46 in Proceedings of IROS '95, the 1995 IEEE/RSJ International Conference on Intelligent Robots and Systems: Human Robot Interaction and Cooperative Robots. Los Alamitos, Calif.: Institute of Electrical and Electronics Engineers, Inc. Computer Society Press.

Thrun, S., D. Fox, and W. Burgard. 1998. Probabilistic mapping of an environment by a mobile robot. Pp. 1546–1551 in Proceedings of the 1998 IEEE International Conference on Robotics and Automation. New York, N.Y.: Institute of Electrical and Electronics Engineers, Inc.

Toscano, M. 2001a. Joint Robotics Program Overview. Briefing by Michael Toscano, Coordinator, Joint Robotics Program Office, to the DoD Robotics Workshop, Zimmerman Associates Inc., Rosslyn, Va., February 8.

Toscano, M. 2001b. Department of Defense Joint Robotics Program. Pp. 313–322 in Unmanned Ground Vehicle Technology III, Proceedings of SPIE Volume 4364. G.R. Gerhart and C.M. Shoemaker, eds. Bellingham, Wash.: The International Society for Optical Engineering.

Toscano, M., 2001c. Program Review of the Joint Robotics Program to the National Research Council Committee on Army Unmanned Ground Vehicle Technology. Briefing by Michael Toscano, Coordinator, Joint Robotics Program Office, to the Committee on Army Unmanned Ground Vehicle Technology, National Research Council, Washington, D.C., August 29.

Toth, J.A. 2002. Briefing on Human Behavior Representations. Briefing by Jozsef A. Toth, IDA, to the Information Systems Technology (IST) Technology Area Review and Assessment (TARA), Fort Monmouth, N.J., April 16.

TRADOC (U.S. Army Training and Doctrine Command). 1997. Future Operational Capability, TRADOC Pamphlet 525-66, May 1. Fort Monroe, Va.: U.S. Army Training and Doctrine Command.

TRADOC. 2001a. Statement of Required Capabilities, Future Combat System of Systems (FCS), November 2. Fort Monroe, Va.: U.S. Army Training and Doctrine Command.

TRADOC. 2001b. TRADOC Mission-Needs Statement. Fort Monroe, Va.: U.S. Army Training and Doctrine Command.

Tunstel, E., A. Howard, and H. Seraji. 2001. Fuzzy rule-based reasoning for rover safety and survivability. Pp. 1413–1420 in Proceedings of the 2001 IEEE International Conference on Robotics and Automation. New York, N.Y.: Institute of Electrical and Electronics Engineers, Inc.

Uchibe, E., M. Asada, and K. Hosoda. 1998. Cooperative behavior acquisition in multi mobile robots environment by reinforcement learning based on state vector estimation. Pp. 1558–1563 in Proceedings of the 1998 IEEE International Conference on Robotics and Automation. New York, N.Y.: Institute of Electrical and Electronics Engineers, Inc.

UMICH (University of Michigan). 2001. Mobile Robotics Lab. Available online at <http://www.engin.umich.edu/research/mrl/index.html> [August 19, 2002].

U.S. Army. 1998. High Mobility Robotic Platform Study, Vol. 1, Contract DAAE07-98-C-L024, CDRL A002. Warren, Mich.: U.S. Army Tank-Automotive Research and Development Engineering Center.

U.S. Army. 2001. Information provided to the Committee on Army Unmanned Ground Vehicle Technology during a Demo III site visit to Ft. Indiantown Gap, Penn., November 14.

U.S. Army. 2002. Advanced Robotics Simulation, STO IV.SN.2002.04. Orlando, Fla.: U.S. Army Simulation, Training, and Instrumentation Command.

VCDS (Vice-Chief of the Defence Staff Group—Canada). 2002. VCDS Joint Experimentation Discussion Paper. Available online at <http://www.vcds.forces.ca/dgsp/dda/je/sec1_e.asp> [June 11, 2002].

Visinsky, M.L., J.R. Cavallaro, and I.D. Walker. 1995. A dynamic fault tolerance framework for remote robots. IEEE Transactions on Robotics and Automation 11(4): 477–490.

Vos, D.W., and B. Motazed. 1999. The application of fault tolerance controls to UAVs. Pp. 69–75 in Navigation and Control Technologies for Unmanned Systems, Proceedings of SPIE Volume 2738. S.A. Speigle, ed. Bellingham, Wash.: The International Society for Optical Engineering.

Wang, J.S., and C.S.G. Lee. 2001. Efficient neuro-fuzzy control systems for autonomous underwater vehicle control. Pp. 2986–2991 in Proceedings of the 2001 IEEE International Conference on Robotics and Automation. New York, N.Y.: Institute of Electrical and Electronics Engineers, Inc.

Washington, R. 2000. On-board real-time state and fault identification for rovers. Available online at <http://ic.arc.nasa.gov/projects/ai-rovers/papers/washington-icra2000.pdf> [August 15, 2002].

Werblin, F., A. Jacobs, and J. Teeters. 1996. The computational eye. IEEE Spectrum 33(5): 30–37.

Williamson, T.A. 1998. A High-Performance Stereo Vision System for Obstacle Detection, Ph.D. Thesis. Pittsburgh, Pa.: Carnegie Mellon University.

Winfield, A.F.T. 2000. Distributed sensing and data collection via broken ad hoc wireless connected networks of mobile robots. Pp. 273–282 in Distributed Autonomous Robotic Systems 4. L.E. Parker, G. Bekey, and J. Barhen, eds. New York, N.Y.: Springer-Verlag.

Yamaguchi, H., and T. Arai. 1994. Distributed and autonomous control method for generating shape of multiple mobile robot group. Pp. 800–807 in Proceedings of the IEEE International Conference on Intelligent Robots and Systems. New York, N.Y.: Institute of Electrical and Electronics Engineers, Inc.

Yamaguchi, H., and J.W. Burdick. 1998. Asymptotic stabilization of multiple nonholonomic mobile robots forming group formations. Pp. 3573–3580 in Proceedings of the 1998 IEEE International Conference on Robotics and Automation. New York, N.Y.: Institute of Electrical and Electronics Engineers, Inc.

Yan, T., J. Ota, A. Nakamura, T. Arai, and N. Kuwahara. 2000. Concept design of remote fault diagnosis system for autonomous mobile robots. Pp. 931–936 in Proceedings of the IEEE/RSJ International Conference on Intelligent Robots and Systems. New York, N.Y., Institute of Electrical and Electronics Engineers, Inc.

Yoshida, E., T. Arai, J. Ota, and T. Miki. 1994. Effect of grouping in local communication system of multiple mobile robots. Pp. 808–815 in Proceedings of the 1994 IEEE/RSJ/GI International Conference on Intelligent Robots and Systems (IROS '94). New York, N.Y.: Institute of Electrical and Electronics Engineers, Inc.

Zadeh, L.A. 1965. Fuzzy sets. Information and Control 8(3): 338–353.

Zadeh, L.A. 1968a. Probability measures of fuzzy events. Journal of Mathematical Analysis and Applications 23(2): 421–427.

Zadeh, L.A. 1968b. Fuzzy algorithm. Information and Control 12(2): 94–102.

Zadeh, L.A. 1971. Toward a theory of fuzzy systems. Pp. 469–490 in Aspects of Network and System Theory. R.E. Kalman and N. DeClaris, eds. New York, N.Y.: Holt, Rinehart & Winston.

Appendixes

A

Committee Member Biographical Sketches

MILLARD F. ROSE, *chair*, is vice-president for research at Radiance Technologies, Inc. He is a former director of the Science Directorate at the National Aeronautics and Space Administration Marshall Space Flight Center, and also a former professor of electrical engineering and director of the Space Power Institute at Auburn University. He is the author of well over 100 technical publications dealing with high-power electromechanics, energy conversion, and environmental effects. Dr. Rose received his B.S. degree in physics from the University of Virginia and his M.S. and Ph.D. in solid-state science from Pennsylvania State University. He has served on the National Research Council's (NRC) Committee on Electric Power for the Dismounted Soldier and is a current member of the National Research Council's Board on Army Science and Technology.

RAJ AGGARWAL is vice-president, Advanced Technology Center, at Rockwell Collins. He is a former director of research and technology for Alliant Techsystems, Inc., and a director of advanced programs for Honeywell, Inc. Dr. Aggarwal received a B.S. degree in physics (with honors) and B.S. and M.S. degrees in electrical and communications engineering from Delhi University in Delhi, India. He received his Ph.D. in electrical engineering from Purdue University.

DAVID E. ASPNES is Distinguished Professor of Physics at North Carolina State University. He has been a department head for Bell Communications Research and a member of the technical staff at Bell Laboratories. He is noted for creative instrumentation, which is widely used in the manufacture of microelectronic devices, and optical control systems. Dr. Aspnes received his B.S. and M.S. degrees in electrical engineering from the University of Wisconsin, Madison, and his Ph.D. in physics from the University of Illinois, Urbana/Champaign.

JOHN T. FEDDEMA is Distinguished Member of Technical Staff at the Sandia National Laboratories. He served as principal investigator for DARPA research projects, including analysis and control software for distributed robotic systems, microassembly, miniature cooperative robotic systems, and microrobots. He is editor of the *Journal of Micromechatronics*. Dr. Feddema received his B.S. degree from Iowa State University and his M.S. and Ph.D. degrees in electrical engineering from Purdue University.

J. WILLIAM GOODWINE, JR. is assistant professor in the Aerospace and Mechanical Engineering Department at the University of Notre Dame. He received his B.S. degree in mechanical engineering from Notre Dame, a J.D. degree from Harvard University, and M.S. and Ph.D. degrees in applied mechanics from the California Institute of Technology.

CLINTON W. KELLY III is senior vice-president at Science Applications International Corporation. He is a former director of the U.S. Strategic Computing Program and former executive director of the DARPA Office for Information Science and Technology. He initiated and directed the DARPA Autonomous Land Vehicle program and currently reviews research on perception for autonomous mobility, planning, and robotic behaviors. Dr. Kelly received his B.S. degree from Duke University and his M.S. and Ph.D. degrees from the University of Michigan, all in electrical engineering.

LARRY LEHOWICZ is vice-president at Quantum Research International. He retired from the U.S. Army as a major general and commander of the U.S. Army Operational Test and Evaluation Command, an organization dedicated to ensuring that warfighting systems, information management systems, and other military equipment are prepared for combat use. Gen. Lehowicz served as Deputy Chief of Staff for Combat Development at the Army Training and Doctrine Command, and he was Assistant Division Commander of the Tenth Mountain Division. He is a graduate of the U.S. Army War College and has an M.B.A. from Syracuse University. He served previously as a member of the National Research Council's Committee on Alternative Technologies for Anti-Personnel Landmines.

ALAN J. McLAUGHLIN is a consultant in the strategic planning and advanced technology fields. He retired as assistant director of the MIT Lincoln Laboratory, where he was responsible for programs in radar and image signal processing, computer networks, and machine intelligence technology. Currently, he is special assistant to the director, MIT Lincoln Laboratory, and a visiting scientist at the Carnegie Mellon University Software Engineering Institute. Mr. McLaughlin received his B.S. and M.S. from Northeastern University, and he saw military service as a lieutenant in the Army Signal Corps. He has served on the Defense Science and Air Force Scientific Advisory Board studies, and he is a past member of the National Research Council's Committees on Future Technologies for Army Multimedia Communications, Modernization of Air Force Computerized Administrative Support System, and Modernization of the Worldwide Military Command and Control System.

ROBIN R. MURPHY is associate professor of computer science and engineering and associate professor of cognitive and neural science at the University of South Florida. She is a consultant to the Institute for Defense Analyses on future unmanned systems and a consultant to the Air Force Scientific Advisory Board. Dr. Murphy received a B.M.E. degree in mechanical engineering and M.S. and Ph.D. degrees in computer science from the Georgia Institute of Technology. She has authored several books on artificial intelligence and robotics.

MALCOLM R. O'NEILL is vice-president and chief technical officer of the Lockheed Martin Corporation. During a distinguished 34-year career in the U.S. Army, Lt. Gen. O'Neill served as director of the DOD Ballistic Missile Defense Organization, director of the Army Acquisition Corps, and commander of the Army Laboratory Command. Additionally, he has managed R&D programs for the Army, DARPA, and NATO. Dr. O'Neill received his Ph.D. in physics from Rice University.

ERNEST N. PETRICK is a consultant on military ground vehicles and propulsion systems. He is a former technical director for the U.S. Army Tank-Automotive Command and retired chief scientist of General Dynamics Land Systems. He received his B.S. degree from Carnegie Institute of Technology and his M.S. and Ph.D. degrees in mechanical engineering from Purdue University. He has served on the Army Science Board and was a member of the National Research Council's Committee to Perform a Technology Assessment Focused on Logistics Support Requirements for Future Army Combat Systems.

AZRIEL ROSENFELD is professor emeritus at the University of Maryland. He is a former director of the Center for Automation Research at the University of Maryland. Dr. Rosenfeld received his Ph.D. in mathematics from Columbia University. He has published over 30 books and over 600 chapters or journal articles. He is an expert in computer image analysis.

ALBERT A. SCIARRETTA is president of CNS Technologies, Inc., consultants on research and development, modeling and simulation, management, and support of advanced information technologies and systems to the Defense Modeling and Simulation Office and other DOD agencies. He is a former manager of advanced information technologies at Quantum Research International, Inc., and a program area manager for the MITRE Corporation. Mr. Sciarretta has a B.S. degree from the U.S. Military Academy and M.S. degrees in mechanical engineering and operations research from Stanford University. He previously served as a member of the National Research Council's Committee on Review of the Department of Defense Air and Space Systems Science and Technology Program and on the National Research Council staff.

STEVEN E. SHLADOVER is deputy director of the California Partners for Advanced Transit and Highways (PATH) program, Institute of Transportation Studies, University of California, Berkeley. He has published extensively on intelligent vehicle and highway systems, advanced vehicle control systems, and vehicle system dynamics. Dr. Shladover received his S.B., S.M., and Sc.D. degrees in mechanical engineering from the Massachusetts Institute of Technology.

B

Meetings and Activities

This appendix lists presentations provided to the committee at meetings and fact-finding sessions conducted during the course of this study.

First Committee Meeting, August 28–29, 2001
National Academy of Sciences, Washington, D.C.

Statement of Task and Army S&T in Robotics
A. Michael Andrews

Army UGV Technology Roadmap
Patrick Eicker
Sandia National Laboratories

Future Combat Systems (FCS) Program
Jim Walbert
Director, Science and Technology
Office of the Project Manager—Objective Force

Lessons Learned from UAV Systems Development
Colleen Devlin
Army Evaluation Center

DoD Robotics Program Overview
Michael Toscano
Office of the Undersecretary of Defense

Second Committee Meeting, September 18–19, 2001
U.S. Army Tank-Automotive & Armaments Command, Warren, Mich.

Welcome/Vetronics Overview
Richard McClelland
U.S. Army Tank-Automotive RD&E Center

FCS Program Overview
Dave Busse
Office of the Program Manager, Future Combat Systems

Robotic Vehicle/Platform Concepts for FCS
Roger Halle
U.S. Army Tank-Automotive RD&E Center

Automotive Research Consortium/Intelligent Vehicle Initiatives
Dave Gorsich
Automotive Research Center, TARDEC Robotics Laboratory

Robotic Follower ATD
Jeffrey Jaczkowski
U.S. Army Tank-Automotive RD&E Center

Vetronics Reference Architecture
Rakesh Patel
U.S. Army Tank-Automotive RD&E Center

Physics Based Simulation of Durability, Mobility and Operations on the Move (planned STO)
Mark Brudnak
TACOM Research Development and Engineering Center

National Automotive Center 8 × 8 Truck
Jeff Ernat
National Automotive Center (TARDEC)

Hybrid Electric Propulsion
Carl Johnson
National Automotive Center (TARDEC)

Intelligent Mobility
Grant Gerhart
U.S. Army Tank-Automotive RD&E Center

Crew Integration & Automation Test Bed ATD
Andrew Orlando
TACOM Research Development and Engineering Center

Foreign Robotics Initiatives
Harold Hutchinson
U.S. Army Tank-Automotive RD&E Center

Obstacle Marking & Vehicle Guidance STO
Andrew Culkin
TACOM Research Development and Engineering Center

Third Committee Meeting, October 25–26, 2001
Wyndham City Center, Washington, D.C.

ASB Study Results
Joe Braddock
Army Science Board

ARL Robotics Developments
Chuck Shoemaker
Army Research Lab

Robotics Developments
Jim Albus
NIST

DARPA UGV and Robotics Overview
Scott Fish
Defense Advanced Research Projects Agency

Organic UAV Program
Sam Wilson
Defense Advanced Research Projects Agency

Fourth Committee Meeting, January 23, 2002
Beckman Center, Irvine, Calif.

Machine Perception
Larry Matthies
Jet Propulsion Laboratory

Mobility, Maneuvering & Tasking
Brian Wilcox
Paul Schenker
Jet Propulsion Laboratory

Cooperative Behaviors
Paul Schenker
Brian Wilcox
Jet Propulsion Laboratory

Cognitive Processes
Homayoun Seraji
Jet Propulsion Laboratory

Interfaces & Mission Operations
Paul Backes
Jet Propulsion Laboratory

Communications
Tom Jedrey
Jet Propulsion Laboratory

Power System Technologies
Rao Surampudi
Jet Propulsion Laboratory

Low Light & Thermal Imaging Technologies
Sarath Gunapala
Jet Propulsion Laboratory

Survivability Design/Fault Protection Technologies
Abdullah Aljabri
Jet Propulsion Laboratory

Test & Evaluation, System Integration
Brian Wilcox
Jet Propulsion Laboratory

Fifth Committee Meeting, April 4–5, 2002
National Academy of Sciences, Washington, D.C.

No presentations were made during this meeting.

Site Visits

U.S. Army Aviation and Missile Command, Huntsville, Ala., November 5, 2001
Ft. Indiantown Gap, Indiantown Gap, Pa., November 14, 2001
CECOM Night Vision and Sensor Lab, Fort Belvoir, Va., January 14, 2002
Jet Propulsion Laboratory, Pasadena, Calif., January 22, 2002
U.S. Air Force Research Laboratory, Tyndall Air Force Base, Fla., March 15, 2002
Naval Research Laboratory, Washington, D.C., March 27, 2002

C

Autonomous Mobility

This appendix provides details on the progress toward achieving autonomous A-to-B mobility through advances in the enabling technology areas of perception, navigation, planning, behaviors, and learning. Except for exclusively teleoperated applications, Army unmanned ground vehicles (UGVs) must be able to move from point A to point B with minimal or no intervention by a human operator. For the foreseeable future, however, soldiers will be needed to control UGVs, even on the battlefield, and the issue will be the number of soldiers required to support UGV operations. The more autonomous the vehicle, the lower the demands on the operator and the higher the degree to which UGVs effectively augment ground forces.

A UGV must be able to use data from on-board sensors, to plan and follow a path[1] through its environment, detecting and avoiding obstacles as required. Perception is a process by which data from sensors are used to develop a representation of the world around the UGV, a world model, sufficient for taking those actions necessary for the UGV to achieve its goals. "Perception is finding out, or coming to know, what the world is like through sensing perception extracts from the sensory input, the information necessary for an intelligent system to understand its situation in the environment so as to act appropriately and respond effectively—to unexpected events in the world" (Albus and Meystel, 2001). The goal of perception is to relate features in the sensor data to those features of the real world that are sufficient, both for the moment-to-moment control of the vehicle and for planning and replanning. Perception by machine[2] is an immensely difficult task in general, and machine perception to meet the needs of a UGV for autonomous mobility is particularly so.

TECHNICAL CHALLENGES

The actions required by a UGV to carry out an A-to-B traverse take place in a perceptually complex environment. It can be assumed that Future Combat Systems (FCS) UGVs will be required to operate in any weather (rain, fog, snow) during the day or night, potentially in the presence of dust or battlefield obscurants and in conjunction with friendly forces likely opposed by an enemy force. The UGV must be able to avoid positive obstacles, such as rocks or trees, and negative obstacles, such as ditches. It must avoid deep mud or swampy regions, where it could be immobilized and must traverse slopes in a stable manner so that it will not turn over. The move from A to B can take place in different terrains and vegetation backgrounds (e.g., desert with rocks and cactus, woodland with varying canopy densities, scrub grassland, on a paved road with sharply defined edges, in an urban area) with different kinds and sizes of obstacles to avoid (rocks in the open, fallen trees masked by grass, collapsed masonry in a street) and in the presence of other features that have tactical significance (e.g., clumps of grass or bushes, tree lines, or ridge crests that could provide cover). Each of these environments imposes its own set of demands on the perception

---

[1]Path planning occurs at two levels: the first is a coarse global plan, A-to-B, produced prior to vehicle movement and based on such map and other data (e.g., overhead imagery, data from the networked environment) as are available. The second is perception based, is developed moment to moment as the vehicle is moving, and consists of a series of local trajectories computed from data provided by the onboard sensors that incrementally refine the global plan. Global replanning may be required subsequent to vehicle movement and will heavily depend on perception.

[2]The phrase "machine perception" or "machine vision" is intended to convey the linkage between perception and action. Machine perception is a subset of the larger image-understanding field, which also includes applications where the real-time linkage to action is absent. Perception as generally used in robotics usually but not exclusively refers to image-forming sensors rather than to all the senses as is found, for example, in the psychological literature. However, there are many examples of robots that use tactile and proprioceptive sensors and a few that use taste and smell.

system, modified additionally by such factors as level of illumination, visibility, and surrounding activity. To do the A-to-B traverse, the robotic vehicle requires perception for the moment-to-moment control of the vehicle and for planning a local trajectory consistent with the global path, detecting, locating, measuring, and classifying any objects[3] that may be on the planned global path so the robot can move to avoid or stop.[4] In addition to obstacles it must detect such features as a road edge, if the path is along a road, or features indicating a more easily traversed or otherwise preferred local trajectory if it is operating off-road.

The perception system must also be able to detect, classify, and locate a variety of natural and manmade features to confirm or refine the UGV's internal estimate of its location (recognize landmarks); to validate assumptions made by the global path planner prior to initiation of the traverse (e.g., whether a region through which the planned path lies is traversable); and to gather information essential for path replanning (e.g., identify potential mobility corridors) and for use in tactical behaviors[5] (e.g., upon reaching B, find and move to a suitable site for an observation post, or move to cover).

Specific perception system objectives for road following, following a planned path cross-country, and obstacle avoidance are derived from the required vehicle speed and the characteristics of the assumed operating environment (e.g., obstacle density, visibility, illumination [day/night], weather [affects visibility and illumination but may also alter feature appearance]). How fast the UGV may need to go for tactical reasons will establish performance targets for road following and cross-country mobility. The principal consideration in road following is the ability to detect and track such features as road edges, which define the road, at the required speed and to detect obstacles at that speed in time to stop or avoid. For the cross-country case, perception system performance will be largely determined by the size of obstacles the vehicle must avoid as a function of speed and the distance ahead those obstacles must be detected in order to stop or turn.

---

[3]Functions: *Detect*—is there a potential feature of interest present or noise? *Locate*—where is it? If it is far from the path, probably there is no need to consider it further for purpose of obstacle avoidance, *Measure*—how large is it? Can the vehicle pass over it or must it be avoided? How far must the vehicle deviate from the planned path to avoid it? *Classify*—what is it? Is it a potential obstacle—a barrier to mobility—or is it obstacle-like based on geometry alone but potentially traversable (e.g., a bush, not a rock)?

[4]An interesting case arises when two obstacles are detected and it is not clear if the vehicle can pass between them. The planner may then choose to neither stop nor avoid, but to move to a different vantage point and reassess.

[5]The tactical behaviors are assumed to also encompass the positioning of the UGV as required by the onboard mission packages (e.g., RSTA, obscurant generation, mine clearance, weapons). The mission packages may also have organic sensors and processing that will not be considered here.

Obstacle detection is complicated by the diversity of the environments in which the obstacles are embedded and by the variety of obstacles themselves. An obstacle is any feature that is a barrier to mobility and could be an isolated object, a slope that could cause a vehicle to roll over, or deep mud. The classification of a feature as an obstacle is therefore dependent both on the mobility characteristics of the vehicle and its path. Obstacle detection has primarily been based on geometric criteria that often fail to differentiate between traversable and intraversable objects or features. This failure can lead to seemingly curious behavior when, for example, a vehicle in an open field with scattered clumps of grass adopts an erratic path as it avoids each clump. The use of more sophisticated criteria to classify objects (for example, by material type) is a relatively recent development and still the subject of research. Table C-1 suggests the scope of obstacles, environments, and other perceptual challenges.

STATE OF THE ART

The state of the art is based primarily on recent research carried out as part of the Army Demo III project, 1998–2002 (e.g., Bornstein et al., 2001); the Defense Advanced Research Projects Agency (DARPA) PerceptOR (Perception Off-Road) project (Fish, 2001) and other research supported by DARPA; the U.S. Department of Transportation, Intelligent Transportation Systems Program (e.g., Masaki, 1998); and through initiatives in Europe, mostly in Germany (e.g., Franke et al., 1998). The foundation for much of the current research was provided by the DARPA Autonomous Land Vehicle (ALV) project, 1984–1989 (Olin and Tseng, 1991) and the DARPA/Army/Office of the Secretary of Defense (OSD) Demo II project, 1992–1996 (Firschein and Strat, 1997). The discussion to follow is divided into three parts: road following, off-road mobility, and sensors, algorithms, and computation.

On-Road Mobility

Army mission profiles show that a significant percentage of movement (70 to 85 percent) is planned for primary or secondary roads. Future robotic systems will presumably have similar mission profiles with significant on-road components. Driving is a complex behavior incorporating many skills. The essential but not sufficient driving skills for on-road mobility are road following or lane tracking and obstacle avoidance (other vehicles and static objects).

On-road mobility has been demonstrated in three environments: (1) on the open road (highways and freeways), (2) in urban "stop and go" setting with substantial structure and (3) following dirt roads, jeep tracks, paths, and trails in less structured environments from rural to undeveloped terrain. In the first two cases there is likely substantial a priori information available, but less in less structured environments. In all on-road environments, the perception system

TABLE C-1 Sample Environments and Challenges

On-Road	Off-Road	Urban
Environments		
Road paved, striped, clear delineations of lanes and edges.	Flat, open terrain, thick, short grass, no trees or rocks, some gullies across planned path, swampy in places.	Low-density construction, two- and three-story buildings, tree-lined, paved streets, rectilinear street patterns, no on-street parking, low two-way traffic density.
Dirt, clear delineation of edges, occasional deep potholes, high crown in places.	Rolling terrain, patches of tall grass, some groves of trees, fallen trees and rocks partially obscured by grass.	High-density construction, two- and three-story mud-brick construction, wandering dirt streets, collapsed buildings, rubble piles partially blocking some streets, abandoned vehicles, refugees in streets.
Jeep track, discontinuous in places, defined by texture and context.	Mountainous, steep slopes partially forested, with huge rocks.	
Challenges		
Broken, faded, or absent lines.	Detect obstacles: • Negative obstacles or partially occluded • Masked or partially occluded obstacles (e.g., rocks, stumps, hidden in grass) • Continuous obstacles: water, swamp, steep slopes, heavy mud • Thin obstacles: posts, poles, wire, fences • Overhanging branches.	Pedestrians, refugees, civilians.
Abrupt changes in curvature.	Differentiate between obstacles and obstacle-like features.	Detect openings in walls, floors, and ceilings.
Low contrast (e.g., brown dirt road embedded in a dried grass background).	Operations in dense obstacle fields (e.g., closely spaced rocks).	Detect furniture, blockades, and materials used as obstacles.
Discontinuities in edges caused by snow, dust, or changes in surface.	Identify tactical features; mobility corridors, tree lines, ridge crests, overhangs providing cover and concealment.	Determine clearance between closely spaced walls and piles of debris.
Glare from water on road.		Avoid low overland wires.
Oncoming traffic.		Avoid telephone poles.
Complex intersections.		Avoid sign poles.
Curbs.		Avoid vehicles.
Read road signs and traffic signals.		

must at a minimum detect and track a lane to provide an input for lateral or lane-steering control (road following); detect and track other vehicles either in the lane or oncoming, to control speed or lateral position; and detect static obstacles in time to stop or avoid them.[6] In the urban environment, in particular, a vehicle must also navigate intersections, detect pedestrians, and detect and recognize traffic signals and signage.

Structured Roads

Substantial research has been carried out using perception to detect and track lanes on structured, open roads (i.e., highways and freeways with known geometries, such as widths and radii of curvature), prominent lane markings, and well-delineated edges (for examples see Bertozzi and Broggi (1997); Masaki (1998); Sato and Cipolla (1996); Pomerleau and Jochem (1996)). Most of the approaches used have been model driven. Knowledge of the road's geometry and other properties is used with features detected by the perception system (e.g., line segments) to define the lane and determine the vehicle's position within it. Sensors used for lane detection and tracking include stereo and monocular color video cameras and forward looking infrared radar (FLIR) for operation at night or under conditions limiting visibility. A representative capability is described in Pomerleau and Jochem (1996). It was called RALPH (rapidly adapting lateral position handler) and used a single video camera. RALPH was independent of particular features as long as the features ran parallel to the road. It could use lane markings, patterns of oil drops, road wear patterns, or road edges. The features did not need to be at any particular position relative to the road and did not need distinct boundaries. A set of features was used to construct a template. Comparisons of current conditions with the template established the vehicle's lateral position and generated steering commands. This system was used in the "No Hands Across America" experiment in 1995,

---

[6]These behaviors are necessary but not sufficient for "driving" behavior, which requires many more skills.

when RALPH drove a commercial van 2,796 miles out of 2,850 miles at speeds up to 60 mph.[7] It worked well at night, at sunset, during rainstorms, and on roads that were poorly marked or with no visible lane markings but with features such as oil drops on the road or pavement wear that could be used to locate the lane. The most challenging situation was when the road was partially obscured by other vehicles. In some of those cases RALPH was able to lock on the vehicle ahead and follow it. When the following vehicle was close to the vehicle ahead of it the prominent vertical edges of that vehicle dominated the scene and RALPH treated it as a lane. RALPH could self-adapt to changing situations by looking far ahead of the vehicle (70 to 100 meters) and using the appearance of the road at that distance to construct a new template. RALPH made assumptions about road curvature between foreground and background to project what the new template should look like when the vehicle was centered in its lane. Comparison between the current image, the current template, and the look-ahead template allowed RALPH to decide if the situation had changed enough to warrant switching to the look-ahead template. RALPH was integrated with obstacle avoidance behavior[8] as part of a demonstration under the U.S. Department of Transportation's Intelligent Transportation Systems program.

Urban Environments

Some of the preceding approaches for lane detection and tracking would work in urban "stop and go" environments; some would not. Parked cars or traffic in the urban environment may intermittently occlude many of the cues used to locate the lane in an open-road environment. Operation in the urban environment is a complex problem; only limited research has been done thus far, most by Franke and his colleagues (Franke et al., 1998). The procedure for urban lane detection and tracking used by Franke et al. was data or feature driven. A geometrical model cannot be easily developed because of the complexity of road topology. A given scene may be an unpredictable, complex combination of curbs, stop lines, pedestrian crossings, and other markings. Franke et al. (1998) first extracted edges and sorted them using specialized feature filters according to such attributes as length, orientation, parallelism, and colinearity. Combinations were created using a rule set and provided to a polynomial classifier trained on example feature sets. The classifier categorized the features as curbs, road markings, or clutter. Vehicles and other objects detected thru stereovision were excluded from consideration as road structure.

Part of road following, particularly in an urban environment, is the detection and navigation of road junctions and intersections. This has not received much emphasis. Early work was done by Crisman (1990) and Kluge and Thorpe (1993). More recently Jochem and Pomerleau (1997) described an approach that used selective image transformations to adapt an existing lane features detector to a wide variety of intersection geometries. They reported successfully detecting each intersection branch in 33 of 35 Y and T intersections. In no case did they report a branch that was not present. This is probably state of the art. Their approach was also notable in its use of active camera control (active vision)[9] to pan the camera and track the detected branch so the vehicle could drive onto it.

As the technology for following structured roads has matured, it has begun to attract serious commercial interest. Carnegie Mellon University (CMU) and AssistWare Technology (Jochem, 2001) have jointly developed the Safe TRAC vision-based lane tracking system under U.S. Department of Transportation (USDOT) funding. A derivative of the RALPH system, it uses a single video camera to measure the vehicle's position in the lane and provides an alarm if the vehicle weaves or drifts. Intended to provide driver warning, it could, like RALPH, be used to control the vehicle. It has undergone 500,000 miles of on-road testing; operating effectively on over 97 percent of all combinations of highways and driving conditions encountered (day, night, rain, snow, various levels of marking quality) with a false alarm rate of one per eight hours of driving. See Jochem (2001) for details.

An optically guided bus system[10] is scheduled to go into service in Las Vegas in 2003. The argument for its use is precision in lane keeping, allowing buses to use a lane that is typically five feet narrower than buses that rely on human drivers. The system, called CIVIS (Eisenberg, 2001), is produced in France by a joint venture of Renault and Fiat. It is already in use in two French cities.

Unstructured Roads

Essentially no work has been done on the related problem of detecting roads embedded in a cross-country environment. This is important when a vehicle is navigating primarily cross-country but where part of the planned path is on a road segment, probably unstructured, that passes through the

---

[7]The operator was responsible for speed control, lane changes, and avoiding other vehicles. RALPH was responsible for maintaining the vehicle in its lane.

[8]ALVINN, a neural-network based road-following predecessor of RALPH, also developed at Carnegie Mellon University, was integrated with obstacle avoidance behavior (stereo-based obstacle detection) for Demo II (see Appendix D.)

[9]Active vision refers to the dynamic control of data sources, field of view (e.g., sensor position and focal length), and processes. It allows sensors and processing to be optimized moment to moment as the environment or requirements change.

[10]The driver controls vehicle speed.

terrain. A related gap exists in the ability to seamlessly switch between cross-country and road-following behaviors. These behaviors have for the most part been developed independently. To switch requires manual intervention by the operator.

Unstructured roads pose a challenge because many of the assumptions behind the approaches described above for structural roads may be invalid: The appearance of the road is likely to be highly variable, making tuning of sensors and algorithms difficult. There are generally no markings, although there may be linear features so a RALPH-like approach might work in some situations. Edges may not be distinct and will likely be discontinuous (e.g., portions of the road or track may be obscured by vegetation or the road may be washed out in places). Lane size and curvature may vary irregularly, as may slope. This suggests that a data-driven (versus model-driven) approach will likely be preferred. The roads may be rough and heavily rutted requiring the vehicle to slow. High crowns may become obstacles and must be measured. Mud is almost guaranteed to be a problem and must be detected.

Because of this variability, the approaches all contain some means for learning from example. Chaturvedi et al. (2001) used a Bayesian classifier to segment roads from background in color imagery. They worked exclusively in a tropical environment with rich color content. The roads were red mud with ill-defined and irregular edges of green vegetation. Variations in light were severe (harsh sun to deep shadows) and visibility was also affected intermittently by rain. Both of these conditions caused the edges to disappear at times. The lack of well-defined edge features motivated the use of color segmentation. They worked in the HIS (hue, intensity, saturation) color space because of the relative invariance of hue to shadows. They were able to successfully segment jungle roads at about 5 Hz under a variety of lighting and weather conditions. Although this specific approach is limited by the constraint that the road be red in color, it does suggest that color segmentation more broadly could be useful in the detection and following of unstructured roads. Because colors change under different illumination, the broad applicability of color segmentation will require finding either color properties that are relatively invariant in shadows or for highlighted surfaces or a means to recover an estimate of the color of the illumination from the scene.

RALPH was a purely reactive system. ROBIN had a deliberative component. Rosenblum (2000) described an improved version of the ROBIN neural-network-based system used in Demo II. This version was used for unstructured road following in the early part of the Demo III program and in other unrelated experiments. Unlike ALVINN, which used a three-layer, feed-forward neural network, ROBIN used a radial basis function (RBF) neural network. An advantage of RBFs is that they smoothly fill gaps in training examples and can be trained very rapidly. A second way that ROBIN differed from ALVINN (or other strictly neural-network solutions) was in the inclusion of a deliberative reasoning module. This monitored performance of the road-following module and could act to improve performance. For example, it could slow the vehicle to obtain multiple looks in an ambiguous situation, change the virtual camera view, or change the parameters used in image preprocessing. ROBIN was able to drive on secondary roads at 25 mph and on ill-defined roads and trails at 10 mph, during daytime, using pseudo black and white video derived from a color camera. Using a FLIR, ROBIN drove secondary roads at night at 15 mph and the ill-defined roads and trails at 10 mph. Rasmussen (2002) described a system that showed the potential of fused laser detection and ranging (LADAR) and color video data in road following. The data were co-registered in space and time. Height and smoothness (height variation in the local vicinity of a point) features were derived from the LADAR. A color histogram was calculated for each color image patch, as was texture. The assumptions were that roads should be locally smooth, be consistent in a mix of brown or gray colors, and exhibit more homogeneous texture than bordering vegetation. The feature data was fused in a three-layer neural-network classifier. The results showed the road was clearly segmented despite shadowing and changes in composition. Training individual neural networks by road type improved performance over a single network. Using data from both sensors produced substantially better performance than any single sensor. This work was done off-line. The approach is currently too computationally demanding for real-time application.

There has been much less research on following unstructured roads than on highways and freeways; systems are not as robust and problems are less well understood. Challenges include roads with sharp curves where the system may lose the road, steep slopes where the slope may be incorrectly classified as an obstacle, judging water depth if the road includes a stream crossing or standing water, and following roads that are defined by texture and context rather than color, changes in contrast, or three-dimensional geometry. Performance in rain is likely to be highly variable, depending on specifics of the road.

On-Road Obstacle Detection

On-road detection includes static obstacles and detecting and tracking other vehicles. Williamson (1998) focused on static obstacles, used a stereo-based approach and could reliably detect objects 14 cm or taller out to distances of 110 meters using narrow field of view, long focal length lenses. He also demonstrated obstacle detection at night using the vehicle's headlights. One obstacle was painted white and was 14 cm tall. It was detected at a range of about 100 meters using the high beams. A similar size black obstacle was detected at 60 meters. Williamson used a three-camera system to reduce the likelihood of false matches. Williamson worked in a structured road environment. There has been little comparable research to detect obstacles on unstructured roads, where for example, abrupt changes in slope may cause false positives with some algorithms. There has been no work specifically to detect on-road negative obstacles. Off-road work is applicable.

Franke et al. (1998) and others worked on vehicle detection and tracking. Franke et al. used a very efficient stereo algorithm that could work in real time; Dellaert and Thorpe (1998) used a two-dimensional approach that also worked in real time. Betke et al. (1996) used edge images to find distant cars. Their approach first did a coarse search to find regions that might contain cars and then did a fine-grained search and match on those regions. Beymer and Malik (1996) used a feature-based technique with such features as a portion of a bumper or prominent corners. They assumed that features that moved together should be grouped together and used Kalman filtering to track the feature groups. Giachetti et al. (1995) used optical flow for detecting and tracking vehicles. This does not work well without good texture and when there is large motion in the image sequence. They developed some multiscale and multiple window algorithms to address these problems. All the above were successful in detecting vehicles both in lane and as oncoming traffic.

Many of the techniques for on-road obstacle detection used video cameras as the sensor. This was driven in part by the desire to put inexpensive systems into private and commercial vehicles. They can be used at night with external illumination but do not work well in fog, smoke, or other situations where visibility is limited. Extensive work was done using FLIR and LADAR for off-road obstacle detection (to be described later) that was equally applicable to the detection of obstacles on road. They provided improved performance at night and under limited visibility but are expensive, and LADAR is range limited. Increasingly it was recognized that no one-sensor type no matter how clever the processing could do everything. Multiple sensor modalities would be required and their results combined to achieve robust obstacle detection under all weather conditions. Langer (1997) developed a system that combined data from a 77-GHz radar with data from a video camera. The radar was used to detect and locate other vehicles. Video provided to the RALPH road-following system sensed road geometry and was used to maintain lateral position. Road geometry information from RALPH was used for clutter rejection and to reduce the number of false positives from the radar. With the addition of radar data RALPH could also autonomously control speed, maintaining a safe driving distance from preceding vehicles. The system was able to track multiple vehicles successfully, both in-lane and in the opposing lane in a cluttered urban environment. Cars could be reliably detected at distances up to 180 meters and trucks up to 200 meters. Langer also detected people at 50 meters. Collision avoidance systems are beginning to find commercial applications (Jones, 2001). Based on radar (77 GHz), LADAR, or stereo from video cameras, these are part of the next generation of adaptive cruise control (ACC) systems, which will maintain a safe distance to the car ahead, braking or accelerating up to the speed preset by the driver.[11] Systems are being sold today by Toyota, Nissan, Jaguar, Mercedes-Benz, and Lexus. GM, Ford, and others plan ACC offerings this year or next. Fujitsu Ten Ltd., in Plymouth Michigan, is developing an ACC for stop-and-go driving. It fuses data from millimeter-wave radar and 640 × 480 stereo video cameras. This takes advantage of the ability of the radar to look far down the road and to provide a wide field of view for tracking cars in turns and using stereo to improve clutter rejection and reduce false alarms caused by stationary objects. So far, no organization has announced that they are developing a commercial system that combines adaptive cruise control and lane tracking.

Leader-Follower Operations

If vehicles can be detected and tracked for collision avoidance, they also can be followed. Of note was the autonomous leader-follower capability demonstrated by Franke et al. (1998). Lead car speed was variable from a stop up to 12 m/s (43 km/h) and was accurately tracked by an autonomous follower vehicle while maintaining a safety distance of 10 meters. More recent perception-based leader-follower work (Bishop, 2000) was intended to enable close-headway convoying of trucks. This project, called CHAUFFEUR, used a pattern of lights on the preceding truck. The distortion of the pattern provided heading correction and the size of the pattern yielded distance. Leader-follower operation was demonstrated in Demo II (three vehicles that also demonstrated formation keeping) and Demo III (two vehicles). In both, the approach was GPS based and not perception based (i.e., the follower vehicle did not make use of perception to track the leader vehicle).

The detection of pedestrians remains a very difficult problem, particularly in cluttered scenes containing many people. Various approaches have been used; Franke et al. (1998) used shape templates and characteristic walking patterns. For detection of walking they used both color clustering on monocular images in a combined color and feature position space and three-dimensional segmentation. Papageorgiou et al. (1998) used a trainable system. Features were encoded at different scales. They used these features to train a support vector machine (SVM) classifier. Without using motion they achieved an 80 percent detection rate with about 10^{-5} false positive rate. The detection rate approached 100 percent for a false positive rate of 10^{-3}. These results were obtained in cluttered urban scenes containing multiple pedestrians at varying distances.

More recent projects include that of Zhao and Thorpe (2000) at Carnegie Mellon University, which used stereo-

---

[11]The driver also steers.

FIGURE C-1 Pedestrian detection. Courtesy of Chuck Thorpe, Carnegie Melon University Robotics Institute, and Liang Zhao, University of Maryland.

vision and a neural-network classifier (see Figure C-1); Broggi et al. (2000) at the University of Pavia, which combined stereovision with template matching for head and shoulder shapes; and Gavrila (2000) at Daimler-Chrysler, which also used stereovision and a time-delay neural network to search across successive frames for temporal pattern characteristics of human gaits. Gavrila (2001) estimates the state of the art at 90 percent to 95 percent detection rate with a false positive rate between 10^{-3} and 10^{-1}.

Similar to vehicle detection and tracking, if a person can be detected and tracked for avoidance in an urban environment then a vehicle could also follow a person in open terrain.

The detection of signage and traffic signals is important in an urban environment. Signage consists of isolated traffic signs on poles and directional or warning symbols painted on the road surface. Franke et al. (1998) used a combination of color segmentation algorithms and gray-scale segmentation (to address situations where illumination or other factors affect color). Segmentation produced a region of interest that served as an input to a radial basis function classifier for signage and a three-layer neural network for traffic light recognition. For signs on roads and on poles they achieved recognition rates of 90 percent with 4 to 6 percent false positives. Recognition rates for traffic lights in a scene were above 90 percent with false positive rates less than 2 percent. Priese et al. (1995) also developed a system to locate and recognize traffic signs. It detected and recognized arrows on the road surface, speed-limit signs, and ideograms. For ideogram classification it used a neural-network-classifier to recognize 37 types of ideograms. They used some image transformation but assumed the signs were essentially viewed directly ahead. Peng and Bhanu (1999) used an adaptive approach to image segmentation in which 14 parameters in the Phoenix color-based segmentation algorithm were adapted to different conditions using reinforcement learning.[12] They were able to achieve about a 70 percent detection rate on stop signs under varying conditions where the sign was prominent in the image (i.e., centered and large). The rate dropped to about 50 percent in more difficult conditions when the sign was smaller and the surrounding clutter greater; note that without adaptation, the rate was about 4 percent. Peng and Bhanu (1999) showed how the performance of a well-understood general-purpose color segmentation algorithm could be improved by using learning to adapt it to changing conditions.[13] In contrast, Franke et al. (1998) developed special purpose classifiers tailored to the sign detection problem.[14] Meyers et al. (1999) considered

---

[12]Learning approaches used in perception fall into two broad categories: supervised learning, or learning by example and reinforcement learning. The neural-network based ALVINN algorithm described in Appendix D is an example of supervised learning. It was trained on examples of typical roads by making a classification guess to which a trainer would respond with the correct result. In reinforcement learning, the system is not given the correct answer but instead is given an evaluation score.

[13]Most image-processing algorithms (e.g., image segmentation, feature extraction, template matching) operate open-loop with fixed parameters. The loop is typically closed by manually tuning the algorithms for a particular operating environment. When a different environment is encountered the use of the initial parameters may lead to degraded performance requiring manual retuning. Instances of this occurred throughout the ALV, Demo II, and Demo III programs. The key contribution of Peng and Bhanu (1999) was to automatically and continuously close the loop using re-enforcement learning. This approach is a way to achieve more robust performance than that provided by a manually tuned system.

[14]Performance is a function of the specific segmentation algorithm chosen. Franke et al. (1998) used a different algorithm than that of Peng and Bhanu and so a direct comparison of the results cannot be made. This points out the general issue of many algorithms for a particular problem but few comparisons among algorithms under controlled conditions.

the problem of reading the characters on a sign viewed from an oblique perspective. They used a transform to rectify and deshear the image using parameters computed from the image itself. They achieved nearly 100-percent recognition accuracy up to azimuth angles of about 50 percent.

Summary

On-road mobility at a minimum requires perception for lane detection to provide lateral control of the vehicle (road following), perception for collision avoidance (i.e., detection and position and velocity estimation for vehicles in lane to maintain a safe distance through adaptive speed control), and perception for the detection of static obstacles.

Perception for lane detection and tracking for structured roads is at the product stage. About 500,000 miles of lane detection and tracking operation has been demonstrated on highways and freeways. Lanes can be tracked at human levels of driving speed performance (e.g., 65 mph) or better under a range of visibility conditions (day, night, rain) and for a variety of structured roads, but none of the systems can match the performance of an alert human driver using context and experience in addition to perception. Lane tracking may function in an advisory capacity providing warning to the driver that the vehicle is drifting out of the lane (Jochem, 2001) or it may be used to directly control steering (Eisenberg, 2001). There are other approaches that have not been as extensively tested; most (e.g., ALVINN, RALPH, ROBIN) have been used to control steering but only in research settings.

Detection and tracking in an urban environment are very difficult. Many of the perceptual clues used to navigate open roads may be available only intermittently because of traffic or parked cars, but these, in turn, can also serve to help define the road. Road following, intersection detection, and traffic avoidance cannot be done in any realistic situation. Signs and traffic signals can be segmented and read only if they conform to rigidly defined specifications and if they occupy a sufficiently large portion of the image. Pedestrian detection remains a problem. A high probability of detection (e.g., 98 percent) is accompanied by a high rate of false positives. This can be addressed by using multiple cues from different sensor modalities. Much research remains to be done.

Although the research has shown the ability of automated vehicles to follow structured roads with performance that appears similar to that of human drivers, there are many situations in which performance is not at the level of a human driver (e.g., complex interchanges, construction zones, driving on a snow-covered road [nearly impossible], driving into the sun at low sun angles, driving in precipitation [heavy rain, snow, or fog], and dust). The systems are almost exclusively sensor driven and are very limited in their ability to use all the context and experience available to a human driver to augment or interpret perceptual cues.

Road following assumes that the vehicle is on the road. A special case is detecting a road, particularly in a cross-country traverse, where part of the planned path may include a road segment. Work done on detecting intersections or forks in paved roads is applicable, but very little research has specifically addressed the detection of dirt roads or trails in open terrain. The level of performance on this task is essentially unknown.

A number of means, both active and passive, have been demonstrated for detecting and tracking other vehicles for collision avoidance, but only in research vehicles. Some have been used to control vehicle speed (e.g., Langer, 1997; Franke et al., 1998). Others have demonstrated the capacity to make the position and velocity estimates necessary to control vehicle speed but were not integrated into the control system. There have been limited demonstrations (e.g., Langer, 1997; Franke et al., 1998) that integrate both lane detection and tracking with collision avoidance for vehicle control. Avoidance of the moving targets represented by animals and pedestrians is another extremely challenging problem that has barely been touched by the research community.

The potential exists (stereovideo or stereo FLIR) to detect static, positive obstacles (e.g., 15 cm) on the road in time to avoid them or stop while traveling at high speed (e.g., 120 km/h with 120 meters look ahead). A very narrow field of view is required, the approach is computationally demanding, and the sensors must be actively controlled. Radar has not been shown to detect small objects reliably (much less than car size) at these distances, and LADAR does not have the range or instantaneous field of view (IFOV).

Obstacle detection and avoidance behavior was integrated with lane-tracking behavior for vehicle control in the ALV program (see Appendix D) for Demo II (ALVINN and color stereo) and for a Department of Transportation demonstration (RALPH and color stereo); however, these demonstrations were staged under conditions much less demanding than real-world operations.

The existing technology is extremely poor at reliably detecting road obstacles smaller than vehicles in time to stop or avoid them. This is an inherently very difficult problem that no existing sensor or perception technology can address adequately. For example, an object 30 cm^3 in size could cause serious problems if struck by a vehicle. A vehicle traveling at highway speed (30 m/s) would need to detect this object at greater than 100 meters to respond in time.

The reported research on reading road markings and road signs represents an extremely primitive capability at this time. It depends on those markings and signs being very carefully placed and designed, and none of the systems can deal with imperfect conditions on either. Even under good conditions, the error rate remains significant for these functions.

A variety of sensors can be used in various combinations for on-road mobility, depending on specific requirements. These include 77-GHz radar for long-range obstacle and vehicle detection and for use under low-visibility conditions; stereo color video or FLIR for lane following, vehicle and obstacle detection, and longer-range pedestrian detection; and LADAR or light-stripers for rapid, close-in collision avoidance, curb detection, and pedestrian detection.

Considerable effort is being invested within the automotive industry and related transportation organizations to develop systems that will enhance driving safety, assist drivers in controlling their vehicles, and eventually automate driving. These activities, which are international in scope, offer the potential for technology spin-offs that could eventually benefit the Army's UGVs by lowering costs and accelerating the availability of components and subsystems. These include sensors, actuators, and software.

Off-Road Mobility

Autonomous off-road navigation requires that the vehicle characterize the terrain as necessary to plan a safe path through it and detect and identify features that are required by tactical behaviors. Characterization of the terrain includes describing three-dimensional terrain geometry, terrain cover, and detecting and classifying features that may be obstacles, including rough or muddy terrain, steep slopes and standing water as well as such features as rocks, trees, and ditches.

Terrain characterization has been variously demonstrated beginning with the ALV program but always in known environments and generally in daytime, under good weather conditions. Performance has continued to improve up to the present but measurement of performance in unknown environments[15] and under a range of environmental conditions is still lacking. Most recent work was also done in daylight, during good weather. The DARPA PerceptOR program is addressing performance measurement in unknown terrain, all weather, day, and night.

Most of the research on perception for terrain characterization was in support of the Demo III and PerceptOR programs. The vehicles for Demo III were XUVs (experimental unmanned vehicles). These weighed about 3,400 pounds, had full-time, four-wheel drive and mobility characteristics essentially equivalent to a high-mobility multi-purpose wheeled vehicle (HMMWV) (Figure C-2 shows the XUV and PerceptOR[16]). Experiments were carried out both on XUVs and HMMWVs. The sensors used on the Demo III XUVs were stereo, color video cameras (640×480), stereo FLIR cameras (3-5, 320×256, cooled, 2-msec integration time), and a LADAR (180×32 at 20 Hz, 50-meter maximum range, 20-meter best performance, 7-cm-range resolution, and 9-mrad angular resolution [$22°$ elevation, $90°$ azimuth]). Foliage penetration (1.5 GHz) and obstacle avoidance (77 GHz) radars were planned but have not yet been integrated. Stereo depth maps, including processing for limited terrain classification, were produced at 4 Hz.

Descriptions of how data from the XUV sensors and from the perception system were used to control the vehicle are given in Coombs et al. (2000) and Murphy et al. (2000). They reported cross-country speeds of up to 35 km/h in benign terrain: "rolling grass-covered meadows where the only obstacles were large trees and shrubs." The conditions were daylight and in good weather. The vehicle they used was an HMMWV. They used a LADAR (128×64 pixels) operating at 1 Hz, detected large obstacles out to 40–50 meters, and concluded that this update rate, plus processing and planning latencies of about one second, limited the speed. Note that the XUV LADAR operates at 20 Hz. The vehicle control software (obstacle detection, cartographer, planner, reactive controller) was ported to the XUV.

Shoemaker and Bornstein (2000) reported that the Demo III Alpha experiments at Aberdeen, Md. (September 1999) used stereo obstacle detection at six or less frames per second. Using only geometric criteria, clumps of high grass were classified as obstacles and avoided. Bornstein et al. (2001) noted that the vehicle did not meet the 10-mph off-road goal, was not able to reliably detect negative obstacles, and had only limited capability in darkness. In October 2000, the Demo III Bravo experiments were held at Ft. Knox, Ky. LADAR was integrated into the vehicle for obstacle detection. Stereo obstacle detection performance was improved for both positive and negative obstacles. The vehicle still had difficulty with tall grass, in this case confusing the tops of the grass with the ground plane and causing the vehicle to avoid open, clear terrain and confusing it with a drop-off. The range limitation of the LADAR led to cul-de-sac situations where, for example, it could not see breaks in tree lines (Bornstein et al., 2001).

In demonstrations at Ft. Indiantown Gap, Pa., Murphy et al. (2002) reported that the XUVs were able to navigate

> over difficult terrain including dirt roads, trails, tall grass, weeds, brush and woods. The XUVs were able to detect and avoid both positive obstacles (such as rocks, trees, and walls) and negative obstacles (such as ditches and gullies). The vehicles were able to negotiate tall grass and push through brush and small trees. The Demo III vehicles have repeatedly navigated kilometers of off-road terrain successfully with only high-level mission commands provided by an operator. . . . The vehicles were often commanded at a maximum velocity of 20 km/h and the vehicles would automatically reduce their speed as the terrain warranted.

Murphy et al. (2002) noted that a major limitation was the limited range and resolution of the sensors, particularly the

---

[15]"Unknown" means that the operators have not previously seen or walked the terrain. No tuning of algorithms or selection of paths has been done based on extensive a priori knowledge.

[16]These vehicles were equipped differently by different teams. A typical vehicle might have stereo video, stereo FLIR, multiple LADARs, and foliage penetrating RADAR.

FIGURE C-2 Demo III vehicle and PerceptOR vehicle. Rows A courtesy of Jon Bornstein, U.S. Army Research Laboratory; Row B courtesy of John Spofford, SAIC.

LADAR, which could not reliably image the ground more than 20 meters ahead and had an angular resolution of about 9 mrad or about 0.5 degree.[17] (By comparison, the human eye has a foveal resolution of about 0.3 mrad.) Murphy et al. suggested that the LADAR was the primary sensor used for obstacle detection. That is, without the LADAR the demonstrations would not have succeeded or performance would have been reduced substantially. Why stereo did not feature more prominently was not discussed.

Members of the committee observed the Demo III XUVs at Ft. Indiantown Gap in November 2001. The demonstrations confirmed reliance on LADAR and the fact that the XUVs pushed through brush with prior knowledge that no obstacles were concealed in the brush. The committee observed that the vehicles on occasion confused steep slopes that were within its performance range with intraversable terrain requiring operator intervention (see Figure C-3). The committee also confirmed the Murphy et al. observation about limited field of view after observing one of the vehicles get trapped in a cul-de-sac. On other occasions the committee noted that the vehicles would stop, or stop and backup, with operator intervention required to reestablish autonomous operation. The problem was dust affecting the LADAR performance. A dust cloud looked like a wall or a dense field of obstacles through which the planner could not find a safe path. Possible solutions include additional sensors, active vision, and algorithms that use last-pulse processing. Reliance on a single sensor is risky.

In related research the PRIMUS German research project reported cross-country speeds of 10 km/h to 25 km/h

FIGURE C-3 Perception of traversable slope as an object. Courtesy of Clint Kelley, SAIC.

---

[17] A pixel size of about 18 cm × 18 cm at 20 meters. Assuming 5 pixels vertical for obstacle detection, this limits positive obstacle size to no less than 90 cm, or about 35 in. An implication is that the vehicle could not detect an obstacle that could damage it in sufficient time to stop at 20 km/h. That is, this speed would be very risky in unknown terrain.

FIGURE C-4 Color-based terrain classification. Courtesy of Larry Matthies, Jet Propulsion Laboratory.

in open terrain (Schwartz, 2000). The project used a small tracked vehicle with good cross-country mobility. Obstacles were less of a problem than with a comparably sized wheeled vehicle. Obstacle detection was done with a Dornier 4-Hz LADAR (129 × 64, 60° × 30°) on a stabilized pan and tilt mount. A monocular color camera was used for contour or edge following. Durrant-Whyte (2001) reported cross-country speeds of 30 km/h over 20-km traverses also using LADAR.

Driving through dense brush, even knowing there are no hidden obstacles, is difficult. The perception system must assess the density of the surrounding brush to determine if the vehicle can push through or must detour. The Demo III system counted the number of range points in a LADAR voxel to estimate vegetation density. If the count was less than a threshold number, the vegetation was assumed to be penetrable. The assumption in Demo III was that the range points were generated by returns from vegetation; the system did not do classification. A fast statistical approach for analyzing LADAR data was described in Macedo et al. (2001) to classify terrain cover and to detect obstacles partially occluded by grass. They found statistically significant differences in the measures used between grass, and rocks partially occluded by grass. Castaño et al. (2002) described a classification approach using texture analysis of color video and FLIR images. The data were collected by the XUV operating at Ft. Knox. They classified a scene into the categories of soil, trees, bushes, grass, and sky. They obtained texture measurements from a bank of 12 spatial filters at different scales and orientation. The measurements were combined in both maximum likelihood and histogram-based classifiers. Classification accuracy (percent correct) during the day with the color data was between 74 percent and 99 percent, depending on the category, and at night with the FLIR data it was about the same: between 77 percent and 99 percent. Figure C-4 shows the process and typical results. The results were obtained off-line because of computational requirements. Bellutta et al. (2000) described stereo-based obstacle detection and color-based terrain cover classification using color and FLIR cameras. Potential obstacles were detected using geometric analysis and then a Bayesian classifier was used to assign a material class to the object. Depth maps were produced at 6 Hz on a 320 × 240 image. A rule-based system could then be used to combine information on geometry and material type to assess traversability. A unique aspect of their approach was the use of active vision to point the narrow field-of-view stereo cameras. The cameras were pointed at the path the vehicle was currently commanded to follow with look-ahead distance determined by the vehicle's speed. They demonstrated the system on an HMMWV. The classification results matched ground truth. The active vision software has not yet been ported to the Demo III XUV.

As part of the PerceptOR program, Matthies (2002) experimented with RGB (red, green, blue), near-infrared, and multiband FLIR imagery for terrain classification. He concluded that near-infrared (nonthermal, two-bands 0.65 μm and 0.80 μm) was more reliable and less affected by variable illumination than RGB, that two to three bands of thermal infrared (IR) in the range from 2 μm to 12 μm showed promise for discriminating vegetation from other material at night, and that texture analysis of thermal IR was very promising for discriminating vegetation from soil. This was the first work on possible means to characterize terrain cover at night.

In separate related research, Bhanu et al. (1997) used 12 spectral bands over the range from 0.44 μm to 1.4 μm to classify terrain. They used a hierarchical classification scheme with which regions in the scene were first labeled: road, field, forest, sky, and unknown. Field regions were then further subdivided into grass, scrub brushes, snowberries, and soil. They used a variety of means for classification, including texture and other feature extraction approaches at the lower level and knowledge based at the higher level for fusion and feature interpretation. They demonstrated the potential for multispectral terrain classification but only on a limited data set. The work was done off-line. Multispectral ground cover classification has long been a standard tool for remote sensing using overhead imaging. This potential should be exploited for ground vehicles.

Hong et al. (2000) and Manduchi (2002) both used data fusion for terrain classification. Hong et al. combined data from a LADAR and a single color camera. They described ways to detect standing water (puddles), signs, and roads and showed how fusion could improve performance over a single sensor. They fused the data in the world model using various heuristics, which also supported the fusion of information over time. Manduchi used a Bayesian classifier to fuse a set of local features statistically. This work also suggested improved performance from fusion. The use of texture as a feature precluded real-time operation.

The evaluation and comparison of terrain classification approaches in particular are qualitative and imprecise because of a lack of adequate ground truth and common datasets.

Substantial work was done on obstacle detection. Matthies et al. (1998) described a systematic effort to evaluate the performance of stereovision with camouflage, concealment, and detection (CCD) cameras and FLIR cameras against putative Demo III requirements. They did a series of obstacle detection experiments with stereo data collected at different times of day and night, and with different size obstacles. They concluded from an analysis of the data that an object that subtended 10 vertical pixels[18] could be reliably detected.[19] Given this value, they calculated the fields of view for a sensor to detect obstacles at several look-ahead distances, each corresponding to a specific cross-country speed (e.g., Demo III at 35 km/h). They included assumptions about vehicle dynamics and processing latencies. The calculations showed that there was an inherent conflict between the narrow field of view required to place 10 pixels on an object for reliable detection and the total field of regard necessary to see all terrain into which the vehicle could steer. They concluded active vision was required.

Owens and Matthies (1999) compared cooled and uncooled FLIRs and image intensifiers for stereo-based detection of obstacles at night. They concluded that only cooled FLIRs would meet requirements. The integration time for uncooled FLIRs, about 15 m/sec, was too long for the required speeds (causing excessive motion blur that washed out features) and the image intensifiers did not have an adequate signal-to-noise ratio for stereo matching. The cooled FLIRs worked very well and provided adequate contrast even at thermal crossover.

More recent and extensive research on obstacle detection is described by Matthies (2002). Sensors used included color video, LADAR, radar, and FLIR (operating in three bands, near [2–2.6 μm], medium [3–5 μm], and long wavelengths [8–12 μm]). The conclusions were that state-of-the-art stereo is 320 × 240 at 10 Hz on a Pentium III and at 30 Hz with application-specific hardware. For positive obstacles, the use of 10 pixels as a minimum obstacle height for detection is a good assumption where vegetation does not obscure the obstacles. Required sensor angular field of view can then be calculated from speed requirements. For positive obstacles in vegetation, detection depends on the size of the obstacle and the density of the vegetation. Both LADAR and radar can detect positive obstacles on the order of 45 cm × 65 cm a few meters into tall grass, depending on the density of the grass. LADAR can detect objects in brush out to about 1 meter. Much more research is required to predict performance across a range of conditions. Unobscured negative obstacles could not be reliably detected more than 10–15 meters ahead of the vehicle (sensor height about 2 meters with angular resolution of about 2.5 mrad). Limited work was done on detecting thin objects (a metal pole about 1.5 meter high and 5 cm wide). Detection with stereo was dem-

---

[18]The 10-pixel value has held up across several experiments (see below).

[19]For example: For a 30-cm obstacle at 8 meters, 128 × 128 window, probability detection 0.99, false alarm rate of 0.03 (Matthies and Grandjean, 1994).

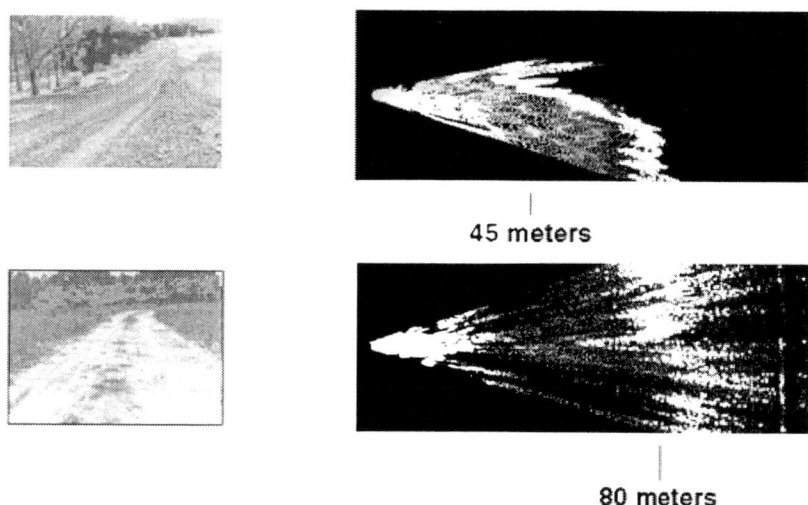

FIGURE C-5 Tree-line detection. Courtesy of Larry Matthies, Jet Propulsion Laboratory.

onstrated at 5 meters but not at 9 meters. Parameters that affect detection of various classes of thin objects were identified. Additional research is needed to be able to model performance. Some work was done on obstacle detection in the presence of obscurants (natural and smokes). The conclusion was that the performance of MWIR (3-5 μm) should be adequate.

Detection of water remained a problem. Matthies (2002) described the use of ratios of colors (green to near infrared), texture analysis of color images (water has little texture), stereo (few matches from the water relative to the surround), LADAR (specular-no return), and polarization. All work to some degree under some situations. For example, if the water reflects sky, then texture works well; if it reflects surrounding trees, then texture provides misleading results. Depending on subsequent research, data fusion may provide the best solution. Green LADAR may be able to measure water depth ahead of the vehicle. Further research is required.

Matthies (2002) and Chang et al. (1999) reported the first work on detection of tactical features. Matthies demonstrated tree-line detection with stereo color imagery at 45 and 80 meters (see Figure C-5). Chang et al. described the use of LADAR to detect overhanging trees and other overhanging features and an algorithm to classify them either as potentially providing cover (a vehicle can go beneath it) or as an obstacle (too low). In one example they were able to detect and classify trees as providing cover or not at a range of 45 meters. If the two were combined, the vehicle could follow a tree line and take cover as required. This capability must be extended well beyond 100 meters.

No quantitative standards metrics or procedures exist for assessing off-road UGV performance. It is difficult to know if progress is being made in off-road navigation and where deficiencies may exist. The assessment problem is exacerbated by the absence of statistically significant published test results reflecting different operating conditions. Emphasis has been placed on meeting demonstration objectives rather than conducting controlled tests. A goal should be data collection sufficient to develop predictive models.

Unlike road following, speed[20] as a metric to gauge progress in off-road mobility is incomplete and may be misleading. No meaningful comparisons can be made without knowing the environmental conditions, the details of the terrain, and, in particular, how much reliance was placed on prior knowledge to achieve demonstrated performance. While it is reasonable to assume that Demo III performance is better than Demo II, there is no way to know from published reports; there are too many uncontrolled variables. Many of the issues identified in the ALV and Demo II programs remain problems today, and many of the capabilities demonstrated in Demo III could be replicated with Demo II algorithms enabled by improved computation. Similarly, there is no way to know how close Demo III performance is to meeting putative FCS or other requirements, since specificity is lacking on both sides. No data exist for performance in an a priori unknown environment. The PerceptOR program is addressing this gap, but results will not be available for about 2 to 3 years.

Because of uncertainty about UGV performance, it is not possible to estimate vehicle operator workload, nor how many vehicle operators might be required for a given force. With statistically valid vehicle performance data, predictive models of UGV performance could be developed, and these issues could be addressed through simulation.

Subjective comparison of UGV cross-country performance with the performance of manned vehicles on comparable terrain suggests that cross-country capability is very

---

[20]Speed is a system-level property but is strongly correlated with perception performance.

immature and limited. Published results and informal communications do not provide evidence that UGVs can drive off-road at speeds equal to those of manned vehicles. Although UGV speeds up to 35 km/h have been reported, the higher speeds have generally been achieved in known benign terrain and under conditions that did not challenge the perception system or the planner. During the ALV and Demo II experiments in similar benign terrain, manned HMMWVs were driven up to 60 km/h. In more challenging and unknown terrain, the top speeds for all vehicles would be lower but the differential likely greater. While off-road performance is limited by sensor range and resolution, it may also be limited by the approach taken. Driving autonomously on structured roads is essentially reactive; surprises are assumed to be unlikely and speeds can be high. Driving off-road, with its inherent uncertainty, is currently treated as a deliberative process, as if surprises are likely. Higher-resolution sensor data is used to continuously produce a detailed three-dimensional reconstruction of the terrain currently limited by sensor capabilities and the way sensors are employed to no farther than 20 to 40 meters ahead of the vehicle. Trajectories are planned within this region. This is unlike the process used by human drivers who look far ahead to roughly characterize terrain. They adjust speed based on expectations derived from experience and context, local terrain properties, and by using higher-resolution foveal vision to continuously test predictions, particularly along the planned path. A UGV could use the same process. Sensors with a wide field of view but lower resolution could look farther ahead. Macro-texture and other features detected at lower-resolution could be used to continuously assess terrain properties for terrain extending out some distance from the vehicle. At the same time, the lower resolution sensors and other data (e.g., the planned path) could be used to continuously cue higher-resolution sensors to examine local regions of interest. Both the predictions and the local data would be used to reactively control speed. This is analogous to road-following systems previously described that look far ahead and judge that the scene still looks like a road and that no obstacle appears to be in the lane ahead. If the terrain ahead is similar to the terrain on which the vehicle is currently driving or can be matched to terrain descriptions in memory using a process like case-based reasoning, then assumptions can be made about likely speeds and verified using active vision. There is predictability in terrain, not as much as on a structured road, but some. The trick is to learn to exploit it to achieve higher speeds.

In principle, LADAR-based perception should be relatively indifferent to illumination and should operate essentially the same in daylight or at night. FLIR also provides good nighttime performance. LADAR does not function well in the presence of obscurants. Radar and FLIR have potential depending on the specifics of the obscurant. There has not been any UGV system-level testing in bad weather or with obscurants, although experiments have been carried out with individual sensors. Much more research and system-level testing under realistic field conditions are required to characterize performance.

The heavy, almost exclusive, dependence of Demo III on an active sensor such as LADAR may be in conflict with tactical needs. Members of the technical staff at the Army NVESD told the committee that LADAR was "like a beacon" to appropriate sensors, making the UGV very easy to detect and vulnerable (U.S. Army, 2002). Strategies to automatically manage the use of active sensors must be developed. Depending on the tactical situation, it may be appropriate to use them extensively, only intermittently, or not at all. Future demonstrations or experiments should acknowledge this vulnerability and move to a more balanced perception capability incorporating passive sensors. RGB (including near IR) provides a good daytime baseline capability for macro terrain classification: green vegetation, dry vegetation, soil and rocks, sky. Material properties can now be used with geometry to classify features as obstacles more accurately. This capability is not yet fully exploited. Two or three bands in the thermal infrared region, $2-12\ \mu$, show promise for terrain classification at night. More detailed levels of classification during the day require multiband cameras (or a standard camera with filters), use of texture and other local features, and more sophisticated classifiers. Detailed characterization of experimental sites (ground truth) is required for progress. More research is required on FLIR and other means for detailed classification at night. Simple counts of LADAR range hits provide a measure of vegetation density once vegetation has been identified. Reliable detection of water remains a problem. Different approaches have been tried with varying degrees of success. Fusion may provide more reliable and consistent results.

Positive obstacles that are not masked by vegetation or obscured for other reasons and are on relatively level ground can be reliably detected by stereo if they subtend 10 or more pixels; LADAR probably requires 5. LADAR, stereo color, and stereo FLIR all work well. Day and night performance should be essentially equivalent, but more testing is required; again, less is known about performance in bad weather or with obscurants. Sufficient data exist to develop limited models for performance prediction (e.g., Matthies and Grandjean, 1994) for some environments.

Obstacle detection performance depends on such factors as the surface properties of the obstacle, level of illumination (for stereo), and the focal length or field of view of the optical system. With a very narrow field of view (e.g., about 4°) a 12-inch obstacle was detected with stereo at about 100 meters (Williamson, 1998). A wider field of view (e.g., 40°) might reduce detection distance for the same obstacle to 20 meters. The width of the obstacle is also important. A wider but shorter object can be detected at a greater distance than an object of the same height but narrower. Although 5–10 pixels vertical is a good criterion, it is more a sufficiency than a necessity.

Little work has been explicitly done to measure the size of obstacles. This bears on the selection of a strategy by the planner. Currently the options are two: stop, and turn to avoid. Others, which are not currently used, are slow and strike or negotiate, and pass over the obstacle if its width is less than the wheel base and its height is less than undercarriage clearance. No proven approach has been demonstrated for the detection of occluded obstacles. LADAR works for short ranges in low-density grass. There have been some promising experiments with fast algorithms for vegetation removal that could extend detection range. Some experiments have been done with foliage penetration (FOLPEN) radar, but the results are inconclusive. Radar works well on some classes of thin obstacles (e.g., wire fences). LADAR can also detect wire fences. Stereo and LADAR can detect other classes of thin obstacles (e.g., thin poles or trees). Radar may not detect nonmetallic objects depending on moisture content. Much more research is required to characterize.

Detection of negative obstacles continues to be limited by geometry (Figure C-6). While performance has improved because of gains in sensor technology (10 pixels can be placed on the far edge at greater distances) sensor height establishes an upper bound on performance. With the desire to reduce vehicle height to improve survivability, the problem will become more difficult. Possible approaches include mast mounted sensors, a tethered lifting body (a virtual mast), or recourse to data from UAVs. Figure C-7 shows the state of the art using stereo video.

Little work has been done on detecting tactical features at ranges of interest. Tree lines and overhangs have been reliably detected, but only at ranges less than 100 meters. Essentially no capability exists for feature detection or situation assessment for ranges from about 100 meters out to 1,000 meters. Requirements for detection of many tactical features to support potential mission packages (e.g., roads and road intersections for RSTA regions of interest) have not been specified.

Sensors

Selection of imaging sensors for a UGV's mobility vision system should be guided by the following: (1) There is no single universal sensor; choose multiple sensor modalities so that the union of the individual sensor's performance encompasses detection of the required features under the required operating conditions; (2) select sensors with overlapping performance to provide redundancy and fault tolerance and as a means for improving signal-to-noise through sensor fusion; and (3) limit the different kinds of sensors employed to reduce problems of supportability, maintainability, and operator training. Concentrate on improving the means for

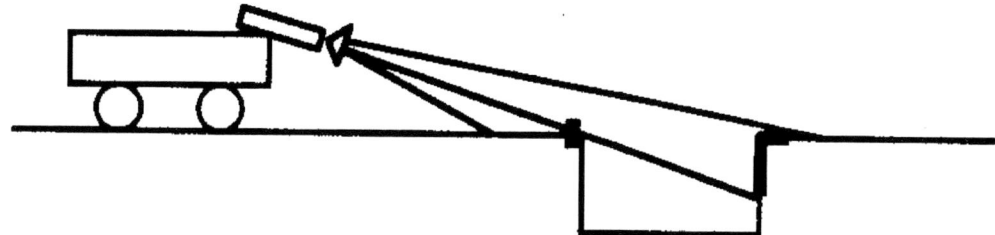

FIGURE C-6 Geometric challenge of negative obstacles. Courtesy of Clint Kelley, SAIC.

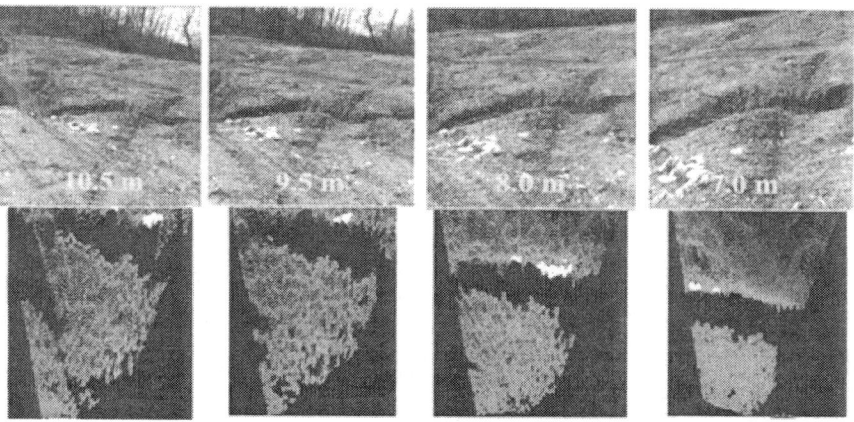

FIGURE C-7 Negative obstacle detection using stereo video. Courtesy of Larry Matthies, Jet Propulsion Laboratory.

extracting from each sensor type all the information each is capable of providing. Resist the tendency to solve perception problems by adding sensors tailored to detecting particular features under specific conditions.

The studies and experiments on sensor phenomenology supporting the ALV, Demo II, Demo III, and the PerceptOR progress, and experiments at the Jet Propulsion Laboratory (JPL) for a Mars rover provide evidence that mobility vision requirements can be met by some combination of color cameras, FLIR, LADAR, and radar. The advantages and disadvantages of each, in a UGV context, are summarized in Table C-2. Environmental sensors (temperature, relative humidity, rain, visibility, and ambient light, including color) complement the imaging sensors and allow automatic tuning of sensors and algorithms under changing conditions.

Table C-3 summarizes sensor improvements since the ALV and Demo II periods. For video, resolution, dynamic range, and low-light capability must be improved. Resolution should be on the order of 2048 × 2048 pixels with a frame rate of at least 10 frames per second. This improved resolution would, for example, more than double the effective range of stereo. A tree line could be detected at 200 to 300 meters compared with today's 80 to 90 meters, assuming a constant stereo base.

Today's cameras with automatic iris control have a dynamic range of about 500:1 shadow to bright; a goal is 10,000:1 with a capability to go to 100,000:1 for selected local regions. This would improve stereo performance and feature classification. The camera should provide a capability to operate "first light to last light." All of these improvements are within reach; no breakthroughs are required. CCD arrays 2084 × 2084 have been fabricated and can be purchased. Data buses based on IEEE 1394, the Firewire standard, support data transfer at 400 Mbps with extensions to 1 Gbps and allow data to be directly transferred to a digital signal processor without intermediate storage in a video buffer. This means that embedded software can do real-time region-of-interest control for locally increased data rates, intensity control, other preprocessing, and local operations such as 3 × 3 correlations for stereo matching. The concept is described in Lee and Blenis (1994).

Although improved resolution would also be useful for FLIR, more desirable would be an uncooled FLIR with an integration time on the order of 2 m/sec, instead of the current 15 m/sec. Uncooled FLIR with 320 × 240 pixels are available today, and 640 × 480 pixel cameras are under development. These operate at 30 fps with a 15-msec integration time[21] (CECOM, 2002). If integration time cannot be reduced, then perhaps a stabilized mount with adaptive optics could be developed, which would allow the FLIR to "stare" for the required integration period. The elimination of the expensive Stirling cooler and the corresponding decrease in cost, weight, and power make this option worth studying.

Although improved over the ALV's scanner, LADAR is still limited in range, angular resolution, and frame rate. LADAR is also affected by dust, smoke, and other obscurants that may be interpreted as obstacles. In addition, the mechanical scanner is heavy, making mounting an issue. Ideally, a LADAR would have a maximum range between 100 to 200 meters, an angular resolution no greater than 3 mrad, and at least a 10-Hz frame rate. Solutions to range, resolution, and frame rate are likely to be found by limiting the wide-angle field of regard of today's systems, the equivalent of foveal vision. Both here and with stereo, such systems require the development of algorithms that can cue the sensors to regions of interest. These are discussed below. It is important to note that most LADAR devices have not been developed with robot vision as an application. They have been designed for other markets, such as aerial surveying or mapping, and adapted for use on UGVs. Various approaches have been tried to eliminate the mechanical scanner. See Hebert (2000) for a survey. Most are in early stages of development or do not meet requirements for a UGV.

There is less to say about desired improvements to radar sensors because of limited experience with autonomous mobility, particularly off-road. The requirements for automotive applications have stimulated research, and production for these applications and for wireless communications has ensured a ready supply of commercial off-the-shelf (COTS) components applicable to UVG requirements. Three areas for improvement important to UGV applications are improved antennas to suppress sidelobes to improve resolution in azimuth and to provide beam steering; use of polarization to reduce multipath reflections and clutter; and improved signal processing to increase resolution in range and azimuth and provide better object classification. A large body of research and practice is available (developed for other applications) that could be adapted to UGV needs.

Algorithms

Improvements in UGV performance have come from more available computation and better algorithms. For example, stereo performance has improved from 64 × 60 pixels at 0.5 Hz in 1990 (Matthies, 1992) to 320 × 240 pixels at 30 Hz in 1998 (Baten et al., 1998). While many, possibly improved algorithms are reported in the literature, there is no systematic process for evaluating them and incorporating them into UGV programs. Many of the algorithms used in Demo III and PerceptOR were also used in Demo II. There is no way to know if they are best of breed.

An approach to software benchmarking (and performance optimization more generally) is to run a "two-boat campaign." In the Americas' Cup and other major competitive sailing events, the tuning of a boat may make the differ-

---

[21] For an uncooled camera, a bolometer, the term "thermal time constant" is more accurate than "integration time."

TABLE C-2 Imaging Sensor Trade-offs

	Advantages	Disadvantages
Stereo color cameras	Provides more pixels than any other sensor. Covers the visible spectrum through the near IR. Binocular stereo provides a daylight depth map out to about 100 meters; motion stereo, out to 500 meters. All points in the scene are measured simultaneously, eliminating the need to correct for vehicle motion. Depth maps can be developed at 10 Hz or faster. Cameras are less expensive ($3,000 to $15,000) than LADAR and very reliable. Red, green, blue (RGB) appropriately processed can provide simultaneous and registered feature classification.	Difficult to exploit full dynamic range. Limited to operation from about 10:00 a.m. to 4:00 p.m. Degrades in presence of obscurants. Requires contrast and texture for stereo matches. Depth calculations may be unreliable in environment cluttered with vegetation. Computationally intensive and very sensitive to calibration. May need to use more than two cameras to improve stereo matches that introduce additional computation, calibration, and mounting issues. Cameras lack region-of-interest control and range operation instructions coupled to dynamic range control. Depth measurements limited by stereo baseline. Motion stereo requires precise state variable data.
Stereo—forward looking infrared radar (FLIR)	Provides depth maps at night and with most obscurants. Multiband FLIR may provide terrain classification capability at night. During the day, FLIR can make use of thermal differences to select correspondences and augment color stereo. Provides additional wavelengths for daytime terrain classification, including detection of standing water.	Fewer pixels than RGB cameras. Very expensive ($15,000 to $125,000). Must use mechanical Stirling-cycle cooler. Reliability issues. Less expensive uncooled FLIR cannot be used because of long integration times.
LADAR	Precise depth measurements independent of external illumination and without extensive computation. Requires fewer pixels than stereo for obstacle detection. Fast means to acquire texture information. Typical scanner will operate with an instantaneous field of view (IFOV) about 3 mrads, with a 5- to 10-Hz frame rate. Numbers of range points per second equivalent to stereo. Some scanners provide coregistered RGB data in daylight.	Poor vertical resolution for negative obstacles. Degrades in the presence of obscurants. Range limited to about 40 to 60 meters. Heavy compared with cameras. Requires comparably heavy pan-and-tilt mount. Expensive, about $100,000. Trade-offs between scan rate and IFOV. The more rapid the scan rate, the larger the IFOV to maintain signal-to-noise level. This places a limit on size of obstacle that can be detected. Requires correction for vehicle motion during scan. Certain tactical situations may limit its use.
Radar	Long-range. Good in presence of obscurants. Relatively inexpensive due to automotive use. Reliable. Can provide some detection of obstacles in foliage with appropriate choice of frequencies and processing. Can detect foliage and estimates of foliage density. Limited classification of material properties can be made. Can sense fencing, signposts, guardrails, and wires. Good detection of moving objects, vehicles, and pedestrians.	Lacks resolution. High levels of false positives unless data are combined with those from other sensors. Performance very sensitive to conductivity and water content of objects. Certain tactical situations may restrict its use.

TABLE C-3 Sensor Improvements

	ALV/Demo II	Present
Video	512 × 485 @30 fps	640 × 480 @30 fps
		1280 × 960 @7.5 fps
	256:1 dynamic range	1000:1 dynamic range
	Slow data transfer: RS 170 (data to frame grabber, then to processor)	Faster data transfer: IEEE 1394-Firewire (data direct to DSP)
	Color (RGB)	Multiband
FLIR	160 × 128, 320 × 256	640 × 512
LADAR	256 × 64 @1/2 Hz	600 × 40 @1/2-5 Hz
	20-meter range, ±8 cm	150-meter range, ±25 cm
	480° × 30°, 5-8.5 mrad	330° × 80°, 3 mrad
		180 × 32 @20 Hz
		50-meter range
		90° × 20°, 9 mrad

ence between winning and losing. The problem is how to know if a tuning modification is for the better. Because sailing conditions are so variable, it is difficult to know what may have caused performance to change. One approach, for those who can afford it, is to use two identical boats. One serves as the standard against which changes to the other are measured. If the changes improve performance, they are incorporated into the benchmark boat and the process is repeated. A similar approach could be used with UGVs.

Algorithms for obstacle detection often fail to differentiate between objects or features that are traversable and those that are obstacles. Detection of obstacles or other features cannot be reliably done much beyond 80 meters. As a practical matter, most is done at no more than 40 to 50 meters.

Active vision can address this problem, but the development and integration of algorithms to cue sensors have lagged. Terrain classification is limited to about four categories if real-time performance is required, and can be done only in daylight. Most features are detected and classified by independently processing data from individual sensors; there is little data fusion. Most algorithms require some tuning to local conditions. There is little self-tuning or adaptation.

The largest gains in UGV performance from algorithms are likely to be found in five areas: active vision, data fusion, texture (from color, FLIR, and LADAR data), color segmentation or classification (from multiband video and multiband FLIR), and machine learning.

Active Vision

References to the need for "active vision"—essentially the dynamic control of field of view, data sources, and process—with respect to regions of interest (Reece and Shafer, 1995), but without limiting consideration to the visual part of the electromagnetic spectrum, have been made throughout this report. Instances of its use in the major programs described have been relatively few. The major issue is selecting regions of interest. The most obvious is to look at where the vehicle plans to go. This was essentially the strategy used by Sukthankar et al. (1993) in developing Panacea for the ALVINN. Elaborations on this are easy to invent. Complementary approaches representative of the state of the art are described in Privitera and Stark (2000) and Backer et al. (2001). Both also contain extensive references to other work. Privitera and Stark used a variety of algorithms to identify regions using such features as high local contrast, color contrast, symmetrical elements, areas with high edge density, varying texture compared with the surroundings, or other characteristics. They compared regions identified with various combinations of these features with those identified by human subjects as a criterion. They found statistically significant agreement. Backer et al. (2001) described a multilevel process for determining regions of interest.

Many of the cues described above are first used by a recurrent neural network to establish the saliency of regions in the scene. Different behaviors used the saliency values assigned to regions to satisfy more complex goals. These behaviors include searching and tracking (trying to find a prespecified target and tracking it), exploring (used when no other task is specified and all areas are of essentially equivalent importance), and detecting changes (scanning the scene for changes). The planner, using data from the world model, invoked these behaviors. The behaviors in turn used different data and algorithms depending on the task and the context, as well as environmental properties. Kelly and Stentz (1997) described how active vision could be used to increase the speed of a UGV by restricting vision processing to regions of interest. Their system controlled look-ahead distance, the size of the region of attention, and resolution.

Active vision raises a number of issues, for example, which sensors should be used together and which should be used independently; when multiple sensors should be focused on the same region of interest; and how processing algorithms should be selected and computational resources assigned. There are also mechanical issues associated with sensor mount design and stabilization. Ideas exist for all of these, but research is required. However, enough work has been done to routinely experiment with some version of active vision. It is clearly what is required to address many of the issues identified earlier in this appendix, the most important of which is the inherent conflict between increased field of view and increased resolution. It offers a means to look for features in the tactical region from 100 meters to 1,000 meters, increase vehicle speed, and improve obstacle identification and terrain classification.

Table C-4 was initially developed for the Demo III program and was subsequently refined. It summarized the judgment of robotic vision researchers about those techniques that could potentially lead to the greatest improvement in feature detection and classification. This table first shows that the use of data fusion could make the largest difference in capability and that fusing texture and spectral-based features in particular was important.

Data Fusion

A properly selected suite of sensors has complementary strengths. The way to capitalize on that complementarity is through data fusion. A fusion system consists of two parts. First, sensor or data source models translate a sensor reading or a data element into a measure of uncertainty (e.g., what is the likelihood this specific reading or value would be obtained given, or conditional upon, hypotheses describing a grid cell?). The hypotheses are tags for the represented elements and are chosen to be relevant to the selection of actions available to the vehicle. Since the vehicle will have multiple behaviors available, perhaps some executing simultaneously, the system may carry multiple sets of hypotheses reflecting the needs of different behaviors. One behavior may reflect concern only for whether a feature is an obstacle or not; another, for whether the feature provides concealment or cover (i.e., is it a tree, a ditch, or a boulder?). Additional conditioning variables must also describe the circumstances or environment surrounding the reading. Knowledge of sensor phenomenology is required to develop these models.

The second part of a fusion system is a rule for evidence accumulation; for example, a Bayesian rule (there are multiple ones depending on the simplifying assumptions made), Dempster's rule of combination, or a number of heuristic approaches, not formally grounded, but that may be useful depending on requirements and constraints (e.g., available computing resources). Multiple levels of hypotheses and multiple ways of accumulating evidence might be employed under the control of an executive who would consider the moment-to-moment importance of the requirements for information and the competing needs for computational resources and would use a satisfying approach to select the appropriate set of hypotheses, sensor model, and means of accumulating evidence at that time.

There is a large body of theoretical work on data fusion (see, for example, Rao, 2001) but few examples of its use for UGV applications. The approaches fall roughly into two categories: (1) those in which features from individual sensors are combined into a supra-feature-vector that is then classified (Rosenblum and Gothard, 1999) and (2) those in which the features from each sensor are classified first (the results may be probabilistic) and the individual classifications are then merged.

The first approach is conceptually simpler but may be computationally demanding for large feature vectors. The second approach allows the use of different classification strategies optimized for particular sensors and features. It is also computationally easier.

Two techniques are popular for merging results: those based on Bayesian probability theory and those that use the Dempster-Shafer algorithm. Manduchi (2002) used a Bayesian framework to merge texture-based and color-based terrain classifications.

Murphy (1996) describes an application based on Dempster-Shafer evidence accumulation. The advantage of both techniques is that they explicitly acknowledge uncertainty in the relationship between sensor output and the derivative classification, and uncertainty in the merged result. Such explicit treatment of uncertainty can be explicitly used by the planner. This should be preferable to acting as if the classification is known with certainty. The difficulty with these techniques is that they require detailed knowledge of sensor phenomenology. However, much of the basis for calculating the required relationship is available in the spectra libraries and predictive sensor models described earlier. It is not clear that the UGV perception community is aware of this work, which was done in support of other programs.

Sensor fusion, correctly done, can always produce results superior to those obtained with any single-sensor (Rao, 2001). The research cited above reported gains over single sensor performance. Algorithms for doing sensor fusion exist and have a sound theoretical basis. The impediment to their use appears to be the requirement for data-linking properties of the scene to sensor output. Much of the required data has been developed to support other programs and should be used by the UGV community to accelerate the application of sensor fusion.

Texture

Texture refers to the spatial frequency of a region or more generally, the abstraction of various statistical homogeneities from that region. It is scale dependent (e.g., a surface that appears rough at one scale may appear smooth at another). A good summary of current texture research is found in Mirmehdi and Petrov (2000) and in Rushing et al. (2001). Mirmehdi and Petrov use a probabilistic approach for the segmentation of color texture. Rushing et al. use a rule-based clustering approach and claim they distinguished textures that other methods could not, and that were difficult for humans. In both cases and in much of the literature, texture analysis has been applied at small scales (e.g., to differentiate between two types of fabric). In most cases the approaches are computationally demanding and cannot be used in real time. The one used by Rushing et al., for example, required 172 seconds for a 512 × 512 image on a 400-MHz processor. The technique of Mirmehdi and Petrov required 60 seconds for a 128 × 128 image. Rosenblum and Gothard (1999) argued that the texture differences that were important in off-road navigation were at larger scales and that methods designed specifically to operate at those scales could operate in real time. They presented such a method and illustrated its application. The value of texture analysis has been suggested in preceding discussions, particularly for terrain classification. Manduchi (2002) showed how it was used in classification experiments to complement color classification, succeeding in some areas where color alone was unsatisfactory. Manduchi used an approach also designed to operate at larger scales and one that could also be implemented in real time. The key to the successful use of texture for off-road navigation is the specific design of algorithms to meet off-road requirements. There is a good research base to draw on.

Spectral Segmentation

Color segmentation (typically based on functions of RGB) is coming into widespread use for off-road navigation, particularly for material classification, to augment geometry in obstacle identification and for terrain classification. The difficulty with color-based methods is that colors change under different illumination. Geusebroek et al. (2001) identified color properties that were more nearly invariant under changes in illumination, including shadows or surfaces

TABLE C-4 Impact of Feature Use on Classification

Environmental feature	Does not meet requirements	Performance meets requirements for optimal conditions	Meets requirements for FCS-level conditions	Failure mode	Benefit
Negative obstacles/ravines				Grazing Angle Equal Illumination No Surface Texture	Close Up Detection Spectral Discontinuities Textures Discontinuities Strong Shape Cues
Positive obstacles				Avoids Traversables	Obstacles by Geometry Vehicle/Obstacle Traversibility
Rough terrain				Poor Resolution	Surface Roughness Filters Rough Traversibles Texture Discontinuities Improved Traversibiltiy Detection Stability From 3D Shape
Roads and trails				Minimal Geom. Cues	Spectral Discontinuities Texture Discontinuities Improved Classification Strong Linear Shape
Traversable vs. non-traversable (rock vs. bush)				Similar Geometries Similar Shapes	Spectral Discontinuities Texture Discontinuities Improved Classification
Water/mud				Geometry-Less	Spectral Uniqueness Texture Uniqueness Improved Classification
Vehicles				Lacks Discrimination	Spectral Properties Separation From Background Improved Segmentation Strong Shape Cues
Humans				Lacks Discrimination	Spectral Properties Separation From Background Improved Segmentation Strong Shape Cues
Trees and tree lines					Spectral Uniqueness Strong Texture Cues Improved Segmentation Strong Shape Cues
Hills/ridge lines				Spectral Independent Texture Independent	Strong Geometry Cues Geographic Classification

continues

TABLE C-4 continued

Environmental feature	Does not meet requirements	Performance meets requirements for optimal conditions	Meets requirements for FCS-level conditions	Failure mode	Benefit
Structures					Spectral Segmentation Periodic Textures Strong Shape Cues

Notes: (1) We will still need sensors to detect soil traversability, see through soft vegetation, and detect standing water depth.
(2) FCS conditions include day/night, low visibility (fog, smoke, airborne precipitation), and adverse weather.

Legend

Geometry only (radar, stereo)
Geometry + classification (radars, stereo, color, polarization)
+ Texture
+ Special constancy
+ Shape/structure

SOURCE: Courtesy of Benny Gothard, SAIC.

with highlights. Finlayson et al. (2001) gave a method for recovering an estimate of the color of the scene illumination and in turn for estimating the reference color of a surface. Both of these could improve current color segmentation algorithms.

A logical extension of color-based segmentation is to add spectral bands to the traditional three of R, G, and B. Although multiband cameras provide up to five bands out to the near (nonthermal) IR available, little experimentation has thus far been carried out for real-time UGV applications. There is, however, a vast remote-sensing literature that documents the utility of multiple bands for a variety of applications (for example, see Iverson and Shen, 1997). The approaches may use multiple bands in the visible (up to 6 to 10 for multispectral, hundreds for hyperspectral) and several in the near to far thermal IR (2.0 μm to 14 μm), depending upon the requirement. Terrain classification is most often done using either unsupervised clustering or by a supervised pattern recognition approach, such as a neural network or other statistical classifier. Extensive libraries of material spectra exist, as do excellent predictive models that can be used to design systems and evaluate algorithms for specific requirements. These models include such factors as time of day, latitude, time of year, cloud cover, and visibility. While work has been done primarily for overhead imagery, and generally without a requirement for real-time processing, much could be adopted for UGV use. The use of multiple spectral bands, combined with texture, could essentially solve most of the terrain classification problems for UGVs, both in daylight and at night, although much more research is required to develop methods based exclusively on thermal IR for use at night. The predictive models could be used to accelerate progress.

Learning

Learning is used extensively in feature classifiers, for example, to classify regions as road or nonroad or for terrain classification. Classifiers typically are based on neural-network or other statistical techniques. The learning is usually supervised or is learning from example, where the classifier modifies its weights or other parameters in response to receiving the correct response from the human operator (Poggio and Sung, 1996). A less well-established application of learning is to adjust parameters in algorithms or to control sensor settings to automatically adapt to changing conditions. The traditional approach to perception has been open loop. Algorithms are hand tuned to adapt them to changes. Many parameters will by necessity be left at their default settings. The Phoenix segmentation algorithm, for example, has 14 adjustable parameters. Often the algorithms will fail if they are not tuned to current conditions; they may be very brittle (i.e., their response to change is very nonlinear). Ideally, a system would monitor its own performance and learn how to improve performance when the environment changes. Peng and Bhanu (1998, 1999) described such a system that detected traffic signs in clutter under varying outdoor conditions. Feedback was provided by the degree of confidence obtained when a model was matched to the segmented region. They assumed that models of the objects to be recognized were known, but the number of objects, their scale, and their location in the image were unknown. Reinforcement learning was used to modify the segmentation algorithm to improve the confidence score. Unlike supervised learning, reinforcement learning requires only a measure of the quality of the performance (the "goodness of fit" of the model to the region), not the correct answer. Peng and Bhanu reported about a 17-fold improvement in performance with learning (confidence levels increased from an average 0.04 to 0.71). The work was done off-line, although they claimed it could be done on the UGV in real time. An issue is the extent to which models appropriate to off-road scenes could be developed. More generally the issue is how to provide performance evaluation functions so the system can self-assess its performance.

Computation

Onboard computation has increased from about 100 MOPS for the ALV(10^8) to about 4.4 giga operations per second (GOPS) (10^9) for Demo II and is about 10^{10} ops for Demo III. Although these are rough order-of-magnitude estimates, they show that embedded computing power has increased about 10^2 to 10^3 over about 14 years. This compares with a Moore's law prediction of about 10 times every 5 years. While computing resources are allocated across all UGV functions, perception accounts for about 85 percent of the total computational load. How much might be required to meet off-road performance objectives?

Gothard et al. (1999) provide estimates of computational load for perception. They concluded that about 40 GOPS was required as a lower bound. The estimates did not include data fusion or active vision. It was also based on the assumption that all processing would be restricted to 200×200 pixel windows. Replacing this last assumption alone with something on the order of 320×240 (current stereo performance) would almost double the computational load. Making allowances for image-processing functions not included and for the increased resolution expected from next-generation sensors, an upper bound of 150 to 200 GOPS may be reasonable. Where could this be obtained? First, much image processing is parallelizable; so one way to obtain the required processing is through a parallel architecture. As an example, an Altivec™ G4 processor delivers 4 to 12 GOPS depending on the specific calculation. An array of these could potentially meet requirements. Two cautions: The code must be carefully parallelized and optimized for the particular architecture selected, and interprocessor bandwidth and sensor-to-processor buses must be designed for expected loads. A second option is to take advantage of special-purpose signal-processing boards specifically designed for image processing. These implement many lower-level image-processing functions in hardware. An example is the Acadia board from Pyramid Vision Technologies, a spin-off from the Sarnoff Corporation. Earlier versions of this board were described in Baten et al. (1998). It can, for example, do correlation-based stereo on a 320×240 window at 30 feet per second. A third option is to employ mixed digital and analog processing. An example of a hybrid MIMD (multiple-instruction, multiple-data) array processor with image processing application was described in Martin et al. (1998). The use of analog processing elements increased speed and reduced size and power; however, analog circuits are limited to about 1 percent accuracy. This is likely acceptable for many image-processing applications.

The challenge is not the availability of raw computing power. It is designing an integrated image processing architecture. The research community has focused on very narrow areas: edge detectors, region segmentation algorithms, texture analysis algorithms, color classification algorithms, stereo algorithms, and algorithms for analyzing range data. The research has also focused typically on single-sensor modalities. Much of the processing has been done off-line. There has been little research on how best to bring all of these components, sensors, algorithms, and processors together in a real-time architecture. Because much of the research is conducted independently, each organization is free to choose programming languages, operating systems, and processors. When these are brought together in a UGV, the integration may involve code running under as many as five different operating systems with variants of each running on different processors. Because of pressure to meet demonstration milestones, code and hardware are typically patched together. There is no system optimization. So, in addition to the uncertainty previously discussed about whether best-of-breed algorithms are being used, there is the issue of system optimization and its effect on performance. Are the architectures for Demo III, PerceptOR, and other programs best-of-breed? If not, how much performance is lost? No one knows.

A related issue is software quality. Perception software will consist of about 750,000 to 1 million source lines of code (SLOC). It has been developed by multiple organizations and is of highly variable quality. It lacks documentation. To go from the present state to optimized code is a major effort. An estimate is 3 to 6 months to document requirements, 3 to 6 months for architecture design, 18 to 24 months to reimplement the code, and 6 months for system and performance testing. The effect of software quality on system performance and reliability is unknown.

SUMMARY

In the 18 years since the beginning of the DARPA ALV program, there has been significant progress in the canonical areas of perception for UGVs: road following, obstacle detection and avoidance, and terrain classification and traversability analysis. There has not been comparable progress at the system level in attaining an ability to go from A to B (on-road and off-road) with minimal intervention by a human operator.

REFERENCES

Albus, J.S., and A.M. Meystel. 2001. Engineering of Mind: An Introduction to the Science of Intelligent Systems. New York, N.Y.: Wiley.

Backer, G., B. Mertsching, and M. Bollmann. 2001. Data and model-driven gaze control for an active-vision system. IEEE Transactions on Pattern Analysis and Machine Intelligence 23(12): 1415–1429.

Baten, S., M. Lützeler, E.D. Dickmanns, R. Mandelbaum, and P.J. Burt 1998. Techniques for autonomous, off-road navigation. IEEE Intelligent Systems and Their Applications 13(6): 57–65.

Bellutta, P., R. Manduchi, L. Matthies, K. Owens, and A. Rankin. 2000. Terrain perception for Demo III. Pp. 326–331 in Proceedings of the IEEE Intelligent Vehicles Symposium 2000. New York, N.Y.: Institute of Electrical and Electronics Engineers, Inc.

Bertozzi, M., and A. Broggi. 1997. Vision-based vehicle guidance. Computer 30(7): 49–55.

Betke, M., E. Haritaoglu, and L.S. Davis. 1996. Multiple vehicle detection and tracking in hard real-time. Pp. 351–356 in Proceedings of the 1996 IEEE Intelligent Vehicles Symposium. New York, N.Y.: Institute of Electrical and Electronics Engineers, Inc.

Beymer, D., and J. Malik. 1996. Tracking vehicles in congested traffic. Pp. 130–135 in Proceedings of the 1996 IEEE Intelligent Vehicles Symposium. New York, N.Y.: Institute of Electrical and Electronics Engineers, Inc.

Bhanu, B., P. Symosek, and S. Das. 1997. Analysis of terrain using multispectral images. Pattern Recognition 30(2): 197–215.

Bishop, R. 2000. Intelligent vehicle applications worldwide. IEEE Intelligent Systems and Their Applications 15(1): 78–81.

Bornstein, J.A., B.E. Brendle, and C.M. Shoemaker. 2001. Army ground robotics technology development experimentation program. Pp. 333–340 in Unmanned Ground Vehicle Technology III, Proceedings of SPIE Volume 4364. G.R. Gerhart and C.M. Shoemaker, eds. Bellingham, Wash.: The International Society for Optical Engineering.

Broggi, A., M. Bertozzi, A. Fascioli, and M. Sechi. 2000. Shape-based pedestrian detection. Pp. 215–220 in Proceedings of the IEEE Intelligent Vehicles Symposium 2000. New York, N.Y.: Institute of Electrical and Electronics Engineers, Inc.

Castaño, R., R. Manduchi, and J. Fox. 2002. Classification experiments on real-world textures. Proceedings of the 2001 IEEE Computer Society Conference on Computer Vision and Pattern Recognition. CD-ROM. Available Los Alamitos, Calif.: Institute of Electrical and Electronics Engineers, Inc. Computer Society.

CECOM (U.S. Army Communications-Electronics Command). 2002. Advanced Sensor Technology. Briefing by Gene Klager, CECOM Night Vision and Electronic Sensors Directorate, to the Committee on Unmanned Ground Vehicle Technology, CECOM Night Vision and Electronic Sensors Directorate, Ft. Belvoir, Va., January 14.

Chang, T., T. Hong, S. Legowik, and M.N. Abrams. 1999. Concealment and obstacle detection for autonomous driving. Pp. 147–152 in Proceedings of the IASTED International Conference, Robotics and Applications. M.H. Hamza, ed. Anaheim, Calif.: ACTA Press.

Chaturvedi, P., E. Sung, A.A. Malcolm, and J. Ibanez-Guzman. 2001. Real-time identification of drivable areas in a semistructured terrain for an autonomous ground vehicle. Pp. 302–312 in Unmanned Ground Vehicle Technology III, Proceedings of SPIE Volume 4364. G.R. Gerhart and C.M. Shoemaker, eds. Bellingham, Wash.: The International Society for Optical Engineering.

Coombs, D., K. Murphy, A. Lacaze, and S. Legowik. 2000. Driving autonomously offroad up to 35 km/hr Pp. 186–191 in Proceedings of the IEEE Intelligent Vehicles Symposium 2000. New York, N.Y.: Institute of Electrical and Electronics Engineers, Inc.

Crisman, J.D. 1990. Color Vision for the Detection of Unstructured Roads and Intersections, Ph.D. dissertation. Pittsburgh, Pa.: Carnegie Mellon University.

Dellaert, F., and C.E. Thorpe. 1998. Robust car tracking using Kalman filtering and Bayesian templates. Pp. 72–85 in Intelligent Transportation Systems, Proceedings of SPIE Volume 3207. M.J. de Vries, P. Kachroo, K. Ozbay, and A.C. Chachich, eds. Bellingham, Wash.: The International Society for Optical Engineering.

Durrant-Whyte, H. 2001. A Critical Review of the State-of-the-Art in Autonomous Land Vehicle Systems and Technology, Sandia Report SAND2001-3685. Albuquerque, N. M.: Sandia National Laboratories.

Eisenberg, A. 2001. What's Next It's a Trolley! It's a Rail Car! No, It's an Optically Guided Bus. New York Times, July 26: G9.

Finlayson, G.D., S.D. Hordley, and P.M. Hubel. 2001. Color by correlation: A simple, unifying framework for color constancy. IEEE Transactions on Pattern Analysis and Machine Intelligence 23(11): 1209–1221.

Firschein, O., and T. Stratt. 1997. Reconnaissance, Surveillance, and Target Acquisition for the Unmanned Ground Vehicle: Providing Surveillance "Eyes" for an Autonomous Vehicle. San Francisco, Calif.: Morgan Kaufmann Publishers.

Fish, S. 2001. Unmanned Ground Combat Vehicles and Associated Perception in Off-Road Environments. Briefing by Scott Fish, Defense Advanced Research Projects Agency, Tactical Technology Office to the Committee on Army Unmanned Ground Vehicle Technology, Wyndham City Center Hotel, Washington, D.C., October 25.

Franke, U., D. Gavrila, S. Görzig, F. Lindner, F. Paetzold, and C. Wöhler. 1998. Autonomous driving goes downtown. IEEE Intelligent Systems & Their Applications 13(6): 40–48.

Gavrila, D. 2000. Pedestrian detection from a moving vehicle. Pp. 37–49 in Computer Vision—ECCV 2000, Proceedings of the 6th European Conference on Computer Vision. D. Vernon, ed. New York, N.Y.: Springer.

Gavrila, D. 2001. Sensor-based pedestrian protection. IEEE Intelligent Systems 16(6): 77–81.

Geusebroek, J.M., R. van den Boomgaard, A.W. M. Smeulders, and H. Geerts. 2001. Color invariance. IEEE Transactions on Pattern Analysis and Machine Intelligence 23(12): 1338–1350.

Giachetti, A., M. Cappello, and V. Torre. 1995. Dynamic segmentation of traffic scenes. Pp. 258–263 in Proceedings of the 1995 IEEE Intelligent Vehicles Symposium. New York, N.Y.: Institute of Electrical and Electronics Engineers, Inc.

Gothard, B.M., P. Cory, and P. Peterman. 1999. Demo III processing architecture trades and preliminary design. Pp. 13–24 in Mobile Robots XIII and Intelligent Transportation Systems, Proceedings of SPIE Volume 3525. H.M. Choset, D.W. Gage, P. Kachroo, M.A. Kourjanski, and M.J. de Vries, eds. Bellingham, Wash.: The International Society for Optical Engineering.

Hebert, M. 2000. Active and passive range sensing for robotics. Pp. 102–110 in Proceedings of the 2000 IEEE International Conference on Robotics and Automation (ICRA '00). New York, N.Y.: Institute of Electrical and Electronics Engineers, Inc.

Hong, T., C. Rasmussen, T. Chang, and M. Shneier. 2000. Fusing LADAR and Color Image Information for Mobile Robot Feature Detection and Tracking. Available online at <http://www.cs.yale.edu/homes/rasmussen/lib/papers/ias2002.pdf> [August 17, 2002].

Iverson, A.E., and S.S. Shen. 1997. Algorithms for Multispectral and Hyperspectral Imagery III, Proceedings of the SPIE Volume 3071. A.E. Iverson and S.S. Shen, eds. Bellingham, Wash.: The International Society for Optical Engineering.

Jochem, T. 2001. Safe-TRAC Technical Brief. Available online at <http://www.assistware.com/Tech_Brief.PDF> [August 8, 2002].

Jochem, T., and D. Pomerleau. 1997. Vision-based neural network road and intersection detection. Pp. 73–86 in Intelligent Unmanned Ground Vehicles. M.H. Hebert, C. Thorpe, and A. Stentz, eds. Boston, Mass.: Kluwer Academic Publishers.

Jones, W.D. 2001. Keeping cars from crashing. IEEE Spectrum 38(9): 40–45.

Kelly, A., and A. Stentz. 1997. Minimum throughput adaptive perception for high-speed mobility. Pp. 215–223 in IROS '97, Proceedings of the 1997 IEEE/RSJ International Conference on Intelligent Robot and Systems: Innovative Robotics for Real-World Applications. New York, N.Y.: Institute of Electrical and Electronics Engineers, Inc.

Kluge, K., and C. Thorpe. 1993. Intersection detection in the YARF road following system. Pp. 145–154 in Intelligent Autonomous Systems IAS-3. F.C.A. Groen, S. Hirose, and C.E. Thorpe, eds. Amsterdam, The Netherlands: IOS Press.

Langer, D.T. 1997. An Integrated MMW Radar System for Outdoor Navigation, Ph.D. dissertation. Pittsburgh, Penn.: Carnegie Mellon University.

Lee, K.M., and R. Blenis. 1994. Design concept and prototype development of a flexible integrated vision system. Journal of Robotic Systems 11(5): 387–398.

Macedo, J., R. Manduchi, and L.H. Matthies. 2001. Ladar-based terrain cover classification. Pp. 274–280 in Unmanned Ground Vehicle Technology III, Proceedings of SPIE Volume 4364. G.R. Gerhart and C.M. Shoemaker, eds. Bellingham, Wash.: The International Society for Optical Engineering.

Manduchi, R. 2002 (in press). Bayesian Feature Fusion for Visual Classification. Pasadena, Calif.: California Institute of Technology Jet Propulsion Laboratory.

Martin, D.A., H.S. Lee, and I. Masaki. 1998. A mixed-signal array processor with early vision applications. IEEE Journal of Solid-State Circuits 33(3): 497–502.

Masaki, I. 1998. Machine-vision systems for intelligent transportation systems. IEEE Intelligent Systems and Their Applications 13(6): 24–31.

Matthies, L. 1992. Stereo vision for planetary rover: Stochastic modeling to near real-time implementation. International Journal of Computer Vision, Vol. 8, No. 1.

Matthies, L. 2002. Perception for Autonomous Navigation. Briefing by Larry Matthies, Supervisor, Machine Vision Group, Jet Propulsion Laboratory, to the Committee on Army Unmanned Ground Vehicle Technology, Jet Propulsion Laboratory, Pasadena, Calif., January 22.

Matthies, L., and P. Grandjean. 1994. Stochastic performance modeling and evaluation of obstacle detectability with imaging range sensors. IEEE Transactions on Robotics and Automation 10(6): 783–792.

Matthies, L., T. Litwin, K. Owens, A. Rankin, K. Murphy, D. Coombs, J. Gilsinn, and T. Hong. 1998. Performance evaluation of UGV obstacle detection with CCD/FLIR stereo vision and LADAR. Pp. 658–670 in Proceedings of the Joint Conference on the Science and Technology of Intelligent Systems. New York, N.Y.: Institute of Electrical and Electronics Engineers, Inc.

Meyers, G.K, R.C. Bolles, Q-T. Luong, and J.A. Herson. 1999. Recognition of Text in 3-D Scenes. Available online at <http://www.erg.sri.com/publications/SDIUTMyers2.pdf> [August 6, 2002].

Mirmehdi, M., and M. Petrov. 2000. Segmentation of color textures. IEEE Transactions on Pattern Analysis and Machine Intelligence 22(2): 142–159.

Murphy, K., M. Abrams, S. Balakirsky, D. Coombs, T. Hong, and S. Legowik, T. Chang, and A. Lacaze. 2000. Intelligent control for unmanned vehicles. Pp. 191–196 in Proceedings of the 8th International Symposium on Robotics and Applications. Albuquerque, N. M.: TSI Press.

Murphy, K.N., M. Abrams, S. Balakirsky, T. Chang, T. Hong, A. Lacaze, and S. Legowik. 2002. Intelligent control for off-road driving. Proceedings of the First International NAISO Congress on Autonomous Intelligent Systems (CD ROM). Sliedrecht, The Netherlands: NAISO Academic Press.

Murphy, R.R. 1996. Biological and cognitive foundations of intelligent sensor fusion. IEEE Transactions on Systems, Man, and Cybernetics Part A—Systems and Humans 25(1): 42–51.

Olin, K.E., and D.Y. Tseng. 1991. Autonomous cross-country navigation—an integrated perception and planning system. IEEE Expert—Intelligent Systems and Their Applications 6(4): 16–32.

Owens, K., and L. Matthies. 1999. Passive night vision sensor comparison for unmanned ground vehicle stereo vision navigation. Pp. 59–68 in Proceedings of the IEEE Workshop on Computer Vision Beyond the Visible Spectrum—Methods and Applications (CVBVS '99). Los Alamitos, Calif.: Institute of Electrical and Electronics Engineers, Inc. Computer Society Press.

Papageorgiou, C., T. Evgeniou, and T. Poggio. 1998. A trainable object detection system. Pp. 1019–1024 in Proceedings of the 1998 DARPA Image Understanding Workshop. G. Lukes, ed. San Francisco, Calif.: Morgan Kaufmann Publishers.

Peng, J., and B. Bhanu. 1998. Closed-loop object recognition using reinforcement learning. IEEE Transactions on Pattern Analysis and Machine Intelligence 20(2): 139–154.

Peng, J., and B. Bhanu. 1999. Learning to perceive objects for autonomous navigation. Autonomous Robots 6(2): 187–201.

Poggio, T., and K-K. Sung. 1996. Networks that learn for image understanding. Pp. 226–240 in Advances in Image Understanding. K. Boyer and N. Abuja, eds. Los Alamitos, Calif.: Institute of Electrical and Electronics Engineers, Inc. Computer Society Press.

Pomerleau, D., and T. Jochem. 1996. Rapidly adapting machine vision for automated vehicle steering. IEEE Expert-Intelligent Systems & Their Applications 11(2): 19–27.

Priese, L., R. Lakmann, and V. Rehrmann. 1995. Ideogram identification in realtime traffic sign recognition system. Pp. 310–314 in Proceedings of the Intelligent Vehicles '95 Symposium. Piscataway, N.J.: Institute of Electrical and Electronics Engineers, Inc.

Privitera, C.M., and L.W. Stark. 2000. Algorithms for defining visual regions-of-interest: Comparisons with eye fixations. IEEE Transactions on Pattern Analysis and Machine Intelligence 22(9): 970–982.

Rao, N.S.V. 2001. On fusers that perform better than best sensor. IEEE Transactions on Pattern Analysis and Machine Intelligence 23(8): 904–909.

Rasmussen, C. 2002. Combining laser range, color and texture cues for autonomous road following. Available online at <http://www.cs.yale.edu/homes/rasmussen/lib/papers/icra2002.pdf> [August 7, 2002].

Reece, D., and Shafer, S. 1995. Control of Perceptual Attention in Robot Driving. Artificial Intelligence 78. pp. 397–430.

Rosenblum, M. 2000. Neurons That Know How to Drive. Available online at <http://www.cis.saic.com/projects/mars/pubs/crivs081.pdf> [August 6, 2002].

Rosenblum, M., and B. Gothard. 1999. Getting more from the scene for autonomous navigation: UGV Demo III program. Paper 3838-21 in Mobile Robots XIV, Proceedings of SPIE Volume 3838. D.W. Gage and H.M. Choset, eds. Bellingham, Wash.: The International Society for Optical Engineering.

Rushing, J.A., H.S. Ranganath, T.H. Hinke, and C.J. Graves. 2001. Using association rules as texture features. IEEE Transactions Pattern Analysis and Machine Intelligence 23(8): 845–858.

Sato, J., and R. Cipolla. 1996. Obstacle detection from image divergence and deformation. Pp. 165–170 in Proceedings of the 1996 IEEE Intelligent Vehicles Symposium. New York, N.Y.: Institute of Electrical and Electronics Engineers, Inc.

Schwartz, I. 2000. PRIMUS autonomous driving robot for military applications. Pp. 313–325 in Unmanned Ground Vehicle Technology II Proceedings of SPIE Volume 4024. G.R. Gerhart, R.W. Gunderson, C.M. Shoemaker, eds. Bellingham, Wash.: The International Society for Optical Engineering.

Shoemaker, C.M., and J.A. Bornstein. 2000. Overview and update of the DEMO III Experimental Unmanned Vehicle Program. Pp. 212–220 in Unmanned Ground Vehicle Technology II Proceedings of SPIE Volume 4024. G.R. Gerhart, R.W. Gunderson, C.M. Shoemaker, eds. Bellingham, Wash.: The International Society for Optical Engineering.

Sukthankar, R., D. Pomerleau, and C. Thorpe. 1993. Panacea: An active controller for the ALVINN autonomous driving system, CMU-RI-TR-93-09. Pittsburgh, Pa.: The Robotics Institute, Carnegie Mellon University.

U.S. Army. 2002. Information presented to the Committee on Army Unmanned Ground Vehicle Technology during site visit to the CECOM Night Vision and Electronic Sensors Directorate, Fort Belvoir, Va., January 14.

Williamson, T.A. 1998. A High-Performance Stereo Vision System for Obstacle Detection, Ph.D. Thesis. Pittsburgh, Pa.: Carnegie Mellon University.

Zhao, L., and C.E. Thorpe. 2000. Stereo- and neural network-based pedestrian detection. IEEE Transactions on Intelligent Transportation Systems 1(3): 148–154.

D

Historical Perspective

This appendix provides historical background for assessing the current state of the art in unmanned ground vehicle (UGV) technologies. The focus is on specific perception research that has directly contributed to autonomous mobility. It is representative, not exhaustive.[1] It traces developments in machine perception relevant to a mobile robot operating outdoors, and it identifies the major challenges in machine perception and disadvantages of different approaches to these challenges.

THE EARLY DAYS 1959–1981

The first mobile robot to use perception, called "Shakey," was developed at the Stanford Research Institute (Nilson, 1984) beginning in 1969 and continuing until 1980. Shakey operated in a "blocks world": an indoor environment containing isolated prismatic objects. Its sensors were a video camera and a laser range finder, and it used the MIT blocks-world image-understanding algorithms (Roberts, 1965) adapted for use with video images.

Each forward move of the robot was a few meters and took about 1 hour, mostly for vision processing. Shakey emerged during the first wave of artificial intelligence that emphasized logic and problem solving. For example, Shakey could be commanded to push a block that rested on top of a larger block. The robot's planner, a theorem prover, would construct a plan that included finding a wedge that could serve as a ramp, pushing it against the larger block, rolling up the ramp, and then pushing the smaller block off. Perception was considered of secondary importance and was left to research assistants. It would become very clear later that what a robot can do is driven primarily by what it can see, not what it can plan.

In 1971, Rod Schmidt of Stanford University published the first Ph.D. thesis on vision and control for a robot operating outdoors. The Stanford Cart used a single black and white camera with a 1-Hz frame rate. It could follow an unbroken white line on a road for about 15 meters before breaking track. Unlike Shakey, the Cart moved continuously at "a slow walking pace" (Moravec, 1999). The Cart's vision algorithm, like Shakey's, first reduced images to a set of edges using operations similarly derived from the blocks-world research. While adequate for Shakey operating indoors, these proved inappropriate for outdoor scenes containing few straight edges and many complicated shapes and color patterns. The Cart was successful only if the line was unbroken, did not curve abruptly, and was clean and uniformly lit. The Cart's vision system[2] used only about 10 percent of the image and relied on frame-to-frame predictions of the likely position of the line to simplify image processing.

Over the period from 1973 to 1980, Hans Moravec of Stanford University (Moravec, 1983) developed the first stereo[3] vision system for a mobile robot. Using a modified Cart, he obtained stereo images by moving a black and white video camera side to side to create a stereo baseline. Like Shakey, Moravec's Cart operated mostly indoors in a room with simple polygonal objects, painted in contrasting black and white, uniformly lit, with two or three objects spaced over 20 meters.

---

[1] Good reviews of the image-understanding literature can be found in Rosenfeld (1984), Fischler and Firschein (1987), Ballard and Brown (1982), Nalwa (1993), and IEEE Transactions on Pattern Analysis and Machine Intelligence.

[2] A scaled-up version of this approach was demonstrated in 1977 by Japan's Mechanical Engineering Laboratory at Tsuba. It could follow roads at up to 30 km/h using very specialized hardware that tracked white lines and other features.

[3] Stereo is the extraction of depth information from multiple images, often assumed to be images acquired simultaneously from two cameras separated by a fixed distance, the stereo baseline. Stereo can use more than two cameras to improve false stereo matches or can be derived from a moving camera that yields a larger baseline and improved depth resolution.

The robot had a planner for path planning and obstacle avoidance. The perception system used an interest algorithm (Moravec, 1977) that selected features that were highly distinctive from their surroundings. It could track and match about 30 image features, creating a sparse range map for navigation. The robot required about 10 minutes for each 1-meter move, or about 5 hours to go 30 meters, using a 1-MIP processor. Outdoors, using the same algorithms, the Cart could only go about 15 meters before becoming confused, most frequently defeated by shadows that created false edges to track. Moravec's work demonstrated the utility of a three-dimensional representation of a robot's environment and further highlighted the difficulties in perception for outdoor mobility.

By the end of 1981, the limited ventures into the real world had highlighted the difficulties posed by complex shapes, curved surfaces, clutter, uneven illumination, and shadows. With some understanding of the issues, research had begun to develop along several lines and laid the groundwork for the next generation of perception systems. This work included extensive research on image segmentation[4] and classification with particular emphasis on the use of color (Laws, 1982; Ohlander et al., 1978; Ohta, 1985), improved algorithms for edge detection[5] (Canny, 1983), and the use of models for road following (Dickmanns and Zapp, 1986). At this time there was less work on the use of three-dimensional data. While Shakey (rangefinder) and Moravec (stereo) demonstrated stereo's value to path planning (more dense range data), stereo remained computationally prohibitive and the range finders of the time, which required much less computation, were too slow.[6]

THE ALV ERA 1984–1991

In 1983, DARPA started an ambitious computer research program, the Strategic Computing program. The program featured several integrating applications to focus the research projects and demonstrate their value to DARPA's military customers. One of these applications was the Autonomous Land Vehicle (ALV),[7] and it would achieve the next major advance in outdoor autonomous mobility.

The ALV had two qualitative objectives: (1) It had to operate exclusively outdoors, both on-road and off-road across a spectrum of conditions: improved and unimproved roads, variable illumination and shadows, obstacles both on- and off-road, and in terrain with varying roughness and vegetative background. (2) It had to be entirely self-contained and all computing was to be done onboard. Quantitatively, the Strategic Computing Plan (DARPA, 1983) called for the vehicle to operate on-road beginning in 1985 at speeds up to 10 km/h, increasing to 80 km/h by 1990, and off-road beginning in 1987 at speeds up to 5 km/h, increasing to 10 km/h by 1989. On-road obstacle avoidance was to begin in 1986 with fixed polyhedral objects spaced about 100 meters apart.

The ALV was an eight-wheeled all-terrain vehicle with an 18-inch ground clearance and good off-road mobility (Figure D-1). A fiberglass shelter sufficient to enclose generators, computers, and other equipment was mounted on the chassis. The navigation system included an inertial land navigation system that provided heading and distance traveled, ultrasonic sensors to provide information about the pose of the vehicle with respect to the ground for correcting the scene model geometry, and a Doppler radar to provide ground speed from both sides of the ALV to correct for wheel slippage errors in odometry.

The ALV had two sensors for perception: a color video camera (512[h] by 485[v], 8 bits of red, blue, green data) and a two-mirror custom laser scanner that had a 30°-vertical by 80°-horizontal field of view, produced a 256(h) × 64(v) range image with about 3-inch resolution, had a 20-meter range, a 2-Hz frame rate,[8] and an instantaneous field of view of 5-8.5 mrad. The scanner also yielded a 256 × 64 reflectance image. The laser scanner used optics and signal processing to produce a real-time depth map suitable for obstacle avoidance that was impractical to obtain from stereo by calculation. The ALV was the first mobile robot to use both these sensors.

The first algorithm used for road following (Turk et al., 1988) projected a red, green, blue (RGB) color space into two dimensions and used a linear discriminant function to differentiate road from nonroad in the image. This was a fast algorithm, but it had to be tuned by hand for different roads and different conditions. Using this algorithm, the ALV made a 1-km traverse in 1985 at an average speed of 3 km/h,

---

[4]One of the fundamental steps in image understanding is to segment or partition a scene into a set of regions, each of which is homogeneous with respect to a set of properties that correspond to semantically meaningful entities in the scene. Classification (sometimes called interpretation) can be thought of as labeling the regions. The division is fuzzy. Depending on how segmentation is done, the desired degree of classification might be obtained as part of segmentation or additional processing may be required. For example, three-dimensional information could be used with geometric criteria to segment objects into obstacle-like or not. Classification as to their characteristics (e.g., a rock or bush) is required to determine if the objects are traversable. Segmentation and classification can be based on colors (all the way to spectra), on ratios of colors, edges, geometry, texture, or other features or combinations of features.

[5]Edges worked well to segment the prismatic objects in the "blocks world" and were useful, but not sufficient, for natural scenes, nor were the edge-finding algorithms developed for the straight edges and controlled illumination of the stylized, less complex "blocks world" appropriate for the natural scenes that contained curved shapes and where there were many opportunities for false alarms because of changes in illumination. For example, the edges of shadows and sun-lit spots are typically much stronger than those of roads.

[6]The range finder was repositioned to acquire each point in turn.

[7]Martin-Marietta was the integrator. Participants included SRI, CMU, University of Maryland, Advanced Decision Systems, Hughes Research Laboratories, U.S. Army Engineering Topographic Laboratory, and ERIM.

[8]This solved the slow data acquisition problem with the range finders. A laser scanner, sometimes referred to as a LADAR (laser detection and ranging), is a scanning laser range finder.

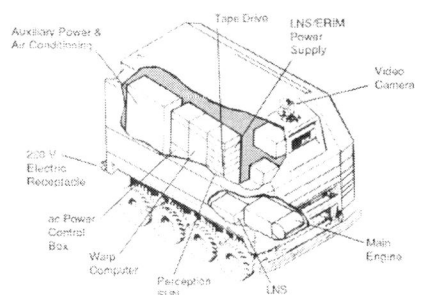
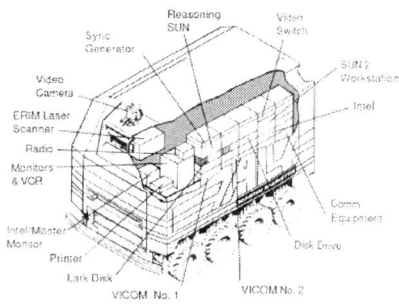

FIGURE D-1 Autonomous land vehicle (ALV). Courtesy of John Spofford, SAIC.

a 10^3 gain over the prior state of the art. This increased to 10 km/h over a 4-km traverse in 1986. In 1987, the ALV reached a top speed of 20 km/h with an improved color segmentation algorithm using a Bayes' classifier (Olin et al., 1987) and used the laser scanner to avoid obstacles placed on the road. The obstacle detection algorithm (Dunlay and Morgenthaler, 1986) made assumptions about the position of ground plane and concluded that those range points that formed clusters above some threshold height over the assumed ground plane represented obstacles. This worked well for relatively flat roads but was inadequate in open terrain, where assumptions about the location of the ground plane were more likely to be in error. Work on road following continued, experimenting with algorithms primarily from Carnegie Mellon University (CMU) (e.g., Hebert, 1990) and the University of Maryland (e.g., Davis et al., 1987; Sharma and Davis, 1986), but the emphasis in the program shifted in 1987 to cross-country navigation.

In doing cross-country navigation, the ALV became the first robotic vehicle to be intentionally driven off-road, and the principle goal for perception became obstacle detection. The ALV relied primarily on the laser range scanner to identify obstacles. Much of the early work on obstacle detection used geometric criteria and mostly considered isolated, simple prismatic objects.

However, using such geometric criteria in cross-country terrain may cause traversable objects, such as dense bushes that can return a cloud of range points, to be classified as obstacles even though the vehicle can push through them. This signaled the need to introduce such additional criteria as material type to better classify the object. As the ALV began to operate cross-country, it also became clear that the concept of an obstacle should be generalized to explicitly include any terrain feature that leads to unsafe vehicle operation (e.g., a steep slope), taking into account vehicle dynamics and kinematics. This expanded definition acknowledged the dependence of object or feature classification on vehicle speed, trajectory, and other variables, extending the concept of an obstacle to the more inclusive notion of traversability.

In August 1987, the ALV performed the first autonomous cross-country traverse based on sensor data (Olin and Tseng, 1991). During this and subsequent trials extending for about a year, the ALV navigated around various kinds of isolated positive obstacles over traverses of several kilometers. The terrain had steep slopes (some over 15 degrees), ravines and gulleys, large scrub oaks, brushes, and rock outcrops (see Figure D-2). Some manmade obstacles were emplaced for experiments. The smallest obstacles that could be reliably detected were on the order of 2 feet to 3 feet in height. On occasion, the vehicle would approach and detect team members and maneuver to avoid them. The vehicle reached speeds of 3.5 to 5 km/h[9] and completed about 75 percent of the traverses.[10] These speeds were enabled by the

---

[9]Speed was limited by look-ahead distance, a function of sensor range and the time required for acquiring and processing the range data.

[10]By comparison a human driver in a pickup truck drove the course at 10 km/h.

FIGURE D-2 ALV and Demo II operating areas. Courtesy of Benny Gothard, SAIC.

Hughes architecture, which tightly coupled the path planner[11] to perception and allowed the vehicles to be more reactive in avoiding nontraversable terrain.

This architecture was the first for a vehicle operating off-road that was influenced by the reactive paradigm of Arkin (1987), Brooks (1986), and Payton (1986). The reactive paradigm was a response to the difficulties in machine vision and in planning that caused robots to operate slowly. Brooks reasoned that insects and some small animals did very well without much sensor data processing or planning. He was able to create some small robots in which sensor outputs directly controlled simple behaviors;[12] there was no planning and the robot maintained no external representation of its environment. More complex behaviors arose as a consequence of the robot's interaction with its environment. While these reactive approaches produced rapid response, because they eliminated both planning and most sensor processing, they did not easily scale to more complex goals. A third paradigm then arose that borrowed from both the deliberative and reactive paradigm. Called the hybrid deliberative or reactive paradigm, it retained planning but used the simple reflexive behaviors as plan primitives. The planner scheduled sequences of these to achieve higher-level goals. Sensor data was provided directly to these behaviors as in the reactive architecture but was also used to construct a global representation for the planner to use in reasoning about higher-level goals.

In a hybrid architecture, the highest level of planning may operate on a cycle time of minutes while the lowest-level sensor-behavior loops may cycle at several hertz. These lower-level loops are likely hard-wired to ensure high-speed operation. At the lowest level, multiple behaviors will be active simultaneously, for example, road-edge following and obstacle avoidance. An arbitration mechanism is used to control execution precedence when behaviors conflict.

The ALV program demonstrated that architectures do matter, that they are a key to performance, and that they affect how perception is done. Without the hybrid deliberative and reactive architecture UGVs would not perform at today's level. There is still room for improvement within the paradigm.

In the ALV program, the cross-country routes were carefully selected to exclude obstacles smaller than those that could be detected. There were no surprises that could endanger the vehicle or operators. Demo II, to be described subsequently, and Demo III (see Appendix C) were also scenario driven with obstacles selected to be detectable. The emphasis in these programs was on tuning performance to meet the demonstration objectives of the scenarios that describe plausible military missions (see, for example, Spofford et al., 1997). Hence, cross-country speeds reported must be viewed as performance obtained in those specific scenarios under controlled conditions not reflective of performance that might be obtained in unknown terrain. For the most part, such experiments remain to be done.

Several other organizations were operating robotic vehicles outdoors during roughly the same period as the ALV (i.e., mid-1980s to about 1990). These included CMU, the Jet Propulsion Laboratory (JPL), and the Bundeswehr University in Munich. Much of the CMU work, by design, was incorporated into the ALV's software. There was however, one significant advance that was not incorporated into the ALV because of timing. It was to have a most singular impact on perception, initially for on-road navigation and

---

[11]The planner did not use data from perception for global path replanning. The sensors had very limited fields of views and the perception system did not have the computational resources needed to recognize landmarks and other features useful in replanning.

[12]Examples of simple behavior: Go forward or in some direction; wandering: go in a random direction for some period; avoiding: turn to right if obstacle is on left, then go, etc.; homing: turn toward home location and go forward.

later for off-road operation. This was the re-appearance, after nearly a 20-year absence from mainstream perception research, of the artificial neural network—not the two-layer Perceptron for the reasons detailed by Minsky and Papert (1969) but a more capable three-layer system.[13] The reason for the renaissance was the invention of a practical training procedure for these multilayer networks. This opened the way for a surge of applications; one of them was ALVINN (autonomous land vehicle in a neural network), and it would increase on-road speeds from 20 km/h in 1998 to over 80 km/h in 1991.[14] ALVINN (Pomerleau, 1992) was motivated by the desire to reduce the hand tuning of perception algorithms and create a system which could robustly handle a wide variety of situations. It was a perception system that learned to control a vehicle by watching a person drive, and learned for itself what features were important in different driving situations.

The two other organizations also made noticeable advances: JPL developed the first system to drive autonomously across outdoor terrain using a range map derived from stereo rather than a laser scanner (Wilcox et al., 1987). The system required about 2.5 hours of computation for every 6 meters of travel. The Bundeswehr University operated at the other extreme, achieving speeds of 100 km/h using simple detectors for road edges and other features and taking advantage of the constraints of the autobahn to simplify the perception problem (Dickmanns and Zapp, 1987). It was highly tuned and could be confused by road imperfections, shadows, and puddles or dirt that obscured the road edge or road markings. It did not do obstacle detection. When the program ended in 1989, the system was using 12 special purpose computers (10 MIPS each), each tracking a single image patch.[15]

By 1990, JPL had developed near real-time stereo algorithms (Matthies, 1992) producing range data from 64 × 60 pixel images at 0.5-Hz frame rate. This vision system was demonstrated as part of the Army Demo I program[16] in 1992 showing automatic obstacle detection and safe stopping (not avoidance) at speeds up to 10 mph. This stereo system used area correlation and produced range data at every pixel in the image, analogous to a LADAR. It was therefore appropriate for use in generic off-road terrain. This was a new paradigm for stereo vision algorithms. Up to that time the standard wisdom in the computer vision community was that area-based algorithms were too slow and too unreliable for real-time operation; most work up to that point had used edge detection, which gave relatively sparse depth maps not well suited to off-road navigation. This development spurred the use of stereo vision at other institutions and in the Demo II program, which followed shortly thereafter.

By the early 1990s, the state-of-the-art perception systems used primarily a single color video camera and a laser scanner. Computing power had risen from about the 1/2 MIPS of the Stanford Cart to about 100 MIPS at the end of the ALV program in 1989. Algorithms had increased in numbers and complexity. The ALV used about 1,000,000 executable source lines of code; about 500,000 were allocated to mobility. Road-following speeds increased by about a factor of 10^4 from 1980 to 1989 (meters per hour to tens of kilometers per hour); this correlated with the increase in computing power. Road following was restricted to daylight, dry conditions, no abrupt changes in curvature, and good edge definition. ALVINN showed how to obtain the more general capability that had been lacking, and the Bundeswehr's work hinted at further gains that might be made through distributed processing.

Cross-country mobility was mostly limited to traverses of relatively flat, open terrain with large, isolated obstacles and in daylight, dry conditions. Feature detection was primarily limited to detecting isolated obstacles on the order of 2 to 3 feet high or larger. There was a limited capability to classify terrain (e.g., into traversable or intraversable), but performance was strongly situation dependent and required considerable hand tuning. The detection of negative obstacles (e.g., ditches or ravines) was shown to be a major problem, and no work had been done on identifying features of tactical significance (e.g., tree lines). The image understanding community was vigorously working on approaches to obtain depth maps from stereo imagery (e.g., Barnard, 1989; Horaud and Skordas, 1989; Lim, 1987; Medioni and Nevatia, 1985; Ohta and Kanade, 1985). However, the computational requirements of most of these approaches placed severe limits on the speed of vehicles using stereo for cross-country navigation. The simpler JPL area-correlation algorithm provided a breakthrough and would enable the use of stereo vision in subsequent UGV programs. Robust performance was emerging as an important goal as vehicles began to operate under more demanding conditions; a limit was being reached on how much hand tuning could be done to cope with each change in the environment (e.g., changes in visibility or vegetation background).

THE DEMO II ERA 1992–1998

The next major advance in the state of the art was driven by the DARPA/OSD-sponsored UGV/Demo II program[17]

---

[13] Neural networks are function approximators, creating a mapping from input features to output. There are many other approaches that can be used.

[14] In fact, on occasion it reached speeds of 112 km/h (70 mph).

[15] This is a good example of a reactive architecture.

[16] The Demo I program was sponsored by the Army Research Laboratory, 1990–1992. The emphasis was mostly on teleoperation. The JPL stereo system was used for obstacle detection and safe vehicle stopping as part of an enhanced teleoperation focus.

[17] There was a Demo I program supported by the Army Research Laboratory, 1990–1992. The emphasis was mostly on teleoperation. JPL did research on using stereovision to produce a three-dimensional representation of the operating environment for the vehicle operator to use in establishing waypoints.

FIGURE D-3 The Demo II vehicle and environment. Courtesy of John Spofford, SAIC.

that began in 1992. The emphasis in the Demo II program was on cross-country A-to-B traverses that might be typical of those required by a military scout mission, including some road following on unimproved roads. The program integrated research from a number of organizations. For a detailed description of the program and the technical contributions of the participants see Firschein and Strat (1997). The Demo II vehicles (see Figure D-3) were HMMWVs equipped with stereo black-and-white video cameras on a fixed mount for obstacle avoidance,[18] and a single color camera on a pan-and-tilt mount for the road following. Experiments were conducted with stereo FLIR (3–5 μm) to demonstrate the feasibility of nighttime obstacle detection, but the results were not used to control the vehicle. Some target detection and classification experiments were conducted with LADAR, but it was not used to control the vehicle. The program manager gave first priority to improving the state of the art of vehicle navigation using passive sensors. The ALVINN road-following system was shown to be robust in Demo II. Neural networks trained on paved roads worked well on dirt roads. Speeds up to 15 mph were achieved on dirt roads. ALVINN was also extended to detect road intersections (Jochem and Pomerleau, 1997) but encountered difficulties at speeds much greater than 5 mph. The researchers concluded that active vision was likely required. The value of active vision was shown in a subsequent experiment (Panacea) by Sukthankar et al. (1993). In this experiment, ALVINN's output was used to pan the camera to keep the road in view and maintain important features within the field of view. The planner could also provide an input. Active vision allowed the CMU Navlab to negotiate a sharp fork it could not otherwise negotiate with a fixed camera.

Significant road curvature and vertical changes were problems, particularly with a fixed camera. There was a problem with a network trained on a road with tall grass on either side when the grass was cut and had turned brown. Low contrast was also a problem (e.g., a brown dirt road against brown grass). A solution for low-contrast situations is to use other features such as texture.

A radial basis function neural network road follower (ROBIN) was developed by Martin-Marietta (Rosenblum, 2000). It was trained using exemplar or template road images. Unlike ALVINN, the input to the neural network was not a real-time road image but a measure of how close the actual images were to each of the templates. It could drive on a diverse set of roads and reached speeds of up to 10 mph on ill-defined dirt roads with extreme geometrics.

Three obstacle detection and avoidance systems were evaluated: GANESHA, an earlier project (Langer and Thorpe, 1997), Ranger (Kelly and Stentz, 1997), and SMARTY (Stentz and Hebert, 1997). Each used somewhat different criteria for obstacle classification and each used the resultant obstacle map in different ways to generate steering

---

[18]Black-and-white cameras were selected because of their higher resolution. Their field of view was about 85° in the horizontal plane. They provided 512 × 480 pixels but only one field, 256 × 240, was processed to avoid the blur from waiting 1/60 second to integrate the second field. Today's cameras provide the capability for full frame integration.

 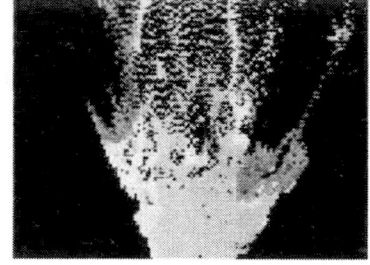 

Dirt road with obstacle   Elevation map of road   Detected obstacles

FIGURE D-4 Stereo obstacle detection results. Courtesy of Larry Matthies, Jet Propulsion Laboratory.

commands. SMARTY, with a simpler classification scheme, was selected.

SMARTY was a point-based system for rapidly processing range data (either from a laser scanner or stereo). It updated obstacle maps, on-the-fly, as soon as data became available, rather than waiting until an entire frame or image was acquired and then batch processing all the data at once. In tests with the ERIM scanner it reduced obstacle detection time from about 700 msec to about 100 msec (Hebert, 1997).

The real significance of SMARTY was the introduction of a new paradigm for data processing: time stamp each data point as it is generated and use vehicle state data (e.g., x, y, z, roll, pitch and yaw)[19] to locate that point in space. This provides a natural way to fuse data from alternative sensors or from a single sensor over time.

Perhaps the greatest change in the perception system for Demo II was the use of stereo images to provide range data. Stereo vision was used in Demo I for obstacle detection at up to 10 mph. Demo II added obstacle avoidance off-board demonstrations of night stereovision with 3–5 μm FLIR, and increased the speed and resolution of the system to 256 × 80 pixel range images at the 1.4-Hz frame rate. This was accomplished through refinement of the algorithms and advances in processor speed.

The stereo algorithms are described in Matthies et al. (1996). They worked particularly on ways to reduce false positives or false matches (i.e., the appearance of false obstacles), which had limited the usefulness of stereo obstacle detection. A typical result from Matthies et al. (1996) is shown in Figure D-4. Note that the obstacle detection algorithm used only geometric criteria; hence, bushes on the left and right were also classified as obstacles. They reported speeds of 5 km/h to 10 km/h over relatively flat terrain with scattered positive obstacles 0.5 meters or higher embedded in 12-inch grass cover. During Demo II(B) the stereo system was integrated with the SMARTY obstacle avoidance system and reached speeds up to 10.8 km/h on grassy terrain with 1-meter or greater positive obstacles.

Obstacle detection performance for Demo II is described in a paper by Spofford and Gothard (1994). Their analysis assumes 128 × 120 stereo depth information from the JPL system used in SMARTY. The system could detect an isolated, positive 8-inch obstacle at about 6 meters and a 16-inch obstacle at about 10 meters. These distances correspond to about 3 mph and 8 mph respectively for the HMMWV. Obstacles less than 8 inches could not be reliably detected. In some experiments 1.5-meter-high fake rocks constructed of foam were used. These could be detected at about 40 meters and allowed for a speed of about 20 mph.[20]

Demo II was remarkable in its demonstration of stereovision as a practical means for obstacle detection, but it came at a cost. The perception system required higher resolution to detect smaller obstacles: the same number of vertical pixels (5-10) must be placed on a smaller surface. The field of view was therefore limited[21] and there was a real possibility that the vehicle might end up in cul-de-sacs; the vehicle lacked perspective. It avoided local obstacles in a way that might have been globally sub-optimal. There are two solutions: Use multiple cameras with different fields of view or use active vision where orientation is controlled to scan over a wider area.

Negative obstacles were successfully detected with algorithms that looked for large discontinuities in range. Any solution was still limited by viewing geometry. No capability was demonstrated in Demo II for detecting thin obstacles such as fences, or for use in other situations where geometry-based methods fail.

There was no on-vehicle demonstration of terrain classification as part of the Demo II program. However, some

---

[19]This requires a very good navigation system and highlights again the role of navigation in perception.

[20]Note again that Demo II was scenario driven. The 1.5-m obstacles were the only obstacles present for these experiments.

[21]The field of view of the Demo II system was further limited to ensure that the computation could be completed in time. This was done using windows whose size was controlled by vehicle speed.

research on doing terrain classification in real time was done by Matthies et al. (1996). Prior work had been successful in doing classification but it took on the order of tens of minutes per scene. Matthies et al. looked for observable properties that could be used with very simple and fast algorithms. They used ratios of spectral bands in the red, green and near infrared (0.08 μm) provided by a color camera with a rotating filter wheel that was not sufficiently robust for on-vehicle use. At that time, no camera was available that could get the required bands simultaneously. This has since changed. With this approach, they were able to label regions as vegetation or nonvegetation at 10 frames per second. Near infrared and RGB was better than RGB alone. They did not extend the classification further (e.g., for vegetation was it grass, trees, bushes; for nonvegetation was it soil, rocks, roads?). Similar work was reported by Davis et al. (1995). They used six spectral bands in the red to the near infrared and obtained classification rates of 10 fields per second. Their conclusions were the same as that of Matties et al., namely, that using near infrared and RGB gave better results than RGB alone.

By the end of the Demo II era, road following had been demonstrated on a wide variety of roads in daylight and in good weather. The potential to use FLIR for road following at night had been demonstrated. There were still issues with complex intersections and abrupt changes in curvature but approaches to both had been demonstrated. Obstacle avoidance was not solved, but performance had improved on isolated, positive obstacles. Smaller objects could be detected

TABLE D-1 Performance Trends for ALV and Demo II

Capability	ALV (1989)	Demo II (1996)
Speed		
Roads		
Paved (day)	13 mph (20 km/h)	44 mph
Dirt roads (day)	NCD	15 mph
Ill defined	NCD	10 mph
Cross country (day)		
Obstacle avoidance		
Isolated positive objects		20 mph
5 feet (1.5 meters)	3 mph (5 km/h)	10–12 mph
2–3 feet	3 mph (5 km/h)	8 mph
16 inches	NCD	3 mph
8 inches	NCD	NCD
4 inches	NCD	NCD
Other obstacles	NCD	NCD
Thin mud, water	NCD	Ridge crest detection and tracking
	NCD	
Weather		
Dry	Yes	Yes
Wet	NCD	NCD
Visibility		
Day	Yes	Yes
Night	NCD	Experiments with FLIR (5 mph on paved road)
Obscurants	NCD	NCD
Behaviors	Road following, cross-country navigation	Road following, cross-country navigation, leader-follower, ridge cresting

Means	ALV (1989)	Demo II (1996)
Architecture	Hybrid DRP	Hybrid DRP
Computing (ROM)	100 MIPS (10^8)	4.4 GIPS (10^9)
SLOC for mobility (ROM)	500 K "spaghetti code"	750 K "spaghetti code"
Sensors	Monocolor video (road following)	Monocolor video (road following)
	LADAR (obstacle avoidance)	Stereo-B&W video (obstacle avoidance)
		Stereo FLIR (night)
Navigation	Inertial land navigation system	IMU, GPS, odometry
	Doppler odometry, tilt	
Data fusion	NCD	NCD
Machine learning	NCD	Limited: neural network for road following

Note: ALV = autonomous land vehicle; NCD = no capacity demonstrated; FLIR = forward looking infrared radar; DRP = dynamic route planning ; ROM = read-only memory; SLOC = source lines of code; LADAR = laser detection and ranging; IMU = inertial measurement unit, GPS = global positioning system.

faster and at greater distance. Requirements were better understood as were the strengths and weaknesses of different sensors and algorithms. The detection of negative obstacles had improved but was still a problem. The use of simple geometric criteria to classify objects as obstacles had been fully exploited. It resulted in many false positives as such traversable features as clumps of grass were classified as obstacles. Additional criteria such as material type would have to be used to improve performance. The goal of classifying terrain on the move into traversable and intraversable regions considering vehicle geometry and kinematics was still in the future although much fundamental work had been accomplished. There was modest progress on real time classification of terrain into categories such as vegetated or not but it fell short of what was needed both to support obstacle avoidance and tactical behaviors. Progress was made on characterizing surface roughness based on inertial measurement data. To obtain a predictive capability would require texture analysis of imaging sensor data. This was difficult to do in real-time. There was no progress on identifying tactically significant features, such as tree lines that fell beyond the nominal 50-meter to 100-meter perception horizon. There was, however, explicit acknowledgment that extending the perception horizon was an important problem. Most of the work continued to be under daylight, good weather conditions with no significant experiments at night or with obscurants. No research was focused on problems unique to operations in urban terrain, arguably the most complicated environment. Leader-follower operation was demonstrated in Demo II, but it was based on GPS, not perception.

Perception continued to rely primarily on video cameras. Some experiments were done with FLIR. LADAR continued in use elsewhere but was not used to enable autonomous mobility in Demo II. Stereo was shown to be a practical way to detect obstacles. There were situations, such as maneuvering along tree lines or in dense obstacle fields, where LADAR was preferred. There were experiments using simple color ratios for terrain classification but there was no work to more fully exploit spectral differences for material classification (e.g., multiband cameras). There was no work in sensor data fusion to improve performance or robustness. Major gains in robustness in road following were achieved with machine learning, but it was not used more generally.

Table D-1 summarizes the state of the art at the end of this period of research and contrasts it with the end of the ALV program.

REFERENCES

Arkin, R.C. 1987. Motor schema-based navigation for a mobile robot: An approach to programming by behavior. Pp. 264–271 in Proceedings of the 1987 IEEE International Conference on Robotics and Automation. New York, N.Y.: Institute of Electrical and Electronics Engineers, Inc.

Ballard, D., and C. Brown. 1982. Computer Vision. Englewood Cliffs, N.J.: Prentice Hall, Inc.

Barnard, S.T. 1989. Stochastic stereo matching over scale. International Journal of Computer Vision 3(1): 17–32.

Brooks, R.A. 1986. A robust layered control system for a mobile robot. IEEE Journal of Robotics and Automation 2(1): 14–23.

Canny, J.F. 1983. Finding Edges and Lines in Images, TR-720. Cambridge, Mass.: Massachusetts Institute of Technology Artificial Intelligence Laboratory.

DARPA (Defense Advanced Research Projects Agency). 1983. Strategic Computing—New Generation Technology: A Strategic Plan for Its Development and Application to Critical Problems in Defense, October 28. Arlington, Va.: Defense Advanced Research Projects Agency.

Davis, I.L., A. Kelly, A. Stentz, and L. Matthies. 1995. Terrain typing for real robots. Pp. 400–405 in Proceedings of the Intelligent Vehicles '95 Symposium. Piscataway, N.J.: Institute of Electrical and Electronics Engineers, Inc.

Davis, L.S., D. Dementhon, R. Gajulapalli, T.R. Kushner, J. LeMoigne, and P. Veatch. 1987. Vision-based navigation: A status report. Pp. 153–169 in Proceedings of the Image Understanding Workshop. Los Altos, Calif.: Morgan Kaufmann Publishers.

Dickmanns, E.D., and A. Zapp. 1986. A curvature-based scheme for improving road vehicle guidance by computer vision. Pp. 161–168 in Mobile Robots, Proceedings of SPIE Volume 727. W.J. Wolfe and N. Marquina, eds. Bellingham, Wash.: The International Society for Optical Engineering.

Dickmanns, E.D., and A. Zapp. 1987. Autonomous high speed road vehicle guidance by computer vision. Pp. 221–226 in Automatic Control—World Congress, 1987: Selected Papers from the 10th Triennial World Congress of the International Federation of Automatic Control. Munich, Germany: Pergamon.

Dunlay, R.T., and D.G. Morgenthaler. 1986. Obstacle avoidance on roadways using range data. Pp. 110–116 in Mobile Robots, Proceedings of SPIE Volume 727. W.J. Wolfe and N. Marquina, eds. Bellingham, Wash.: The International Society for Optical Engineering.

Firschein, O., and T. Stratt. 1997. Reconnaissance, Surveillance, and Target Acquisition for the Unmanned Ground Vehicle: Providing Surveillance "Eyes" for an Autonomous Vehicle. San Francisco, Calif.: Morgan Kaufmann Publishers.

Fischler, M., and O. Firschein. 1987. Readings in Computer Vision. San Francisco, Calif.: Morgan Kaufmann Publishers.

Hebert, M. 1990. Building and navigating maps of road scenes using active range and reflectance data. Pp. 117–129 in Vision and Navigation: The Carnegie Mellon Navlab. C.E. Thorpe, ed. Boston, Mass.: Kluwer Academic Publishers.

Hebert, M.H. 1997. SMARTY: Point-based range processing for autonomous driving. Pp. 87–104 in Intelligent Unmanned Ground Vehicles. M.H. Hebert, C. Thorpe, and A. Stentz, eds. New York, N.Y.: Kluwer Academic Publishers.

Horaud, R., and T. Skordas. 1989. Stereo correspondence through feature grouping and maximal cliques. IEEE Transactions on Pattern Analysis and Machine Intelligence 11(11): 1168–1180.

Jochem, T., and D. Pomerleau. 1997. Vision-based neural network road and intersection detection. Pp. 73–86 in Intelligent Unmanned Ground Vehicles. M.H. Hebert, C. Thorpe, and A. Stentz, eds. Boston, Mass.: Kluwer Academic Publishers.

Kelly, A., and A. Stentz. 1997. Minimum throughput adaptive perception for high-speed mobility. Pp. 215–223 in IROS '97, Proceedings of the 1997 IEEE/RSJ International Conference on Intelligent Robot and Systems: Innovative Robotics for Real-World Applications. New York, N.Y.: Institute of Electrical and Electronics Engineers, Inc.

Langer, D., and C.E. Thorpe. 1997. Sonar-based outdoor vehicle navigation. Pp. 159–186 in Intelligent Unmanned Ground Vehicles: Autonomous Navigation Research at Carnegie Mellon. M. Hebert, C.E Thorpe, and A. Stentz, eds. Boston, Mass.: Kluwer Academic Publishers.

Laws, K.I. 1982. The Phoenix Image Segmentation System: Description and Evaluation, Technical Report 289, December. Menlo Park, Calif.: SRI International.

Lim, H.S. 1987. Stereo Vision: Structural Correspondence and Curved Surface Reconstruction, Ph.D. Dissertation. Stanford, Calif.: Stanford University.

Matthies, L. 1992. Stereo vision for planetary rover: Stochastic modeling to near real-time implementation. International Journal of Computer Vision, Vol. 8, No. 1.

Matthies, L., A. Kelly, T. Litwin, and G. Tharp. 1996. Obstacle detection for unmanned ground vehicles: A progress report. Pp. 475–486 in Robotics Research: Proceedings of the 7th International Symposium. G. Giralt and G. Hirzinger, eds. Heidelberg, Germany: Springer-Verlag.

Medioni, G., and R. Nevatia. 1985. Segment based stereo matching. Computer Vision, Graphics, and Image Processing 31(1): 2–18.

Minsky, M.L., and S. Papert. 1969. Perceptrons: An Introduction to Computational Geometry. Cambridge, Mass.: The MIT Press.

Moravec, H.P. 1977. Towards Automatic Visual Obstacle Avoidance. Proceedings of the 5th International Joint Conference on Artificial Intelligence.

Moravec, H.P. 1983. The Stanford CART and the CMU Rover. Proceedings of the IEEE 71(7): 872–884.

Moravec, H.P. 1999. Robot: Mere Machine to Transcendent Mind. New York, N.Y.: Oxford University Press.

Nalwa, V. 1993. A Guided Tour of Computer Vision. Chicago, Ill.: R.R. Donnelley & Sons Company.

Nilson, N.J. 1984. Shakey the Robot, Technical Report 323, April. Menlo Park, Calif.: SRI International.

Ohlander, R., K. Price, and D.R. Reddy. 1978. Picture segmentation using a recursive region splitting method. Computer Graphics and Image Processing 8(3): 313–333.

Ohta, Y. 1985. Knowledge-Based Interpretation of Outdoor Color Scenes. Boston, Mass.: Pitman.

Ohta, Y., and T. Kanade. 1985. Stereo by intra-scanline and inter-scanline search using dynamic programming. IEEE Transactions on Pattern Analysis and Machine Intelligence 7(2): 139–154.

Olin, K.E., F.M. Vilnrotter, M.J. Daily, and K. Reiser. 1987. Developments in knowledge-based vision for obstacle detection and avoidance. Pp. 78–86 in Proceedings of the Image Understanding Workshop. Los Altos, Calif.: Morgan Kaufmann Publishers.

Olin, K.E., and D.Y. Tseng. 1991. Autonomous cross-country navigation—an integrated perception and planning system. IEEE Expert—Intelligent Systems and Their Applications 6(4): 16–32.

Payton, D.W. 1986. An architecture for reflexive autonomous vehicle control. Pp. 1838–1845 in Proceedings of the IEEE Robotics and Automation Conference. New York, N.Y.: Institute of Electrical and Electronics Engineers, Inc.

Pomerleau, D.A. 1992. Neural Network Perception for Mobile Robot Guidance, Ph.D. thesis. Pittsburgh, PA.: Carnegie Mellon University.

Roberts, L. 1965. Machine perception of three-dimensional solids. Pp. 159–197 in Optical and Electro-optical Information Processing. J. Tippett and coworkers, eds. Cambridge, Mass.: MIT Press.

Rosenblum, M. 2000. Neurons That Know How To Drive. Available online at <http://www.cis.saic.com/projects/mars/pubs/crivs081.pdf> [August 6, 2002].

Rosenfeld, A. 1984. Image analysis: Problems, progress, and prospects. Pattern Recognition 17(1): 3–12.

Sharma, U.K., and L.S. Davis. 1986. Road following by an autonomous vehicle using range data. Pp. 169–179 in Mobile Robots, Proceedings of SPIE Volume 727. W.J. Wolfe and N. Marquina, eds. Bellingham, Wash.: The International Society for Optical Engineering.

Spofford, J.R., and B.M. Gothard. 1994. Stopping distance analysis of ladar and stereo for unmanned ground vehicles. Pp. 215–227 in Mobile Robots IX, Proceedings of SPIE Volume 2352. W.J. Wolfe and W.H. Chun, eds. Bellingham, Wash.: The International Society for Optical Engineering.

Spofford, J.R., S.H. Munkeby, and R.D. Rimey. 1997. The annual demonstrations of the UGV/Demo II program. Pp. 51–91 in Reconnaissance, Surveillance, and Target Acquisition for the Unmanned Ground Vehicle: Providing Surveillance "Eyes" for an Autonomous Vehicle. O. Firschein and T.M. Strat, eds. San Francisco, Calif.: Morgan Kaufman Publishers.

Stentz, A., and M. Hebert. 1997. A navigation system for goal acquisition in unknown environments. Pp. 227–306 in Intelligent Unmanned Ground Vehicles: Autonomous Navigation Research at Carnegie Mellon. M. Hebert, C.E Thorpe, and A. Stentz, eds. Boston, Mass.: Kluwer Academic Publishers.

Sukthankar, R., D. Pomerleau, and C. Thorpe. 1993. Panacea: An active controller for the ALVINN autonomous driving system, CMU-RI-TR-93-09. Pittsburgh, Pa.: The Robotics Institute, Carnegie Mellon University.

Turk, M.A., D.G. Morgenthaler, K.D. Gremban, and M. Marra. 1988. VITS—a vision system for autonomous land vehicle navigation. IEEE Transactions on Pattern Analysis and Machine Intelligence 10(3): 342–361.

Wilcox, B.H., D.B. Gennery, A.H. Mishkin, B.K. Cooper, T.B. Lawton, N.K. Lay, and S.P. Katzmann. 1987. A vision system for a Mars rover. Pp. 172–179 in Mobile Robots II, Proceedings of SPIE Volume 852. W.H. Chun and W.J. Wolfe, eds. Bellingham, Wash.: The International Society for Optical Engineering.